SETTING THE MOULD

SETTING THE MOULD

THE UNITED STATES AND BRITAIN
1945–1950

ROBIN EDMONDS

W · W · NORTON & COMPANY
New York   *London*

First American Edition

Library of Congress Cataloging-in-Publication Data
Edmonds, Robin.
Setting the mould.
Bibliography: p. 327.
Includes index.
1. United States—Foreign relations—Great Britain.
2. Great Britain—Foreign relations—United States.
3. United States—Foreign relations—1945–1953.
4. Great Britain—Foreign relations—1945–.
I. Title.
E183.8.G7   1986b   327.73041   86–12883

ISBN 0-393-02382-6

W. W. Norton & Company, Inc., 500 Fifth Avenue, New York, N. Y. 10110
W. W. Norton & Company Ltd., 37 Great Russell Street, London WC1B 3NU

1 2 3 4 5 6 7 8 9 0

FOR ENID

Preface

ELEVEN years ago, one evening in Washington, my host threw out the suggestion that the Anglo-American relationship was due for a fresh study. Why not write it? I was surprised. The fascination of the subject was obvious, but so were the objections: the amount that had already been written—and spoken—about it by others, the nostalgic trap that a British author writing a book on it must avoid, the absence at that time of British official documents, under the United Kingdom's thirty-year rule, and—most of all—the vast breadth of the historical canvas.

Of the first and the last of these objections I remain keenly aware. I can only hope that the first may be to some extent counter-balanced by the fact that this is a study by someone who observed most of the events described at not too great a remove and was close to some of them.* In the face of the last objection, I have drawn encouragement from the words of another writer approaching an equally daunting theme, who wrote in his preface that he believed that there could be advantage in a 'unitary, though partial, vision of a grand theme by a single pair of eyes'.[1] That the eyes are those of a British observer will be clear to the reader, especially perhaps in the final chapter, but throughout the book I have sought to minimize the risk of succumbing to nostalgia by trying, so far as is possible for an Englishman, to see international events from the American point of view as much as from the British.

The remaining problem—the British thirty-year rule—has been solved by the passage of time. A first shot at research, made in 1977, convinced me that a study of the Anglo-American relationship must be based on a comparison of the official records of both countries; however valuable the records of one of the two alone may be, they cannot be enough for the study of the interaction between the two countries. It was only in 1981 that this comparison became possible for the whole of the post-war quinquennium; since January 1981 British documents up to the end of 1950 have been available in the United Kingdom Public Record Office. I therefore set the project aside until these documents—though still with some important exceptions[3]—had been released.

At first I intended this book to cover a longer period, and to take as

---

* The author was a member of the British Foreign (later Diplomatic) Service from 1946–77—effectively from 1945[2]. His later appointments included Head of the American Department and Assistant Under-Secretary for American Affairs at the Foreign and Commonwealth Office.

its starting-point the end of the Second World War. On reflection, I decided to concentrate its focus on the formative years 1945–50 and to include in the book the final months of the war. The reason why the opening date of 1 January 1945 has been preferred to VE or VJ Day is that without Potsdam nothing that followed is intelligible; Potsdam itself is barely intelligible without Yalta; and the post-war relationship between the United States and Britain cannot be fully understood if it is isolated from the relationship that had been built up between them during the war. Moreover, without the atomic dimension it is hard to make sense either of the 1945 conferences or of the post-war evolution of the Anglo-American relationship. The choice of the earlier starting-point has also made it possible to include both a brief account of the domestic political changes that occurred in both countries in 1945 and a discussion of the two countries' attitudes towards the problems of the post-war international economic order, including the Bretton Woods agreements. Here again, without consideration of these linkages, what followed is easily misinterpreted.

Parts I–V of the book are based almost entirely on American and British primary sources of the years 1945–50. These are the only sure guide to what really happened during these extraordinary years. Diaries of all kinds have been used sparingly. The accuracy of the observation that, because 'no human memory is so arranged as to recollect everything in continuous sequence', diaries 'often turn out to be bad assistants'[4] has been illustrated by historical studies of this period which have relied on the evidence of diaries and memoirs, only to be revealed as inaccurate—sometimes seriously so—by documents published after the studies were written.

Official documents, however, sometimes omit altogether, or only partially reveal, the human factor. For this essential supplementary evidence I have relied, as well as on my own recollection, on that of a wide range of former colleagues and historians—primarily, but not exclusively, in the United States and Britain—whose names will be found among those listed at the end of the Preface. In order to provide a further dimension, some Soviet cross-bearings have been included in Part II. There is still no full Soviet documentation covering these years, but Soviet collections of documents relating to the 1945 conferences are now available, although—unlike western volumes—these do not include internal minutes or memoranda.

To paraphrase a document may easily mislead, however much care is taken to avoid this. In Parts I–V, therefore, I have preferred quotation to paraphrase, even at the cost of lengthening the book. By contrast, Part VI (covering the years 1951–62) has been written with a broad brush. At the time of writing, British documents available in the

Public Record Office do not extend beyond 1954; and the *Foreign Relations of the United States* series has also come to a halt in that year. The quotations from official documents are, therefore, much fewer in Part VI, whose purpose is in any case quite different from that of the rest of the book: being to relate the evolution of the Anglo-American relationship in the early post-war years to the modern relationship between the two countries, with signposts clear enough to enable a balance sheet to be drawn up in the final chapter.

In Part VI, too, the question of the time-frame has not been easy to decide. Arguably, the definitive political watershed for the post-war Anglo-American relationship was the Suez crisis of 1956. On balance, however, the end of 1962 has been preferred as a cut-off date, because it was in the autumn of that year that John Kennedy and Harold Macmillan concluded the Nassau atomic weapons agreement, critically important both for Britain at the time and for the contemporary Anglo-American relationship; and this agreement was signed shortly after the United States had handled the greatest crisis since the Second World War—the Cuban missile crisis—as a superpower, acting entirely on its own.

1945–50 were not only the years by the end of which the mould was set, but also during which the relationship between the United States and Britain was unquestionably of global importance. This universality presents a problem of choice of method: the shot-gun or the rifled approach, the chronological or the thematic treatment of events. The first and the last of the six years 1945–50 call for a considerable measure of chronological treatment, if only because both were years of war. The Second World War was still being waged for the greater part of 1945; in 1950 a major war was launched in Asia; and 1950 was in several respects the pivotal year in post-war history. But the intervening four years do not easily lend themselves to chronological treatment of the year-by-year kind which, for example, a biographer covering this period can scarcely avoid. It is not historically sound to divide the Anglo-American relationship during this period into neat sections. True, by the end of 1945 the United States and Britain found themselves estranged, by comparison with the years of war; and—at any rate so far as their European policy was concerned—their relationship began to become intimate again in 1947. But there were also important overlaps, both of discord and of harmony, which make it awkward to construct a chronological schema for the four years 1946–9; and a clear understanding of particular issues—that of Palestine, for example—can be obtained only if they are pursued straight through, with as few breaks as possible. This book, therefore, seeks to combine both the chronological and the thematic approach.

Whatever the method adopted, it is bound to have some disadvantages. In preferring the rifled to the shot-gun approach in most of this book, I am conscious that it does scant justice to some major political developments of the period: for example, the post-war transformation of Japan, which—seen from an Anglo-American standpoint—was almost entirely an American responsibility, and the transfer of power in South Asia in 1947, which was almost entirely a British one. Moreover, discussion of two areas of the world, both of which were to become of critical importance in the 1950s—South-East Asia and Latin America—has suffered from the constraints of space. And one (deliberate) omission will be evident: the 'missing dimension of diplomatic history'[5] finds no place in this book, except for brief references in Chapters 2 and 21.

Three other points may require a word of explanation: names and the transliteration of cyrillic and Chinese. First names are given on first mention in the text, and occasionally thereafter, except in the case of lesser known Russians, for whom the Russian practice of the initials of name and patronymic has been followed; nicknames have generally been avoided in all languages; and all titles, other than military, are omitted in the text, but may be found in the index or sometimes in the notes. Since many of the names mentioned in the text will be unfamiliar to a later generation, other than to specialists in the post-war period, and in order to reduce the need for frequent recourse to the index, a separate list of names has been provided.

For cyrillic, the system followed by the journal *Soviet Studies* has been followed, with the single exception of the word *nyet*,[6] which became anglicized during this period. As for Chinese, rather than convert to pinyin the spelling in documents of the time, which used the older system of Chinese transliteration, and in maps relating to the period, it has seemed preferable to retain the older system both in quotation and in the text.

Finally, there is the problem of the use of hindsight. Ideally, I would have preferred to confine value judgements, which by definition oblige the historian to look back and to take a view, to the end, leaving the account of the sweep of events offered by the rest of the book to speak for itself. In practice, it would have been pedantic to adhere rigidly to this rule. Instead, in addition to the assessment offered at the end of the book, at certain places in the narrative (where indeed it might even have appeared evasive not to have done so) I have stood back momentarily, in order to point out a consequence unobserved at the time, or to suggest a conclusion that would not then have been generally accepted. Having at first considered including, in every case of an observation written with hindsight, a reminder to that effect, I came to the conclusion that this ran the risk of sounding intolerably repetitious. In each such instance in Parts I–V it will, I hope, be clear when the use of hindsight is involved.

My first opportunity to observe, at close quarters, the interaction between the Americans and the British, came well before the opening date of this book. Quite unexpectedly, I spent the early autumn of 1943 living the life of an American in every respect except my British uniform, as a liaison officer with a US Army unit in South Italy. Whatever I may have learned about the relationship between the Americans and the British—both its potential and its limitations—was derived in the first instance from that memorable episode in my life. Across the span of forty-two years, I would like to thank my American colleagues of those days both for their understanding and for what they taught me.

It goes without saying that this book does not express the views of the British Government. The opinions expressed in the book, like the conclusions that it reaches, are entirely my own. But my 'partial vision' has been greatly expanded through the eyes of others, for whose help I am very grateful. The typescript has been read by my wife, Lord Bullock, Mr Michael Cullis (who also kindly read the proofs), Mr Joseph Harsch, Professor Roger Louis, Mr Thomas Sorensen, and Dr Stephen White. (Dr White also rendered invaluable assistance both in running elusive references to earth and by instilling discipline into a body of notes that at first looked rather mutinous.) Individual chapters of the book have been read by Professor Margaret Gowing (both she and Lord Bullock have been particularly generous with their time), Mr Austen Albu, Sir John Barnes, Sir Ashley Clarke, Monsieur Yves Delahaye, Professor Michael Howard, Mr Keith Kyle, Monsieur Jean Laloy, Sir Martin Le Quesne, Dr Michael Leifer, Dr Robert O'Neill, Mr Derek Thomas, Dr Gregory Treverton, and Mr Samuel Wells Jr. All of them will, I hope, find at any rate some of their wisdom reflected in changes that I have made in the light of their advice. My especial thanks are due to Professor Louis, both for his penetrating comments and for kindly allowing me to read in typescript his preface to the collection of essays on *The Special Relationship: Anglo-American Relations since 1945*, to be published by the Clarendon Press in 1986.

In addition, the number of people whom I have consulted on specific points, both before putting pen to paper and while writing the book, is considerable. They include Mr George Ball, Lord Beloff, Mr Robert Bowie, the late Professor Hedley Bull, Sir Julian Bullard, Sir James Cable, Sir Alan Campbell, Mr Joseph Dobbs, Professor Jean-Baptiste Duroselle, Professor John Erickson, Lord Franks, Professor Qiu Ke'an, Professor George Kennan, the late Mr Henryk Krzeczkowski, the late Professor Hugh Seton-Watson, Lord Shackleton, and the late Sir Duncan Wilson. If there is anyone whose name I have inadvertently omitted in this list or in the preceding paragraph, I offer my apologies. To all of them I am gratefully indebted.

My thanks also go to Miss Laura Tatham and Mrs Pauline Kingsley-Ducane for bearing the brunt of deciphering my handwriting; and to Mrs Julie Willbourn and Ms Terry Webb, each of whom typed one chapter. Although the greater part of the book has been written in Somerset, both the initial research and the writing of the final chapter benefited from an American environment; for making this possible I am grateful to the Woodrow Wilson Center for Scholars, Washington, DC and to the Center for International Affairs, Harvard University, respectively. I would also like to express especial thanks to Mrs Nicole Gallimore, the Librarian at Chatham House, and her staff for their kind assistance, unstintingly given over a long period; as well as to Mr Duncan Chalmers and his staff at the Public Record Office and to Ms Sarah Marks, at the National Archives.

Sir Edgar Williams once again cast his discerning eye over the proofs. Both Mr Ivon Asquith, at the Oxford University Press, and Mr Donald Lamm, President of W. W. Norton, New York, have been towers of strength. And—last, but by no stretch of the imagination least—my wife has encouraged and sustained me through two exacting years.

Street,
Somerset
November, 1985

Acknowledgements

I EXPRESS my thanks to Mr Raymond Hyatt and Mr Peter Grove, for the maps on which those included in this book are based; to the Borthwick Institute of Historical Research, York, for permission to quote from the entries of Lord Halifax's Secret Diary of 1 and 2 December 1941; to the Library of Churchill College, Cambridge, for permission to quote from Lord Halifax's letter to Winston Churchill of 3 August 1945; to Messrs William Heinemann and W. W. Norton, for permission to quote the passage from Lord Bullock's *Ernest Bevin: Foreign Secretary, 1945–51*, pp. 182–3; and to the holders of the copyrights of the photographs reproduced in this book.

All quotations from documents at the Public Record Office, Kew in Crown copyright appear by permission of the Controller of Her Majesty's Stationery Office.

Contents

Illustrations

Maps

Abbreviations

ANZUS	Australia, New Zealand, and United States Pact
AK	*Armja Krajowa* (the Polish Home Army)
CEEC	Committee of European Economic Co-operation
CFM	Council of Foreign Ministers
CIA	Central Intelligence Agency
CPC	(Anglo-American atomic) Combined Policy Committee
CPR	Chinese People's Republic
CPSU	Communist Party of the Soviet Union
CRO	Commonwealth Relations Office
ECA	Economic Cooperation Administration
EDC	European Defence Community
EEC	European Economic Community
ERP	European Recovery Plan
FCO	Foreign and Commonwealth Office
FO	Foreign Office
FRG	Federal Republic of Germany
GATT	General Agreement on Tariffs and Trade
GHQ	General Headquarters
GNP	Gross National Product
GOC	General Officer Commanding
HC	House of Commons
HL	House of Lords
HMSO	His/Her Majesty's Stationery Office
IBRD	International Bank for Reconstruction and Development
ICBM	Intercontinental ballistic missile
IDA	International Development Agency
(I)ISS	(International) Institute for Strategic Studies
IMF	International Monetary Fund
IRBM	Intermediate range ballistic missile
NA	(US) National Archives
NATO	North Atlantic Treaty Organization
NKVD	Soviet Committee State Security (now the KGB)
NSC	National Security Council
OEEC	Organization for European Economic Cooperation (afterwards became Organization for Economic Cooperation and Development—OECD)

PRO	(United Kingdom) Public Record Office
RAF	Royal Air Force
RIIA	Royal Institute of International Affairs
SEATO	South-East Asia Treaty Organization
SLBM	Submarine-launched ballistic missile
TA	Tube Alloys
UK	United Kingdom of Great Britain and Northern Ireland
UKAEA	United Kingdom Atomic Energy Authority
UN(O)	United Nations (Organization)
UNCOK	United Nations Commission on Korea
UNRRA	United Nations Relief and Rehabilitation Administration
UNSCOP	United Nations Special Commission on Palestine
US(A)	United States (of America)
USAEC	United States Atomic Energy Commission
USSR	Union of Soviet Socialist Republics
VE	Victory in Europe (Day)
VJ	Victory in Japan (Day

NOTE

1. For a further list of abbreviations used in this book, for sources to which frequent reference is made, see *Note on Documentary Sources*.
2. The word 'billion' (bn.) is used to mean one thousand million.
3. In order to convert to modern value amounts expressed (as they are throughout Parts I–V of this book) in prices current during the years 1945–50, an average US$ multiplier of 5 and an average £ sterling multiplier of 12 should be applied (source: US Department of Commerce consumer price index and UK retail price index).

A Selective Anglo-American Chronology, 1945–62

1945

January–February	Malta Conference	
February	Yalta Conference	
April		Death of Roosevelt; Truman becomes President
May	End of war in Europe	
June	UN Charter signed	
July		Churchill loses general election, replaced by Attlee
July-August	Potsdam Conference	
August	Atomic bombs dropped on Hiroshima and Nagasaki; end of war with Japan	
September–October	First (London) CFM	
December	Anglo-American Financial Agreement	
December	Meeting of three Foreign Ministers in Moscow	

1946

March	Churchill's address at Fulton, Missouri	
April	Anglo-American Committee of Enquiry reports on Palestine	
October	Truman's Yom Kippur statement on Palestine	
December	Anglo-American agreement on formation of German 'Bi-Zone'	

1947

February	Peace Treaties signed with Italy and with Germany's wartime allies	
March	Truman Doctrine (US assumes British financial commitments to Greece and Turkey)	

March–April	Moscow CFM	
June	Marshall's Harvard speech	
June–July	Bevin's response to Marshall Plan	Molotov walks out of Paris Conference
August	Convertibility of sterling suspended	
October	'Pentagon Talks' on Middle East	
December	Lancaster House CFM	

1948

January	British Government gives up right of veto on US use of atomic bomb	
February		Prague *coup d'état*
March	Anglo-American talks in Washington on western defence system	Treaty of Brussels signed
April	British mandate in Palestine ends; Israel recognized *de facto* by US	
June	Berlin blockaded by Soviet Union	
July	US strategic bombers stationed in Britain	
November		Truman elected President

1949

April	North Atlantic Treaty signed	
May	Palais Rose meeting of CFM	Berlin blockade lifted
August–September	Anglo-American financial talks in Washington; devaluation of sterling	First Soviet atomic explosion
October		Chinese People's Republic proclaimed

1950

	(Year of hydrogen bomb decisions)	
January–February	British recognition of Chinese People's Republic; US continues recognition of Chinese Nationalist Government (Taiwan); failure of Anglo-American atomic negotiations	Sino-Soviet Treaty of Alliance signed
February		Labour Party wins British general election

May	Tripartite (Anglo-Franco-US) Declaration on Middle East	
June	Outbreak of Korean War	Britain refuses French invitation to attend Schuman Plan Conference in Paris
October	UN forces cross 38th Parallel in Korea	
November– December		Chinese troops drive UN forces out of North Korea
December	Anglo-American summit conference in Washington; North Atlantic Council agrees to rearm German Federal Republic and to Eisenhower's appointment as Supreme Allied Commander Europe	

1951

April		Death of Bevin; MacArthur relieved of (Far East) command
September	Japanese Peace Treaty signed	
October		Churchill wins general election

1952

January	Churchill meets Truman in Washington	
October	First British atomic explosion	
November	First US hydrogen bomb explosion	Eisenhower elected President
December	Churchill meets Eisenhower in New York	

1953

March		Death of Stalin
May	Churchill proposes summit meeting	
July	Korean Armistice agreed	
August		First Soviet hydrogen bomb explosion
December	Bermuda Three-Power Conference (US, Britain, and France)	

Selective Chronology

1954

April–July	Geneva Conference on Far East; Indo-China partitioned	
September	South-East Asia Treaty Organization formed	
October		Western European Union agreement signed (permits West German membership of NATO)

1955

April		Churchill resigns; Eden Prime Minister
May	Austrian State Treaty signed	Warsaw Pact signed
July	First Four-Power summit meeting (Geneva)	

1956

February		Khrushchev denounces Stalin at 20th CPSU Congress
July		Nasser nationalizes Suez Canal
October–November	Suez Crisis	Hungarian revolt; Polish 'October revolution'
November		Eisenhower re-elected President

1957

January		Eden resigns; Macmillan Prime Minister
March	Eisenhower and Macmillan meet in Bermuda	
March		Treaty of Rome (EEC) signed
October	First British hydrogen bomb exploded	First Soviet *sputnik* launched in space
October	Eisenhower and Macmillan meet in Washington; 'Declaration of Common Purpose' signed	

1958

June		de Gaulle returns to power
July	McMahon Atomic Energy Act amended; Anglo-American Atomic Agreement signed	

October		Macmillan wins general election
1959		
January		Castro enters Havana
February–March		Macmillan's visit to Soviet Union
June		Soviet Government rescinds Sino-Soviet nuclear agreement
September		Khrushchev's visit to US
1960		
May	Second Four-Power summit meeting (abortive), in Paris	
November	Agreement on Polaris base in Britain announced	
November		Kennedy elected President
1961		
March	First meeting between Kennedy and Macmillan	
April		Invasion of Bay of Pigs, Cuba
June		Kennedy and Khrushchev meet in Vienna
August	Beginning of Berlin Wall	Britain applies to join EEC
1962		
October–November		Cuban missile crisis
December	Nassau meeting between Macmillan and Kennedy; 'Statement on Nuclear Defense Systems' (Polaris) agreed.	

Selective List of Names appearing in the narrative

In this list, which is intended as an aid for readers unfamiliar with the period (not as an addition to the Index, which will be found at the end of the book), the appointment shown opposite each name is the one held at the time when the name occurs for the first time. In a few cases, other appointments or offices relevant to the years covered by the book have been added in brackets.

Acheson, Dean G., Secretary of State
Adenauer, Konrad, Chancellor of the Federal German Republic
Anderson, John, Chancellor of the Exchequer
Antonov, Alexei, Chief of Soviet General Staff
Attlee, Clement, Prime Minister of Great Britain

Baruch, Bernard M., American financier; US Representative, UN Atomic
    Energy Commission, 1946
Berlin, Isaiah, Chichele Professor of Social and Political Theory, Oxford
    University
Bevan, Aneurin, Minister of Health
Beveridge, William, Chairman of British Official Committee on Social
    Insurance and Allied Services, 1941–2
Bevin, Ernest, Foreign Secretary
Bidault, Georges, French Foreign Minister
Bohlen, Charles E., Minister (later Ambassador), US Embassy, Paris, tempor-
    arily recalled to State Department, 1950
Bruce, David K., US Ambassador, Paris (later Ambassador, London)
Buchan, Alastair, Professor of International Relations, Oxford University
Bullard, Reader, British Ambassador, Teheran
Bush, Vannevar, Director (US) Office of Scientific Research and Develop-
    ment.
Byrnes, James F., Secretary of State

Cadogan, Alexander, Permanent Under-Secretary, Foreign Office
Campbell, Ronald, British Minister, Washington (later Ambassador, Cairo)
Chadwick, James, British scientific adviser to (atomic) Combined Policy
    Committee
Chamberlain, Neville, Prime Minister of Great Britain
Cherwell (Lindemann), Frederick, Paymaster-General
Chiang Kai-shek, Generalissimo, President of China
Chou En-lai, Prime Minister of China
Churchill, Winston, Prime Minister of Great Britain

Clarke, Richard, Under-Secretary, (British) Treasury
Clay, Lucius, D., US Military Governor, Germany
Clayton, William L., (US) Assistant Secretary of State for Economic Affairs
Clifford, Clark M., Special Counsel to President Truman
Cockcroft, John, Chairman, UK Atomic Energy Authority
Conant, James B., President of Harvard University; Chairman, US National
    Defense Research Committee
Creech-Jones, Arthur, Colonial Secretary
Cripps, (Richard) Stafford, President, Board of Trade (afterwards Chancel-
    lor of the Exchequer)
Crossman, Richard, Labour Member of Parliament

Dalton, Hugh, Chancellor of the Exchequer
Davies, Joseph E., Chairman, President's War Relief Control Board, personal
    emissary from Truman to Churchill, May 1945 (formerly Ambassador,
    Moscow)
De Gasperi, Alcide, Prime Minister of Italy
de Gaulle, Charles, President of France
Dening, Esler, Assistant Under-Secretary for Far Eastern Affairs, Foreign
    Office
Dill, John, Head of British Joint Services Mission, Washington
Djilas, Milovan, Yugoslav communist partisan leader
Douglas, Lewis, US Ambassador, London
Dulles, John Foster, Secretary of State

Eden, Anthony, Foreign Secretary (later Prime Minister of Great Britain)
Eisenhower, Dwight D., Supreme Commander, Allied Expeditionary Force in
    Western Europe (later NATO Supreme Commander, Europe, before
    becoming President of the United States)

Forrestal, James V., (US) Secretary of Defense
Franks, Oliver, British Ambassador, Washington
Fuchs, Klaus, atomic scientist

Gaitskell, Hugh, Minister of State, Economic Affairs (later Chancellor of the
    Exchequer and afterwards Leader of the Opposition)
Gorbachev, Mikhail, General Secretary, Soviet Communist Party
Gottwald, Klement, Prime Minister (later President) of Czechoslovakia
Grady, Henry F., Head of US delegation, Anglo-American discussions on
    Palestine, July 1946
Gromyko, Andrei, Soviet Deputy Foreign Minister (later Minister of Foreign
    Affairs)
Groves, Leslie R., Commanding General 'Manhattan Engineering District'
    (war-time codename of US atomic project)

Halifax, Edward, British Ambassador, Washington (formerly Foreign Secretary)

Harriman, W. Averell, US Ambassador, London (formerly Ambassador, Moscow)

Harsch, Joseph C., American columnist, broadcaster, and writer

Healey, Denis, Secretary, Labour Party International Department

Hinton, Christopher, Deputy Controller, Atomic Energy (Production), Ministry of Supply

Hiss, Alger, former State Department official

Hoffman, Paul, US Economic Cooperation Administrator

Hopkins, Harry L., Special Assistant to President Roosevelt

Hull, Cordell, Secretary of State

Inverchapel (Clark Kerr), Archibald, British Ambassador, Washington (formerly Ambassador, Moscow)

Jacobson, Edward, personal friend of Harry Truman (and former business partner)

Johnson, Louis A., (US) Secretary of Defense

Kennan, George F., Director, State Department Policy Planning Staff (formerly Counsellor, US Embassy, Moscow)

Kennedy, John F., President of the United States

Kennedy, Joseph P., US Ambassador, London

Keynes (John) Maynard, British negotiator of Bretton Woods agreements and US loan to Britain

Killearn (Lampson), Miles, British Ambassador, Cairo

King, Mackenzie, Prime Minister of Canada

Kissinger, Henry A., Secretary of State

Khrushchev, Nikita, First Secretary, Soviet Communist Party and Chairman of Council of Ministers

Kurchatov, Igor, Director, Soviet atomic project

Leahy, William D., Chief of Staff to US Commander-in-Chief (the President)

Lilienthal, David E., Chairman, US Atomic Energy Commission

Lippmann, Walter, American columnist and author

Lovett, Robert A., US Under Secretary of State (later Defense Secretary)

Luce, Henry R., co-founder and editor, *Time* Magazine

MacArthur, Douglas A., Supreme Commander, South-West Pacific (subsequently Supreme Commander in Japan)

McCarthy, Joseph R., (Republican) US Senator from Wisconsin

McCloy, John J., US Military Governor and High Commissioner, Germany

McMahon, Brien, Chairman, Senate Atomic Energy Committee

Macmillan (Stockton), Harold, Conservative Member of Parliament, in Opposition (later Prime Minister)

McNeil, Hector, Minister of State, Foreign Office

Makins (Sherfield), Roger, Minister, British Embassy, Washington (later Ambassador there, following his years at the Foreign Office as Assistant, then Deputy, Under-Secretary of State)

Mao Tse-tung, Chairman, Chinese Communist Party

Marshall, George C., Chief of US Army Staff (later Secretary of State)

Massigli, René, French Ambassador, London

Mikołajczyk, Stanisław, Polish Prime Minister (London), subsequently Deputy Prime Minister (Warsaw)

Molotov, Vyacheslav, Soviet Commissar (later Minister) for Foreign Affairs

Monnet, Jean, Head of French Commissariat du Plan, responsible for drafting Schuman Plan

Montgomery, Bernard, GOC 21 Army Group (later Chief of Imperial General Staff and Chairman, Western Union Commanders-in-Chief)

Morgenthau, Henry, Jr., (US) Secretary of the Treasury

Morrison, Herbert, Leader of House of Commons

Nasser, Gamal Abdel, President of Egypt

Nehru, Jawaharlal, Prime Minister of India

Niles, David K., Special Adviser to President Truman

Nitze, Paul H. Director, State Department Policy Planning Staff

Nixon, Richard M., President of the United States

Oppenheimer, J. Robert, Director, Los Alamos Scientific Laboratory, 1943–45; Chairman, Advisory Committee, US Atomic Energy Commission, 1946–52

Ormsby Gore, David, British Ambassador, Washington

Osóbka-Morawski, Edward, Polish Prime Minister and Foreign Minister, Lublin Committee (and later in Warsaw Government)

Potemkin, Vladimir, Soviet Vice-Commissar for Foreign Affairs

Robertson, Brian, British Military Governor, Germany

Roosevelt, Franklin D., President of the United States

Rusk, Dean, Assistant Secretary for Far Eastern Affairs (later US Secretary of State)

Sargent, Orme, Deputy Under-Secretary, Foreign Office (later Permanent Under-Secretary)

Saud, Abdel Aziz Ibn, King of Saudi Arabia

Schuman, Robert, French Foreign Minister

Shtemenko, S. M., Deputy Chief of Soviet General Staff

Sidky, Ibrahim, Prime Minister of Egypt

Smart, Walter, Minister (Oriental), British Embassy, Cairo

Smuts, Jan Christian, Prime Minister of South Africa

Snyder, John W., US Federal Loan Administrator (later Secretary of the Treasury)

Sokolovskii, V. D., Chief of Soviet Military Administration in Germany
Spaak, Paul-Henri, Prime Minister of Belgium
Stalin, Joseph, General Secretary of Soviet Communist Party and Chairman
   of Council of People's Commissars (Council of Ministers from 1946)
Stettinius, Edward R., Secretary of State
Stimson, Henry L., US Secretary of War
Strang, William, Permanent Under-Secretary, Foreign Office

Taft, Robert A., Chairman, Senate Republican Party Policy Committee
Tedder, Arthur, Head of British Joint Services Mission, Washington
Truman, Harry S., President of the United States

Vandenberg, Arthur H., ranking Republican member (later Chairman),
   Senate Foreign Relations Committee
Varga, Evgenii, Head of Institute of Economics, Soviet Academy of Sciences
Vinson, Fred M., Secretary of the Treasury (later Chief Justice)

Waley, David, Under-Secretary, (British) Treasury
Wallace, Henry A., US Vice President (formerly Secretary of Agriculture,
   later Secretary of Commerce)
Weizmann, Chaim, President, Jewish Agency for Palestine and World Jewish
   Organisation (later President of Israel)
White, Harry D., (US) Assistant Secretary of the Treasury
Willkie, Wendell L., Republican Party candidate for US presidency, 1940
Wilson, Harold, President, Board of Trade
Wilson, Henry Maitland, Head of British Joint Services Mission, Washington
Winant, John G., US Ambassador, London
Wolff, Karl, senior SS General in Italy, 1945
Wright, Michael, Assistant-Under-Secretary for Middle Eastern Affairs,
   Foreign Office

Zhukov, Georgii, Commander of First Belorussian Front; captor of Berlin
   (later Chief of Soviet Military Administration in Germany)

CHAPTER I

Introduction

IN the course of the one hundred and thirty-one years between 1814 and 1945, three conferences were held in Europe, at what would now be called summit level, in the immediate aftermath of wars that had convulsed the continent. The three wars—the Napoleonic and the two world wars—had much in common, as did the conferences that followed them: at Vienna in 1814–15, at Versailles in 1919, and at Potsdam in 1945. All these wars were, at least in their origins, European civil wars. At the heart of all three conferences lay the ancient problem of Central Europe. Winston Churchill harangued Joseph Stalin about Poland in language that echoed some of Lord Castlereagh's words* to the Tsar Alexander I at Vienna. Moreover, the concept of compensation, used to justify the mass migrations of 1945, also underlay the 'transference of souls' of 1815. At the end of the Second World War, however, it was not the comparison[1] with the negotiations of the Congress of Vienna that was fashionable, but the anxiety to avoid the mistakes made in the treaty concluded at Versailles only a quarter of a century earlier. Everyone, including Stalin, was talking about the mistakes of Versailles in 1945.[2]

Among the many lessons of the Versailles fiasco generally accepted in 1945, the simplest was that 'after a long war it is impossible to make a quick peace'.[3] The conference at Potsdam, therefore, was not in a formal sense a peace conference, although it did establish peace-making machinery. Secondly, whereas in November 1918, many months earlier than expected, the German Government surrendered in reliance on Woodrow Wilson's 'clearly defined principles which should set up a new order of right and justice'—the source of bitter controversy thereafter—the Allies insisted on the unconditional surrender of their enemies. After May 1945 there simply was no German Government left to argue. A third mistake at Versailles had been widely acknow-ledged soon after the ink was dry on the treaty. Its drafters had ignored

---

* 'The Emperor insinuated that the question [of Poland] could only end in one way, as he was in possession. I observed that it was very true His Imperial Majesty was in possession, and he must know that no one was less disposed than myself hostilely to dispute that possession; but I was sure His Imperial Majesty would not be satisfied to rest his pretensions on a title of conquest in opposition to the general sentiments of Europe.'

what John Maynard Keynes was to call the economic consequences of the peace—so much so that a book published in Britain in 1942 went so far as to argue that this time it would be 'prudent to let the work of economic reconstruction proceed a long way before attempting to create the rigid political forms of a lasting settlement'.[4] A year before the end of the Second World War the framework for a new structure of international economic relations had already taken shape—a brilliant thrust of Anglo-American vision, even though the Bretton Woods agreements failed to foresee that post-war economic problems would require far more radical treatment than could be provided by the institutions that they set up. A fourth defect had already adjusted itself in the course of the Second World War. Whereas Soviet absence from Versailles was bound sooner or later to frustrate much of the settlement agreed there, in 1945 the Soviet Union had regained the position held by Russia at the end of the Napoleonic War—that of the greatest military power in continental Europe—and was playing an international role reminiscent in many respects of Russian imperial policy. Finally, instead of the League of Nations, which was crippled from the outset, three weeks before the conference opened at Potsdam the charter of the United Nations Organization had been signed at San Francisco: an organization which was to be 'policed'[5] by the victors of the Second World War.

These differences between 1919 and 1945 were important. But the greatest difference was the fact that, of the two world wars, it was the second that was truly a world war. The Second World War began in Central Europe; and it was from Potsdam that the final 'triple summons' to the Japanese Government to surrender was issued.[6] But this war engulfed the Mediterranean, the Near East, South-East Asia, and the Far East on a global scale for which the First World War, though not confined to Europe, provided no real precedent. The leaders of the Grand Alliance—the United States, the Soviet Union, and Great Britain—were not being grandiloquent when they described themselves in one of the Yalta agreements as 'the three Great Powers'. In the West, moreover, phrases like 'one world', the title of a best-seller written by Wendell Willkie, and Henry Wallace's 'the century of the common man' did not then have the naive ring that they have today; there was not only the relief that followed years of carnage, but a general hope; and for one American observer of the international scene, 'not since the unity of the ancient world was disrupted' had there been 'so good a prospect of a settled peace'.[7] This time perhaps the backing of the overwhelming authority of the victorious powers might enable a just and enduring peace, in the words of the communiqué issued after the Potsdam Conference, to be established.

The victors did indeed establish peace, of a kind, in the European continent, but in few of the others. Although the scope of this book includes the years during which the great rift between East and West began, and may therefore throw some indirect light on its origins and its consequences, it is not intended to address the question: which side bears the greater responsibility for the cold war? Enough has perhaps been written on this controversy by orthodox historians on both sides and by revisionists. Soviet published documents for the late Stalinist years are still meagre, even by comparison with other periods of Soviet history. Moreover, in order to permit a comprehensive judgement, especially of the pivotal year 1950, the opening of the Chinese archive would be needed as well. By contrast, US documents of this period have long been available—of all kinds and in great quantity. And what is now also available is the British archive. Since 1981 it has been possible for the first time to use the American and the British documents in conjunction, to study the post-war evolution of the Anglo-American relationship. Moreover, taken together, these documents also offer the historian a lens through which he may re-examine, from the viewpoint of two of the principal actors, the formative years which largely determined the political structure of the modern world; and it is with these years that any study of this structure must begin.

However the period 1945–50 is broadly regarded—good or bad, accident or design, an inevitable sequence of events flowing from deep-set trends or the result of a few critical decisions—and whatever the field of observation—political, economic, or military—these were the last years after the end of the Second World War during which major political options were open. For most of this period the mould had not yet set. Apart from the war itself, they were also the years during which the relationship between the United States and Britain mattered most. True, they included moments when the relationship between the two countries became—in the jargon of a later age—semi-adversarial. Nevertheless, the Anglo-American relationship has never been more significant since the war than it was during the period of approximately three years measured from mid-1947. And throughout the whole period the relationship was significant, not only for both countries and for their subsequent history, but also for the course of international events and hence for the international community.

Today, nearly half a century later, most of the achievements of the post-war quinquennium are being called in question. Of the international institutions established at its outset—the International Monetary Fund, the International Bank, and the United Nations Organization—perhaps the first two have best withstood the test of time. The United Nations, though providing an international forum, has hardly fulfilled the intentions of its founding fathers. In 1945 there were also the beginnings of a belief, though not then expressed in these terms, that the North–South relationship was ripe for a fresh start. And if there was one commitment shared in 1945 by all governments, it was the maintenance of full employment.* Both those concepts have been severely battered in recent years.

By the end of the quinquennium, the field of achievement in which western leaders, like their eastern counterparts, felt most certain that they had deserved well of their countries was that of defence. In the event the Sino-Soviet alliance lasted, in effect, little more than eight years; and the political cohesion of Eastern Europe is now at best a wasting asset for the Soviet Union. The transatlantic relationship that underlies the North Atlantic Treaty Organization has also been under great strain in the current decade. Small wonder, therefore, that today the Anglo-American relationship, which formed the hard core of the North Atlantic Alliance at the time of its creation, has not been exempt from the general reassessment of the values of the late 1940s, and in particular of the concepts of defence that date from that period. This reassessment is by no means confined to the Left. In Britain, it was a right-wing Member of Parliament who wrote in 1983 that 'those were the years in which Britain became committed to that axiomatic identification with the philosophy and strategy of the United States which a whole series of current events . . . are conspiring to undermine and discredit'.[8]

No issue in the field of post-war defence policy has attracted more searching scrutiny than that of nuclear weapons. For most people today the central event of August 1945 is not the Potsdam Protocol, but the fact that the world whose destiny the three Heads of Government were trying to decide at Potsdam was, at that very moment, on the eve of entering the nuclear age. As the US President and the British Prime Minister knew, and as Stalin was left to deduce from an informal exchange with Truman on 24 July, the 'prompt and utter destruction', offered two days later by the allied proclamation to Japan as the alternative to unconditional surrender,[9] was to be assured by the atomic bomb. (The first was dropped on Hiroshima on 6 August; the second on Nagasaki three days later; and, but for the Japanese surrender on 14 August, a third might have followed before the month was out.)[10] Later chapters assess the remarks that briefly passed between the President and Stalin after the plenary meeting of the Potsdam Conference on 24 July 1945. Here what matters is not only

---

* The 'maintenance of high levels of employment' was one of the declared purposes even of the International Monetary Fund, included in Article I.

the fact that the three Heads of Government left Potsdam without any joint discussion of the atomic bomb, but also that, even after the two atomic attacks, the number of people in any of the three countries represented at Potsdam who realized the full significance of what had just happened in Japan and what was likely to follow, was very small indeed—in spite of the call for debate with which Henry Smyth concluded his official report on atomic energy, released for publication in the United States only six days after Hiroshima. In Smyth's words, the questions raised by future atomic developments 'in a free country like ours . . . should be debated by the people and decisions made by the people through their representatives'.* As the *History of the United States Atomic Energy Commission* recorded twelve years later, in the decade that followed the attacks on Hiroshima and Nagasaki 'the relatively few persons who were privileged to work behind the security barrier imposed by the Atomic Energy Act of 1946 found themselves ever more isolated in a world their fellow citizens had never seen'.[11] And most of these citizens, both in the United States and in Britain, shared a basic assumption immediately after the war, which subsequent events were to prove spectacularly wrong.

All memory distorts. With the passage of time, it also telescopes events, especially when, as they did in 1945–50, they follow each other thick and fast in every quarter of the globe. However it may seem today, the cold war and the age of the superpowers did not burst on the world suddenly; they developed slowly, in an international environment which, to begin with, was still fluid. The negotiating process between the three† victorious powers continued, on any reckoning, for at least two and a half years after the end of the Second World War; indeed the sixth and, as it turned out, final meeting of the Council of Foreign Ministers was convened in May 1949. The post-war relationship between the United States and Britain was an equally gradual evolution, coloured, but by no means determined, by the relationship built up between the two countries during the war—a relationship that had never been expressed in a written treaty of alliance.

The Anglo-American relationship has attracted many different epithets. By far the most famous, though it was an adjective from which both the governments of the day shied away at the time, is the word 'special'. This was coined by Churchill. In the speech which he delivered at Fulton, Missouri on 5 March 1946, he advocated:

a special relationship between the British Commonwealth and Empire and the

---

* It remains, however, to the credit of the US Government that the Smyth report was published. The British Government agreed only with reluctance.

† Joined by France and China in the Council of Foreign Ministers—an instrument agreed on by the Three at Potsdam.

United States ... the continuance of the intimate relationships between our military advisers, leading to the common study of potential dangers ... the continuance of the present facilities for mutual security by the joint use of all naval and air force bases in the possession of either country all over the world.[12]

In a later passage in this speech Churchill declared that 'from Stettin in the Baltic to Trieste in the Adriatic, an iron curtain has descended across the continent'. It says something about the state both of Anglo-American and of US-Soviet relations at that time that, although the US President was seated at Churchill's side (indeed they had travelled to Missouri together from Florida), the speech got a mixed reception even in the English-speaking press. The *Wall Street Journal*, though not itself unsympathetic to Churchill's approach, wrote that 'the country's reaction to Mr Churchill's Fulton speech must be convincing proof that the United States wants no alliance or anything that resembles an alliance with any other nation ...'.[13] In a speech delivered in New York on 16 March, the US Secretary of State said that the United States was no more interested in an alliance with Britain against the Soviet Union than in an alliance with the Soviet Union against Britain.[14] Stalin called Churchill's speech 'a dangerous act calculated to sow the seeds of discord among the Allied Governments and to make their cooperation difficult'.[15] And many years later Charles de Gaulle repaid a series of Anglo-Saxon injuries, when asked what he thought of the Anglo-American relationship; he replied: 'Existe-t-il?'[16]

Since Churchill's historic definition of the Anglo-American relationship as 'special', it has been called unique, natural, uneasy, ambiguous, ambivalent, and—most recently at the time of writing and perhaps most accurately—sweet and sour.[17] Common to all contemporary assessments, at whichever end of the spectrum, is the recognition of the fact that, as the asymmetry between the power of the two countries grew, for the United States the relationship with Britain became only one strand in its broader transatlantic relationship, whereas for Britain its relationship with the United States has on the whole—the Suez crisis in 1956 is the dramatic exception—remained the centre-piece of every post-war government's foreign policy. This said, there are two reasons why—now that the greater part of the evidence is publicly available—the Anglo-American relationship deserves a fresh look. The first is the one mentioned above: during 1945–50 the way in which the relationship between the United States and Great Britain evolved was a political factor not only of bilateral, but also of global, significance. The second reason lies in the searching questions put by a committed Atlanticist, Alastair Buchan at the beginning of an article written for the US Bicentennial Year just before his death and posthumously published in 1976:

Has this continuous and pervasive contact ... damaged or strengthened the two countries? Has the one society been able to prevent the other from making serious mistakes or to contribute to its learning process? Have the two countries lured each other into needless adventures, inspired a false sense of confidence in each other, distorted the other's perspective; or has the relationship been as benign as much Bicentennial oratory will no doubt maintain?[18]

In order to suggest answers, the final section of this book offers a summary account of the subsequent development of the Anglo-American relationship up to the end of 1962, followed by a balance sheet.

Thus the book has three themes: the evolution of the Anglo-American relationship from the beginning of 1945 to the end of 1950; its impact on world events of that period; and its consequences for the United States and for Britain in the longer term. On the cardinal point—that 1945–50 were the years by the end of which the post-war mould was set—there is general agreement.[19] But there can be no simple answers to the great range of questions that a study of these years throws up today. It has been well said that 'too often people still regard the world either as a closed field in which inspired heroes confront each other, or as the theatre of a disorder that is impossible to disentangle'.[20] We should not be misled by the fact that the chief contestants in the heroic conflicts, military and political, of the post-war quinquennium themselves saw the issues in black and white. It does not diminish the historical stature of those who resolved these issues by their decisions,[21] to recognize, as a dispassionate reassessment must now do, that these were years not only of confrontation, but also of confusion. The confrontation and the confusion were equally memorable. This is an attempt, without over-simplifying either, to illuminate both.

PART I

THE ANGLO-AMERICAN SETTING

CHAPTER 2

Attitudes, 1920–45

THE entire text of Churchill's speech at Fulton, Missouri—well over 4,000 words—was carried by the *St Louis Post-Dispatch* on 6 March 1946. At the top of the page devoted to the speech was an inset entitled 'Highlights of Address', thirteen in all. These did not include Churchill's proposal of 'a special relationship between the British Commonwealth and Empire and the United States', which he had defined in precise military terms and described in his speech as 'the crux of what I have travelled here to say'.[1] The reasons for this omission—the speech owes its fame not to Churchill's proposal of a special relationship, but to the passage defining the Iron Curtain—do not lie only in the political context of that time. They must also be sought in the erratic course followed by Anglo-American relations over the preceding twenty-five years.

In early 1946, barely six months after the end of the Second World War, anyone looking back at the previous quarter of a century with an objective view, from whichever side of the Atlantic, would have agreed with the conclusion reached by an historian long afterwards, that during the war the two countries' 'joint military effort was underpinned by a wholly unprecedented pooling of national resources'. He might also have accepted, as a legitimate hyperbole, the same writer's description of the wartime relationship as 'a genuine merging of national identities' for nearly four years.[2] What he would not have found credible was the creation of a continuous 'hands across the sea' image of the relationship between the two countries; this mid-Atlantic myth was to win widespread acceptance during the next decade and retained its roseate haze well into the 1960s.[3] In reality, measured against the background of 1920–45, the Anglo-American partnership developed during the Second World War, under the leadership of Franklin Roosevelt and Winston Churchill, had been atypical. It was also predominantly Eurocentric. In what Churchill in 1942 called 'those wild lands' of the Far East, the United States, and Britain were, at best, wartime allies of a kind.[4] Moreover, by the time that Churchill delivered his address to Westminster College at Fulton, he had been out of office for eight months and Roosevelt had been dead for nearly a year.

Today, what stands out from the inter-war relationship between the United States and Britain is how chancy it was; and how little each country then knew about the other—this in spite of their shared heritage of language, law, and racial stock (just over half the population of the United States in 1920 has been estimated to have been of British origin).[5] And in Woodrow Wilson's own view, had Germany accepted, and the Anglo-French Alliance rebuffed, his appeal for peace in December 1916, the United States would have found itself fighting on the opposite side in the following year.[6] The Anglo-American treaty guaranteeing the Franco-German frontier after the First World War was defeated by only ten votes in the US Senate. In December 1941, in spite of the remarkable 'common law' alliance already established between the neutral United States and belligerent Britain, it was only at the eleventh hour that the critical US commitment—to give armed support if the Japanese attacked the British or their Dutch allies—was given to the British Ambassador in Washington, and then in a Rooseveltian aside.[7] Even then it was the German Government who, once again, simplified the US decision, this time by declaring war on the United States in the wake of the Japanese attack on Pearl Harbor. What would have happened if, instead of attacking in the Pacific, Japan had acceded to German pressure and attacked the Soviet Union first? No one can say, although one thing is clear: the survival of the Soviet Union in 1941 owed much to Japanese neutrality.

The statesman cannot fairly be blamed by the historian for not being clairvoyant; chance plays its part in all history. None the less, the scale on which the Americans and the British left things to chance and hoped for the best in their relations with each other during the inter-war years can be fully explained only if the depth of their mutual ignorance is taken into account. The 1920s and 1930s were the years of American isolationism. But this was also a period during which Western Europe turned in on itself. The State Visit paid to the United States by King George VI, on the eve of the outbreak of war, was the first ever made to that country by a reigning British monarch.

At governmental level, British shilly-shallying in the face of the approaching storm in 1938–9 was enough to justify, in most American eyes, the United States' reluctance to become involved in Europe; and on 28 September 1938 the standing ovation that greeted Neville Chamberlain's announcement that he was leaving for Munich, was given to him by cheering members from all parties in the House of Commons.[8] In fact, Chamberlain's policy of appeasement divided the country. Nevertheless, when he remarked—in response to Roosevelt's initiative earlier in 1938—that it was always best and safest to count on nothing from the Americans but words,[9] he reflected the general view of his fellow-countrymen at that time.* Most of the British generation that entered university in that year took it for granted that, in the war that they saw approaching, Britain's principal ally would again be France, not the United States. Insular though most British members of Parliament were, they knew American politicians even less well than they did European. As for the press, there were only three British journalists resident in Washington in the late 1930s. The more numerous members of the American press corps based on London were there to cover Europe. (Some of them were later to become famous— above all, Edward Murrow, whose broadcasts from London during the Battle of Britain in 1940 were a significant factor in the change in American opinion about Britain that began in the second half of that year.) The dense network of academic links between the United States and Britain, taken for granted today, then lay far in the future. Scholarships and fellowships operating in both directions were only beginning to bear Anglo-American fruit.[10]

Of the armed services, only the two air forces knew each other well during the inter-war period, with significant results later on; the two navies for the most part disliked each other. In the two foreign services there were honourable exceptions to the general norm of mutual high-hatting; but even the wartime archives of the State Department and the Foreign Office—long after 1941—abound in memoranda and minutes on Anglo-American matters whose polished condescension does no credit to their writers on either side of the Atlantic.[11] As for political appointees, Joseph Kennedy's ambassadorial mission to London was an unhappy episode (it is his three successors—John Winant, Averell Harriman, and Lewis Douglas—who are remembered today). Nor did the incumbent of Lutyens' neo-Virginian mansion in Massachusetts Avenue always enjoy the Washington environment. Edward Halifax, Chamberlain's Foreign Secretary, whom Churchill appointed to succeed Philip Lothian as Ambassador after the latter's death in Washington, found Americans 'very crude and semi-educated';[12] he was fortunate in having the support of Ronald Campbell, a career diplomat whose perceptive comments on the Anglo-American scene are now receiving from historians the attention that they deserve.[13] Of Halifax's post-war successor, Archibald Inverchapel, a close colleague[14] wrote that 'he disliked Washington society, did not exert

---

* Summed up in the quip: 'those who live in White Houses should not throw stones'. However, by 14 May 1940 the Cambridge Union Debating Society had decided that 'neutrality is an anachronism in this war'; this motion was proposed and seconded by two visiting speakers from the Oxford Union—Nicholas Henderson and the author.

himself to become friends with leading Americans, and was bored by his surroundings and his work'. It was not until after the arrival of Oliver Franks at the Washington Embassy that a wholly different relationship developed between the US Secretary of State and the British Ambassador.*

So much for the official world. Commercial relations did not bring the two countries much closer. Although the flow of trade on the eve of the Second World War was substantial, it gave no hint of the massive increase that was to follow (the United States then accounted for 12.8 per cent of British imports and only 3.9 per cent of British exports).[15] Indeed, any search for continuously significant Anglo-American links in the inter-war years has to turn mainly to the world of the arts and of the professions. Two branches of the latter with a long transatlantic tradition were the law and finance. But the most important single transatlantic current of the period was that of show business. The picture that each people received of the other through the medium of theatre, cinema, and gramophone record,† though often inaccurate, was pervasive in a way that nothing else was; and paradoxically the most potent influence exercised on Britain across the Atlantic in those years was largely the work of two American ethnic minorities.

This inter-war state of Anglo-American affairs has been discussed at some length, because it is only by reference to the thinness of the texture of the inter-war relationship that it is possible to gauge both the intensity of the wartime partnership between the United States and Britain and the nature of the reaction that followed in the immediate aftermath of the war. 'So we had won after all!', wrote Churchill of the events of December 1941.[18] They dramatically transformed the war; and within it—for nearly four years—the Anglo-American relationship. This wartime partnership was both epitomized and reinforced by an extraordinary understanding between two extraordinary men: Roosevelt and Churchill. No war has ever been fought by allies whose leaders' correspondence[19] with each other casts so vivid a light both on the conduct of the alliance and on the national characters—the weaknesses as well as the strengths—of the two allies. In what Isaiah Berlin described as an *éloge*, written after Roosevelt's death, but at a time when Churchill was still living, he contrasted Roosevelt's 'half-conscious premonitory awareness of the coming shape of society' with

---

* Of this relationship, Dean Acheson wrote: 'Not long after becoming Secretary of State, I made him an unorthodox proposal ... that we talk regularly, and in complete personal confidence, about any international problems we saw arising. Neither would report or quote the other unless he got the other's consent ... for four years ... we met alone ... at the end of the day ... No one was informed even of the fact of the meeting.'[16]

† To take only one example, any doubt about the impact of American pre-war jazz on my generation is removed by a glance at Philip Larkin's *All What Jazz*.[17]

Churchill's 'strongest sense ... of the past'; and he concluded that 'something of the fundamental unlikeness between America and Europe, and perhaps between the twentieth century and the nineteenth, seemed to be crystallised in this remarkable interplay'.[20]

This unlikeness stood out with particular clarity in the Anglo-American wartime relationship in Asia, which was qualitatively different from the partnership between the two countries in Europe at the same time. In part this difference was political, stemming from sharp and persistent disagreements between the United States and Britain over their policies towards the two major Asian countries in the wartime alliance—China and India—and over their intentions about the future role to be played by them in the post-war world. It was also military. The global balance of the two allies' direction of the war as a whole was radically altered almost at once by an unforeseen military event in South-East Asia:[21] the surrender of Singapore. This disaster, on 15 February 1942, preceded by the loss of the two battleships, *Prince of Wales* and *Repulse*, was the greatest in the entire history of British arms.

The full dimensions of this double catastrophe were perhaps not generally grasped in Britain at the time, although Churchill himself realized what had happened and what it would mean.[22] The political consequences in Asia as a whole, in both the short and the long term, were far-reaching. Militarily, the way lay open for the war to be brought to the gates of India and of Australia. But a further result in practical military terms was that, until the Japanese surrender on board the battleship *Missouri* in Tokyo Bay, the direction of the war in the Pacific lay entirely in American hands.[23] In the China theatre of war, the British were present only on sufferance as late as July 1945.[24] South-East Asia Command was established in 1943 under a British Supreme Commander, Admiral Louis Mountbatten; one of Britain's finest generals, William Slim, conducted a successful campaign in Burma, which culminated in the recapture of Rangoon in May 1945; but the direction of the war in South-East Asia, 'although nominally a British responsibility ... became a combined* British-American effort'.[25]

In Asia as a whole this was predominantly an American war. In the overall logistic context the size of the American economy was bound

---

* A combination that at times produced a situation which, in the words of the British official history of *The War against Japan*, 'became almost Gilbertian. The Americans refused to agree to resources being directed for amphibious operations against Rangoon in order that every effort should be concentrated in opening land communications to China by way of North Burma, and the Chinese refused to cooperate unless there were simultaneous large-scale amphibious operations.'[26]

sooner or later to make the United States the senior partner in the war. This indeed is what happened. But in the European theatre of operations (which—together with the Middle East—had for most of the British loomed larger than the Asian war) the fact that Britain had become the junior partner did not become apparent until the war was nearing its end. Because they had been continuously at war for nearly six years, because for part of this time they had fought alone, and because for most of these years their gaze had been fixed on the English Channel and the Mediterranean, the British had no feeling of being left behind by the march of events. On the contrary, they believed that they had been at the centre of world history and that the world owed them a debt for what they had done.

All this said, beneath the drab nomenclature of the Anglo-American structure established at the turn of 1941/2—the Combined Chiefs of Staff, the Combined Munitions Assignment Board, the Combined Shipping Adjustment Board, the Combined Raw Materials Board, later supplemented by the Combined Food Board and the Combined Production and Resources Board—lay what General Marshall, the greatest staff officer of any country that took part in the Second World War, afterwards described as 'the most complete unification of military effort ever achieved by two Allied nations'.[27] This process was underpinned financially by Lend-Lease, in the secret intelligence field by the sharing of the Ultra product,[28] and in the secret scientific field (at any rate from August 1943 onwards) by the atomic interchange. It was reflected not only by the presence of nine thousand British officials in Washington and by vast inter-allied headquarters round the world, but also by the deep personal friendships that were then formed between individual American and British officers and civilians. One of the most important of these, that between Marshall himself and Field-Marshal Dill,[29] was cut short by Dill's death in Washington in November 1944. Many other wartime friendships—and not only those formed between senior commanders—were to become significant elements in the subsequent development of the post-war transatlantic relationship. Nevertheless, the very closeness of the two countries during the war, including these personal links, made the contrast of the closing months of 1945 even sharper. By the end of that year the transatlantic 'willing suspension of disbelief for the moment'[30] of the war was over.

CHAPTER 3

The Political Scene in the United States and Britain, 1945

It would have been surprising if, once the Second World War ended, there had not been a reaction away from wartime fusion and at least some degree of throw-back towards the Anglo-American attitudes of the inter-war years. The warmth of Churchill's last bilateral meeting with Roosevelt—the second Quebec conference held in 'a blaze of friendship' in September 1944, followed immediately afterwards by their meeting at Roosevelt's house, Hyde Park—soon evaporated.[1] At the turn of the year 1944/5 the personal understanding between Roosevelt, by now a sick man, sitting at the centre of world power, and Churchill, restlessly flying to and fro, was no longer what it had been three years earlier, at the time of Roosevelt's remark: 'it is fun to be in the same decade with you'. For Churchill, however, Roosevelt's re-election in November 1944, for a fourth presidential term, had been a profound relief; and he meant what he said when, after Roosevelt's death on 12 April 1945, he described him to the House of Commons as 'the greatest American friend we have ever known'.[2] It was also not a great exaggeration to describe the American response to the news of Churchill's electoral defeat three months later as 'an astonishment reminiscent of Pearl Harbour'.[3] The sudden and unforeseen caesura in each country's management of its internal affairs in 1945, first in the United States and then in Britain, certainly compounded the post-war reaction.

THE UNITED STATES

The White House was occupied by the Democratic Party for twenty years from 1933 onwards. Roosevelt's death at Warm Springs, Georgia brought an end to an unprecedented tenure of office. Thereafter there were changes of personality, of governmental style, and of policy. Gradually, not dramatically as in Britain after the Labour Party's sweeping victory in July, a new political climate was formed in the United States. The central thread running through this development was the composition of Congress—a thread partly masked during the war years by Roosevelt's personal dominance of the conduct of the war. In fact, even Roosevelt had been unable to get a major domestic bill through Congress since the mid-term elections of 1938 had given the Republican Party eight more seats in the Senate and doubled their strength in the House of Representatives. Four years later, though still a minority, the Republicans won their biggest electoral gains since the 1920s. Thereafter they were strong enough in the Senate to block any measure requiring two-thirds approval; and, taken together, an alignment of Republican and Southern Democrats assured a conservative majority in Congress as a whole. This was the Congress that the President had to deal with at the end of the war. A little over a year later, in the elections to the Eightieth Congress, the Republican Party was to win outright control of both Houses for the first time for eighteen years: 246–188 in the House of Representatives and 51–45 in the Senate.* From the outset, therefore, it was axiomatic that in the United States the President would have to secure a new, bipartisan consensus to support whatever post-war international settlement might be reached.

From 12 April 1945, the man who had to form this consensus in both domestic and foreign policy had no experience whatever of foreign affairs. Roosevelt had not even kept Harry Truman informed, let alone consulted him, on the conduct of foreign policy (at the age of sixty, Truman's only journey abroad had been to France, as an artillery officer in 1918). Even his nomination as vice-presidential candidate by the Democratic Party in 1944 owed more to negative than to positive reasons: he had few enemies; and for Roosevelt he was the running mate least likely to damage the ticket. Elected senator from a border state ten years earlier, he had become a quintessential middle of the roader in the Senate. 'A little left of centre' was how he described himself publicly just before he became President; to his diary a few weeks later he confided his belief that a 'professional' liberal was 'a low form of politician'; and in the early stage of his presidency it was not difficult to make fun of Truman—his poker, his bourbon, and his jaunts, for example. 'To err is Truman' was a Republican jibe.[4] And in Moscow he was called the *galantereishchik*—a reference to the failure in 1922 of the haberdashery which he and his friend Edward Jacobson had set up in Kansas City after the First World War.[5]

On the other hand, Truman brought to the White House some valuable assets. The very fact that his senatorial record had been uncontroversial was politically helpful in the circumstances of 1945. Moreover, in the Senate he had not only become a member of the

* Freshmen in the House of Representatives included John Kennedy and Richard Nixon. Joseph McCarthy was a freshman senator.

Military Affairs Committee: his chairmanship of the Special Committee set up in 1941 to investigate the National Defense Program had given him administrative experience and won him national respect; and it also brought him into close contact with the American whom he most admired, General Marshall. Few presidents have equalled Truman's passionate interest in history, ancient and modern. In his boyhood he had 'pored over Plutarch's *Lives* time and time again', as he recalled in his memoirs, the first volume of which was aptly entitled *Year of Decisions*; decisiveness was perhaps his chief personal characteristic, coupled with the ability to see complex problems in simple terms.[6] The other side of his personal coin, as the course of his presidency proved, was a tendency to over-simplify and to fire from the hip. But he took decisions and, once they had been taken, he accepted responsibility for them. The buck really did stop at the White House.

The contrast between Truman and Roosevelt was the subject of innumerable comments, American and foreign, at the time. So also was the change in the style of government. This applied particularly to the conduct of foreign relations (in which the State Department had played only a minor role during the war), although the foreign policy of the Truman Administration remained, broadly speaking, Rooseveltian throughout the rest of 1945, as will be seen in Part II of this book. Within three days of becoming President, Truman asked James Byrnes to take the place of Edward Stettinius, who had succeeded Cordell Hull as Secretary of State after his resignation in November 1944 (Roosevelt had not even bothered to take Hull with him to the Quebec Conference). Five years older than Truman, Byrnes did not know very much more about foreign affairs than the President, but he had been abroad; he had attended the Yalta Conference; he was a southern senator of long experience; and he had been Roosevelt's Director of War Mobilization. Indeed, he had been one of the two original front-runners for the vice-presidential nomination in the previous year; disappointed, he had resigned from his post in the Roosevelt Administration just before Roosevelt's death. He had also helped Truman in his climb up the greasy pole. Byrnes' appointment was not announced at once. When it was, the choice was applauded; but in the event it did not work.[7] This failure, however, would lead on to the successive appointment by Truman of two of the most eminent Secretaries of State in American history: Marshall and Acheson.*

With Capitol Hill in mind, Truman gave three other cabinet posts to

---

* Although Dean Acheson did not become Secretary of State until January 1949, his role as Under Secretary, first under Byrnes and then for six months under Marshall, was of great importance. Apart from eighteen months in 1947–8, when he returned to his law firm, his governmental career spanned the decade, and beyond, as did Marshall's.

members or former members of Congress, of whom the most important
was another conservative southerner, Fred Vinson. Elected to the
House of Representatives from Kentucky in 1924, at Bretton Woods in
1944 he had been Vice-Chairman of the US delegation; from the
Office of War Mobilization and Reconversion, he was appointed
Secretary of the Treasury in July 1945; and he later became Chief
Justice. By the standards of subsequent presidents, the number of
personal friends that Truman brought to Washington was small.
Perhaps the one closest to him was John Snyder, an even more
conservative banker from Arkansas, whom Truman had first met at
reserve officers' summer training in 1928. First appointed Federal Loan
Administrator, Snyder later followed Vinson's route to the Treasury.
Of the Rooseveltian New Dealers, soon only Harold Ickes remained as
Secretary of the Interior and Henry Wallace as Secretary of Commerce; both resigned in the following year. The Secretary of War,
Henry Stimson, an anglophile Republican, who had already served in
earlier Republican Administrations, both as Secretary of War and as
Secretary of State, and had been called in by Roosevelt in 1940, was
seventy-seven years old and by now a tired man. He had little in
common with the new President except that he too had served in the
field artillery in the First World War (he was usually addressed as
Colonel, rather than Henry). Nevertheless, he stayed at his post until
the autumn of 1945; and during this time he shared with Truman some
of his most critical decisions—those relating to the atomic bomb.

Such then were some of the leading figures of the US Administration, led by a man who had not been elected to the office which he now
held. On the Republican side, given the views of Robert Taft,[8]
Chairman of the Senate Republican Policy Committee, it was small
wonder that the Seventy-Ninth Congress rejected almost every piece of
social and economic legislation put forward by Truman. In the field of
foreign affairs, the speech in the Senate in which Arthur Vandenberg,
then the senior Republican member (later chairman) of the Senate
Foreign Relations Committee, renounced isolationism on 10 January
1945 was a major advance,[9] although he still had to be treated with kid
gloves by the Administration (as Byrnes was to learn, to his cost). What
was common to both sides of Congress was their opposition to 'Santa
Claus philosophy'. In the following month, two days before Truman
became President, it was only his vice-presidential casting vote that
broke a 39–39 tie in the Senate, thus narrowly defeating an amendment which would have prevented even the *sale* of Lend-Lease supplies
to any foreign government for purposes of post-war relief and
reconstruction.[10] One of the earliest pieces of advice that he received as
President came from Bernard Baruch: that the United States must

'resolutely resist British pleas for special consideration'.[11] Neither the depth of Baruch's suspicions nor the magnitude of Truman's task would have been lessened if they had read the sentence in a Foreign Office memorandum on 'The Essentials of an American policy', written in 1944, which suggested that it should be the British aim 'to make use of American power for purposes which we regard as good'.[12]

BRITAIN

Twelve days after the end of the war in Europe the Labour Party members of the wartime coalition government, formed by Churchill five years earlier, resigned. The first general election in Britain for ten years was held on 6 July 1945; Churchill formed a caretaker government in the interval. The spectacular results of this election were not known for nearly three weeks, mainly because of the need to collect servicemen's votes from abroad (Churchill, therefore, had to return to London in the middle of the Potsdam Conference, at which the Labour leader, Clement Attlee, had accompanied the British delegation, at Churchill's invitation). For the first time in British history, the Labour Party was then able to form a government backed by a parliamentary majority (393–222) beyond any conceivable challenge and with a clear popular mandate for a programme of reform; throughout the next six years it lost only one by-election. The defeat of Churchill's party in this election astounded many people at the time, especially in the United States, where the Labour Party's programme was perceived as radical (Stalin had also expected Churchill to win).[13] The electoral results were in the main caused by two related factors: a popular reaction, pent up by the six-year interval of the war, against the mistakes committed, largely under Conservative Party leadership, in both domestic and foreign policy during the 1930s, coupled with a widespread belief that the Labour Party, as well as being the party of full employment, had learned the lessons[14] of their disaster of 1931.*

On the face of it, in the summer of 1945 the domestic political situations of the United States and Britain seemed almost the direct opposites of each other. The United States had lost its wartime leader through death, Britain through the ballot-box. In Washington the inexperienced Democratic president, called to the White House by an act of fate, led a largely new team gingerly forward in the face of a conservative alignment on Capitol Hill. In London a Labour Government, many of whose ministers had held key offices of state for over

---

* Churchill himself had been out of office from 1929 until the outbreak of war. In 1931, under the pressure of the financial crisis, the Labour Prime Minister, Ramsay Macdonald, formed a 'National' Government, thus splitting his own party, most of whose members refused to follow him (and from then on were led by Attlee, in Opposition).

five years, embarked on an intensive programme of reform simultaneously with the task of leading their country in its transition from war to peace. Whereas Truman had hardly three years in which to prove himself before the onset of the next presidential election campaign, Attlee and his colleagues, secure in the House of Commons, could look forward to a full term* of power. By the turn of the year Truman's initial honeymoon period was over. In domestic political terms 1946 was to prove a bad year for him, whereas it was to be described as the *annus mirabilis* of the Labour Government. Both these early prospects proved deceptive. Truman surprised almost everyone by defeating Thomas Dewey in the election of November 1948, thus becoming President for the next four years in his own right.[15] Attlee's leadership ran into deep trouble in 1947; and although he remained at 10 Downing Street until October 1951, his government ran out of steam long before that; most of its reforms were enacted in its first three years.

This said, if Attlee had anything in common with the other world leaders of 1945, it was perhaps with Truman, of whom he himself said '... we talked the same language. We became friends'.[16] Whether or not they really became friends in a politically significant sense is open to question; except at Potsdam, they met only twice during the whole of their terms of office. Both of them were, however, outwardly unassuming men; both had fought in the First World War; both actively enjoyed the despatch of public business—what Attlee called 'polishing off'—and in each of their characters there was a strong didactic streak. Both men were to demonstrate exceptional powers of political resilience.

For its first two years the Labour Government was largely run by a triumvirate consisting of Clement Attlee, Ernest Bevin, and Herbert Morrison (Stafford Cripps joined this inner circle only after the financial crisis of 1947). All three men had been born in the late Victorian era and had in the early years of the twentieth century become the moderate socialists that they were to remain for the rest of their lives. Unlike his two colleagues, Attlee was—in every respect except his socialism—a typical member of the British middle class of his day. The years that he had spent among the poor of London—he became Mayor of Stepney in 1919—had altered neither his dress, as any photograph still shows, nor his speech, which retained the clipped, laconic idiom characteristic of a blimpish colonel in Graham Greene's novels. The public impression conveyed by Attlee was summed up by a British journalist as that of 'a great headmaster, controlled, efficient, and, above all, good'.[17] In his own Cabinet, Attlee's role remained,

* A maximum of five years. In the event Attlee went to the country for the first time in February 1950.

with few exceptions, remarkably like what it had been in Churchill's: that of a master of work in committee, a back-seat co-ordinator. In economic affairs—the area of policy that was soon to become crucial—he was out of his depth, regarding 'dollar trouble' almost as though it were a bout of influenza.[18] The leadership of the House of Commons, like the handling of the government's links with the Labour Party, rested in the hands of Morrison, one of the ablest public administrators of the period. The conduct of foreign policy Attlee largely delegated to Bevin, until—indeed somewhat beyond—the point where Bevin's ill health made this impracticable. The exceptions to this delegation were, however, extremely important: in particular, Attlee took personal responsibility for decisions in the field of atomic energy, including those regarding the atomic bomb, and for the conduct of the policy that gave independence to the Indian sub-continent.*

In the breadth of his vision and the intensity of his feelings, Bevin towered above his colleagues in 1945 (and when he made a mistake, it was liable to be on a correspondingly gigantic scale). Although Bevin and Churchill were in one way chalk and cheese (the former born, illegitimate, in rural poverty, the latter born in Blenheim Palace), the only British politician of comparable strength and stature was Churchill. Had Churchill fallen or failed during the war, Bevin was the only man who could have effectively succeeded him as Prime Minister.[20] Of the members of the post-war Labour triumvirate, he alone had a grasp of economic management. This ante-dated the 1940s by many years; Keynes had long ago (in the Macmillan Committee on Finance and Industry) observed Bevin's capacity for absorbing new economic ideas. But for a last-minute change of mind by Attlee, he would have become Chancellor of the Exchequer (a post which was given instead to Hugh Dalton).[21] In return perhaps for the free hand given him at the foreign Office, Bevin remained consistently loyal to Attlee—"little Clem'— even when other members of the Cabinet were urging him to take over as Prime Minister. His remark about Attlee, made at the critical juncture of September 1947, says it all: 'I love the little man. He is our Campbell-Bannerman'.[22]

Personalities apart, the British governmental style changed little in 1945. Broadly speaking, the same civil servants went on submitting advice to Labour ministers, many of whom were familiar to Whitehall from the coalition years, through the same machinery of government as before. There were, however, two important differences. The first

---

* In relation to the atomic project, Attlee's 'habit of working quietly, swiftly, without the slightest fuss and with the minimum of written direction, masks for a later generation his powers of decision and control'.[19]

was the proliferation of Cabinet committees, which was perhaps Attlee's main legacy to Whitehall: 157 standing committees and 305 *ad hoc* committees (of these one, GEN 163, was the very small committee, then secret, now famous, which took the final decision to manufacture the British atomic bomb). The smooth working of this bureaucratic system—in essence the system which British governments have operated in Whitehall ever since—has been the subject of pride in Britain and respect abroad, not least in Washington, where senior officials come and go with each change of Administration. Although, over the years, it became one of the most potent factors inhibiting change—whether from right or left—in British society, it was at that time an instrument geared to change.

The second difference related to the conduct of Britain's external relations in the broad sense of the term. In one way, with Bevin as Foreign Secretary, the Foreign Office came back into its own. In strong contrast to its experience under both Chamberlain and Churchill, it was now able to operate with the minimum of interference from 10 Downing Street (but with Attlee's backing in Cabinet or in Parliament). In another, even Bevin's immense personal authority—and he spent long periods abroad at conferences and meetings—was not enough to span the whole spectrum of external relations in the period now beginning. (Between April and December 1946, for example, he missed forty-six out of sixty-nine Cabinet meetings, through either absence abroad or illness.) Nor was a network of interlocking committees an adequate substitute. It was not only the fact that international economic relations, for which the responsible Department of State was the Treasury, would soon loom larger than ever before. Besides the Foreign Office itself, in 1945 there were three other Departments engaged in the conduct of what would today be called external* relations: the India Office, the Colonial Office, and the Dominions Office (renamed the Commonwealth Relations Office in 1947). To all three Attlee appointed, in 1945, elderly (and two of them minor) figures as Secretaries of State, all with seats in the Cabinet; in addition the Control Office for Germany and Austria (under the War Office until October 1945) was placed under the supervision of the Chancellor of the Duchy of Lancaster.[23] The three Departments of State had indeed existed before the war; two of them had a long history. Long before becoming Foreign Secretary Bevin had urged that the India Office should be abolished; that responsibility for India should be transferred to the Dominions Office; and that an India Department should be formed in the Foreign Office;[24] the Foreign Office had

---

* Or—in the phrase finally adopted in 1968—'Foreign and Commonwealth relations'.

expected to take over responsibility for relations with India after India became independent; and in 1948 the Foreign Office took, at official level, an unsuccessful initiative which Commonwealth Relations Office officials interpreted as a take-over bid.[25] The conceptual and bureaucratic weakness inherent in these divisions was to continue over a wide field for many years.*

In domestic policy, the post-war system of social security, taken as a whole, was the lineal descendant of the proposals first put forward in William Beveridge's *Report on Social Insurance and Allied Services*, published in November 1942. Beveridge was, like Keynes, a member of the Liberal Party. His 1942 report, prepared by an inter-departmental committee, was accepted by the coalition government. This report, together with his (privately produced) 1944 report on *Full Employment in a Free Society* and the Education Act introduced in the same year (by a Conservative Minister), was inherited by the post-war Labour Government.[27] Within this framework the greatest single achievement of the Labour Government of 1945 was, in the eyes of most of my generation at the time, the establishment of the National Health Service: the one exception, for which Aneurin Bevan was the responsible Minister, where Britain took the lead. Otherwise, the post-war social reforms seemed to most of us then to do little more than bring Britain's practice into line with that of several other advanced industrial countries.

In the economic field, the legislative proposals which the Government presented to Parliament were those listed in its electoral manifesto, *Let us face the future*. With one exception—the nationalization of the iron and steel industry, added to the list only in April 1945 at the Labour Party Conference and the subject of long debate within the Cabinet two years later—this legislation was subsequently accepted by all political parties in Britain. When the Conservative Party returned to power in October 1951—and they held it for thirteen years thereafter—they made no attempt to reverse the other measures: the nationalization of the Bank of England, the coal mines, the railways, electricity, and gas. Over thirty years were to pass before the demarcation of the boundaries between the public and the private sectors of the British economy became the subject of serious political debate. Finally, by a strange paradox, much as the Labour Government talked about central planning, in practice it pursued economic policies far less *dirigiste* than those of the French Government of the same period; instead it relied essentially on the maintenance in being of the British wartime system of controls.[28]

* One of the earliest examples, which was later to have tragic results, was the disagreement in Whitehall over Cyprus in 1945–7.[26]

Foreign affairs having played little part in the general election, there was none of the clarity in the new government's foreign policy in July 1945 that there was in its domestic programme. Within the Labour Party there was a wide spectrum of views,[29] but support for the concept of an independent line in foreign policy was not confined to the Labour Party. For *The Economist*, writing two weeks after the general election, Churchill's 'policy of deference towards the United States' had 'greatly hindered any effort to evolve a separate British approach' to world affairs.[30] Perhaps the best general guide to the Labour Party's broad view of Britain's role in the post-war world is a passage in a book written soon after the war by a socialist novelist:

If Britain is to survive as what is called a 'great' nation, playing an important and useful part in the world's affairs, one must take certain things as assured ... that Britain will remain on good terms with Russia and Europe, will keep its special links with America and the Dominions, and will solve the problem of India in some amicable fashion ... If the savage international struggle of the last twenty years continues, there will only be room in the world for two or three great powers, and in the long run Britain will not be one of them.[31]

Nearly forty years afterwards, some phrases in this passage may provoke a smile: the bracketing of the United States with the British (pre-war) Dominions and the way in which the Soviet future super-power is lumped together with 'Europe'. Nevertheless, the gist of these words was prescient. Their author was George Orwell; the book was published two years after *Animal Farm*. Bevin (who himself despised intellectuals) simply said, in a sentence attributed to him by a witness of his arrival in July 1945 at Berlin airport: 'I am not going to have Britain barged about'.[32]

THE 'THIRD MONROE'

Had this book been written ten years ago, there would not have been much that could have been usefully added, based on documentary evidence, in order to show how Bevin saw the British position in the world and the part that Britain should play in it, once the first flush of Labour's electoral success was over. A comparative newcomer to Parliament,[33] he was seldom at his best in the House of Commons. His speeches on foreign policy delivered in the House are, therefore, not the best quarry for the historian who does better to seek Bevin's views from other sources. Later in his term as Foreign Secretary, it was said of him that 'the more interesting Mr Bevin is, the more obscure and difficult he becomes'.[34] By good fortune, however, a Top Secret memorandum by the Foreign Secretary entitled 'The Foreign Situation' is now available in the Public Record Office, which was almost certainly drafted by the Foreign Secretary himself;[35] no one else in the Foreign Office would have used phrases like 'three Monroes' or 'this sphere of influence business' in a formal document. This memorandum was probably not circulated to Bevin's Cabinet colleagues—indeed it seems more likely to have been intended as a sounding-board for senior members of the Foreign Office—and it was filed among his private papers. Dated 8 November 1945, less than two months before the inaugural session of the United Nations General Assembly (convened at Church House, Westminster on 1 January 1946), and with a perspective markedly different from the Churchillian vision that was to be portrayed in the Fulton, Missouri speech four months later, this summary of Bevin's own view of the world early in his term of office is so illuminating that it warrants quotation at length:

Instead of world cooperation we are rapidly drifting into spheres of influence or what can be better described as three great Monroes. The United States have long held, with our support, to the Monroe Doctrine for the western hemisphere, and there is no doubt now that notwithstanding all the prot- estations that have been made* they are attempting to extend this principle financially and economically to the Far East to include China and Japan, while the Russians seem to me to have made up their mind that their sphere is going to be from Lubeck to the Adriatic in the west and to Port Arthur in the east. Britain therefore stands between the two with the western world all divided up ... If this sphere of influence business does develop it will leave us and France on the outer circle of Europe with our friends, such as Italy, Greece, Turkey, the Middle East, our Dominions and India, and our colonial empire in Africa: a tremendous area to defend and a responsibility that, if it does develop, would make our position extremely difficult. It will be realised that the continental side of this western sphere would also be influenced and to a very large extent dominated by the colossal military power of Russia and by her political power which she can bring to bear through the Communist parties in the various countries. Meanwhile, France would stand in a kind of intermediate position, balancing herself against the east and out of sheer necessity resting upon us. The future too of the German people is going to be a constant source of insecurity, and every sort of political trick will be resorted to in order to control or eliminate this eventual reservoir of power. The French demands on the Ruhr and the Rhineland, and Russia's action in transferring Eastern Germany to Poland are already examples of this tendency, and when the German people recover consciousness we may be sure they themselves will soon be playing an active part in these highly dangerous manoeuvres.

After drawing a comparison between Soviet use of military power to

---

* Byrnes did indeed interject six weeks later in Moscow, when Bevin described the US Monroe Doctrine as being 'extended to the Pacific'.[36]

incorporate Eastern Europe into the Soviet economy and American use of financial and economic power to maintain US preserves in the Far East and South America, Bevin deduced that the British Government was 'dealing with power politics naked and unashamed'. He therefore considered it vital not to deceive the people of the world by leading them to believe that a United Nations Organization was being created that would protect them from future world wars. There were, he went on,

at least two mighty countries in the world which, by the very nature of things, are following the present policy which is bound to see them line up against each other, while we in Great Britain who have had the brunt of two great wars will be left to take sides either with one or the other.

Bevin's conclusion, in the final paragraph of his memorandum, was that, with the present deadlock between the Soviet Union, the United States, and Britain, not very much could be accomplished. The only safe course for Britain was therefore

to rely on our right to maintain the security of the British Commonwealth on the same terms as other countries are maintaining theirs, and to develop, within the conception of the United Nations, good relations with our near neighbours in the same way as the United States have developed their relations on the continent of America.[37]

It was not only Churchill who would have disagreed with much of Bevin's analysis. Halifax, having read a copy sent to the Washington Embassy, replied with a letter dissenting from his assessment of US policy; indeed he even counter-attacked Bevin over British policy in the Middle East.[38] But Bevin's concept of the 'third Monroe' was not a passing fancy. Three months later he circulated to members of the Cabinet Defence Committee a memorandum which included the following sentence:

In the European scene . . . we are the last bastion of social democracy . . . this now represents our way of life as against the red tooth and claw of American capitalism and the Communist dictatorship of Soviet Russia.[39]

Bevin was his own man. While he was at the Foreign Office, British foreign policy was 'my policy'. This said, a clear thread links Bevin's concept of Britain as the third centre of world power, uncomfortably placed between the other two, and some ideas which his predecessor at the Foreign Office had approved two weeks before he left it. On 12 July 1945 Eden had minuted that he agreed 'wholeheartedly' with a memorandum called by its author, Orme Sargent (who soon afterwards became Permanent Under-Secretary at the Foreign Office),

'Stocktaking after VE-Day'.[40] Sargent's point of departure was the feeling

in the minds of our two big partners, especially in that of the United States ... that Great Britain is now a secondary Power and can be treated as such, and that in the long run all will be well if they—the United States and the Soviet Union—as the two supreme World Powers of the future, understand one another. It is this misconception which it must be our policy to combat.

There are parts of this memorandum that date from an older school of diplomacy, of which Sargent was one of the few surviving members in the senior ranks of the post-war Foreign Office:[41] for example, the idea that the governments of the Dominions could be 'enrolled' by the British Government and the absence of any sign of awareness of Britain's economic position, to which, in a gentle minute, Ronald Campbell—now back from Washington—later drew Sargent's attention.[42] Judging by a minute that Sargent wrote on 1 October 1945 he subsequently took both these points. Commenting on a long despatch from the Washington Embassy, addressing the question of the steady increase in the American belief that Great Britain had come to occupy a position inferior to that of the United States and the Soviet Union, he wrote:

This despatch describes the position of Lepidus in the triumvirate with Mark Antony and Augustus ... it behoves us all the more to strengthen our own world position *vis-à-vis* of our two great allied rivals by building ourselves as *the* great European Power. This brings us back to the policy of collaboration with France with a view of our two countries establishing themselves politically as the leaders of all the Western European Powers ... Once we have acquired this position both the United States and the Soviet Union are more likely to respect us and therefore collaborate with us than they are at present.[43]

Had Bevin read Sargent's memorandum* of 11 July, he would also have approved of its robustness, as Eden had done at the time. Indeed, Bevin's own remark in Berlin about not having Britain 'barged about' has much in common with the general thrust of the conclusion reached by Sargent a few days earlier:

We must not be afraid of having a policy independent of our two great partners and not submit to a line of action dictated to us by either Russia or the United States, just because of their superior power or because we despair of being able to maintain ourselves without US support in Europe.[44]

Part II of this book will examine to what extent the British position of Lepidus was sustained and how the United States interpreted the

* Under Whitehall conventions, officials do not submit to the Ministers of an incoming government memoranda prepared for members of the preceding government.

role of Augustus in the trilateral framework, within which the Anglo-American relationship was operating throughout 1945: a year during which there were four meetings of the Grand Alliance—at Yalta, Potsdam, London, and Moscow.*

* Or five, if the United Nations founding conference held at San Francisco is included in the series.

PART II

THE UNITED STATES AND BRITAIN IN THE TRILATERAL FRAMEWORK, 1945

CHAPTER 4

Yalta

THE Yalta Conference (officially called the Crimean Conference) was the first step in a concerted attempt by the three wartime allies to establish the post-war political structure. Today, it is most remembered because of the subsequent history of Eastern Europe, and above all, of Poland. The fact that at Yalta the three leaders exchanged no fewer than eighteen thousand words about Poland, which was discussed at seven of the eight plenary meetings, and that the compromise agreement finally reached on the last day regarding the Polish question was soon to become the focus of inter-allied recrimination, has tended to obscure the other issues, of world-wide scope, which the conference addressed. Moreover, neither Yalta nor the trilateral meetings that followed in the rest of 1945 can be understood except against the backdrop of the kaleidoscopic series of events that accompanied the end of the Second World War* (which, like the end of the First World War, came much earlier than expected). The military context of the Yalta Conference now looks simple enough: the meeting, which lasted from 4–11 February, took place, as it turned out, only three months before VE Day, four months before the first atomic explosion at Alamogordo, New Mexico, and six months before VJ Day. It looked very different at the time. In Western Europe, General Eisenhower's forces had not yet crossed the Rhine (indeed it was only seven weeks since the German counter-offensive had taken them by surprise in the Ardennes). In North Italy, Field-Marshal Alexander's depleted forces were bogged down.[1] In the Pacific, US forces were about to land on Iwojima; the bitter battle for Okinawa lay ahead; and the final assult on the Japanese mainland was expected to cost massive American casualties.[2] In Central Europe, Marshal Zhukov's forces, having captured Warsaw on 17 January, had reached the Oder; the western-most tip of his salient was, at the moment when the Yalta Conference began, thirty-five miles from Berlin.

One other battle formed a critical part of the political backdrop to the scene at Yalta, although it had been fought—and lost—six months earlier: the Warsaw Rising. Of the many capital cities entered by allied troops in 1945, Warsaw was the only one that was not only in ruins but

* A chronology of some of the principal events of 1945 will be found at Appendix 1.

virtually deserted. On 1 August 1944, the day on which Soviet troops reached the outskirts of the city and the *Armja Krajowa** rose against the Germans, the population of Warsaw numbered almost a million. By the time the *AK* capitulated, on 2 October, nearly 200,000 Poles had been killed by the German army, who removed the survivors from the city, which was systematically destroyed; 150,000 were deported for forced labour. What happened during the two traumatic months for which Operation *Burza* lasted, lies outside the timespan of this book, but three points must be noted here. First, although the military object of the operation was to attack the German army in Warsaw, it was also aimed politically against the Soviet Union; the list of objectives sent to London on 22 July 1944 by Tadeusz Bór-Komorowski, the *AK* commander, leaves no room for doubt about this.[3] For Stalin, therefore, in Warsaw as in the rest of Poland, the *AK* was 'a group of criminals'.[4] Secondly, once the rising had failed, the exiled Polish Government in London no longer commanded the allegiance of armed forces capable of exercising a direct influence on the outcome of events in Poland, unlike the Soviet-backed Polish Committee of National Liberation—the 'Lublin Committee'—which the Soviet Government recognized as the Provisional Government of the Polish Republic on 1 January 1945. Finally, because the exiled Government was in London and because over 150,000 'London' Polish troops were fighting, with great distinction, under British command in Italy and in Western Europe, it was on the British Government, whose declaration of war in 1939 had been a response to the German invasion of Poland, that fell the primary obligation to try to rescue what could still be saved from the wreckage of the Second Polish Republic.

THE APPROACH TO THE CONFERENCE

The run-up to Yalta illustrates the fundamental unlikeness to which reference has been made in Chapter 2. To defer decisions until the latest possible moment was one of Roosevelt's characteristics. At the very end of 1944 Roosevelt told Stalin that he was 'more than ever convinced that when the three of us get together we can reach a solution of the Polish problem', whereas ten days later Churchill described the forthcoming meeting to Roosevelt as 'a fateful conference, coming at a moment when the great allies are so divided'.[5] Roosevelt had set his sights on other targets: above all, Soviet

* The *AK* was the Home Army, the Polish Resistance, under the orders of the exiled Polish Government in London. Diplomatic relations between this government and the Soviet Government had been broken off in 1943, following the discovery of the Katyn massacre (of Polish officers, carried out by the NKVD, but ascribed to the Germans by the Soviet Government).

participation in the war against Japan and Soviet agreement on the outstanding issues related to the establishment of the United Nations Organization. For his Chief of Staff in the White House, Admiral Leahy, the essential condition for the prevention of another great war was that 'Britain and Russia cooperate and collaborate in the interests of peace'; the writer of the State Department briefing book paper on liberated countries, prepared for the Yalta Conference, saw 'growing evidence of Anglo-Soviet rivalry on the continent of Europe'; and as late as 17 November 1944 it was declared US policy that the question of Poland's future frontiers was a matter for agreement between the governments of Poland, the Soviet Union, and Britain. On that date, one month after Churchill had exerted 'dire threats'[6] to get the Polish Prime Minister, Stanisław Mikołajczyk, and his colleagues to come from London to Moscow and, once they had arrived, urged them to accept the Curzon Line as Poland's eastern frontier, Roosevelt wrote formally to Mikołajczyk* saying that if the Polish, Soviet, and British governments were to reach a mutual agreement on Poland's frontiers, the US Government 'would offer no objection'.[7]

Quite apart from this essential difference of perspective, the Anglo-American relationship entered the year 1945 uneasily. In December 1944 the two governments had allowed themselves the luxury of a public row over an Italian political crisis.[8] By the end of the month they had patched up their Italian altercation, but much mud was slung across the Atlantic by the press of both countries: a process not helped by an ill-judged press conference given by Field-Marshal Montgomery, on 7 January 1945,[9] immediately after the shock of the Ardennes battle. The new year was brought in by *The Economist* with a leader which began by describing the current American view of the British as spending 'half their time imitating Lord North and the other half of their time aping Dr Schacht' and concluded:

But let an end be put to the policy of appeasement which, at Mr Churchill's personal bidding, has been followed, with all the humiliation and abasements it has brought in its train, ever since Pearl Harbour removed the need for it. Henceforward if British policies and precautions are to be traded against American promises, the only safe terms are cash and delivery. And, if Americans find this attitude too cynical or superior, they should draw the conclusion that they have twisted the lion's tail just once too often.

The Economist's leader was taken with sufficient seriousness in Washington for the Secretary of State to send a memorandum about it to the

---

* The Curzon Line, put forward by the British Delegation at the 1919 Peace Conference, is shown on the map on p. 41. Mikołajczyk, unable to convince his London colleagues on his return from Moscow, resigned on 24 November 1944. Most of the Polish troops fighting in Italy came from the region of Poland lying to the east of the Curzon Line.

President on 2 January 1945. The article, he wrote,

represents what is in the minds of millions of Englishmen ... the underlying
cause is the emotional difficulty which anyone, and especially any Englishman, has in adjusting himself to a secondary role after having always accepted
a leading one as his national right.[10]

In his message to Roosevelt on 1 January, Churchill's doggerel—'No
more let us palter, from Malta to Yalta, let nobody alter'—reflected
the fact that he had had to press the President very hard indeed in
order to get him to agree to any Anglo-American meetings on the way
to Yalta at all.[11] As late as 6 January Roosevelt was telling Churchill
flatly that it would not be possible to have an Anglo-American staff
meeting at Malta, as Churchill had suggested.[12] In the end both the
Chiefs of Staff of both countries and their two Secretaries of State,
Edward Stettinius and Anthony Eden, did meet in Malta for the last
days of January; and on the morning of 2 February the President
arrived in the U.S.S. *Quincy*, in time for a lunch and one meeting with
the Prime Minister before they all took off by air for the Crimea after
dinner. Thus honour was satisfied, but the Malta Conference was not a
happy Anglo-American occasion. Politically, it was Hamlet without
the Prince of Denmark until 2 February. Militarily, the combined
Chiefs of Staff spent their time arguing about Eisenhower's plan for
the forthcoming Rhineland battle.[13] For the Yalta issues themselves
there was little Anglo-American preparation.

THE CONFERENCE

About seven hundred American and British officials arrived in mid-
winter to spend just over a week in the Crimea. (Roosevelt had thought
of spending not more than four or five days; and Churchill had
remarked to Hopkins beforehand: 'if we had spent ten years on
research, we could not have found a worse place in the world'.* Yet the
words of the final toasts exchanged at Yalta and what was written[15]
and said immediately after the conference bear out the sanguine
reports that Churchill submitted to the House of Commons on 27
February and Roosevelt to a joint session of the two Houses of
Congress four days later. Even the extremes of Hopkins' euphoria—
'the Russians have given in so much at this conference' and 'we really
believed ... that this was the dawn of the new day'—had some

---

* The places canvassed as venue in the messages exchanged before the conference included,
among others, Alaska, Scotland, Athens, Salonika, Cyprus, Constantinople, Jerusalem, Alexandria, Rome, Taormina, and Malta. It was Stalin, pleading medical advice, who obliged Roosevelt
and Churchill to come to Yalta, on which agreement was finally reached at the very end of
December 1944.[14]

foundation: both the United States and Britain achieved major aims at Yalta. Indeed, the only reservation recalled by Hopkins afterwards was that 'we could not foretell what the results would be if anything should happen to Stalin'.[16] This fantasy—Stalin the prisoner of the hawks in the Politburo—was a belief that was not confined to Hopkins or to the United States; it persisted long after the war had come to an end.

Of the two central American objectives, that relating to the United Nations Organization was reached early in the conference. On 7 February Vyacheslav Molotov announced that the Soviet Union would not, after all, insist that permanent members of the Security Council should have the right of veto on procedural questions; and the number of Soviet republics to become members of the organization was reduced from sixteen to 'at least two'; in Roosevelt's words, 'a great step forward', which would bring 'joy and relief to the peoples of the world'.[17] The timing of these two Soviet moves was tactically well judged, in the Soviet interest. But the convening of the United Nations Conference at San Francisco decided at Yalta was not simply an act of American idealism. Given the lesson of 1919, the United Nations Organization was the instrument through which bipartisan Congressional support for US post-war foreign policy was to be secured. The speed with which the names of the US delegation to be invited to go to San Francisco, including Vandenberg, were agreed at Yalta (for a press release to be issued by the White House on the day after the conference communiqué), is telling evidence of the significance attached to this instrument by Roosevelt and his advisers, which was rooted in American domestic politics.[18]

So far as the United States and the Soviet Union were concerned, the core of the conference was the agreement on the Far East, signed on the last day of the conference.[19] Although this agreement was tripartite on paper, it was in fact a bilateral US-Soviet deal. The country vitally affected—China—was not consulted about its terms; indeed the Chinese Government was not even aware of the existence of the agreement at the time. This secret Yalta Agreement committed the Soviet Union to enter the war against Japan 'in two or three months after Germany has surrendered'. In return three conditions were to be 'unquestionably fulfilled'. The status quo in Outer Mongolia was to be preserved; 'the former rights of Russia violated by the treacherous attack of Japan in 1904'* were to be restored; and the Kurile islands

---

* The pre-1904 Russian rights in the Far East recovered at Yalta by the Soviet Union were, first, the southern part of Sakhalin, as well as all the islands adjacent to it; and secondly, as further interpreted in the 1945 Sino-Soviet Treaty, a Soviet lease, free of charge, of half of the facilities of the port of Dairen (the port being intended to pass under Chinese administration), joint Sino-Soviet use of the Port Arthur naval base, and joint ownership of the Chinese Eastern and South Manchurian Railway—all three for thirty years.

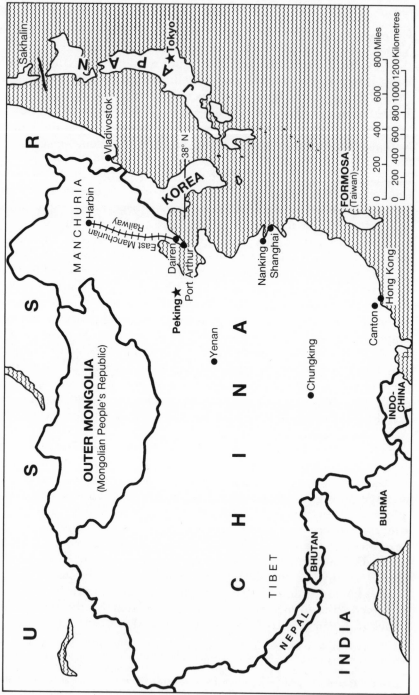

Map 1. East Asia, 1945

were to be handed over to the Soviet Union. A further clause of the agreement recognized that Chiang Kai-shek's concurrence in the fulfilment of the conditions relating to Outer Mongolia and Manchuria was required; the US President was to 'take measures to obtain this concurrence on advice from Marshal Stalin'. Finally, the Soviet Union expressed its readiness to conclude a pact of friendship and alliance with the National Government of China.

In the Sino-Soviet negotiations that followed six months later, the phrase 'status quo', in relation to Outer Mongolia, which for the Soviet Union meant independence, became the chief obstacle to agreement, because the Chinese Government was reluctant to admit that, the Sino-Soviet Treaty of 1924 notwithstanding, the Chinese writ had ceased to run in that territory for over twenty years. In the end, under the terms of the new Sino-Soviet Treaty signed on 14 August 1945,[20] China agreed to recognize the independence of Outer Mongolia (today the Mongolian People's Republic) if a plebiscite after the defeat of Japan confirmed that this was what the people wanted, while the Soviet Union recognized Chinese sovereignty in Manchuria. The first article of this treaty committed Soviet 'moral support and aid in military supplies and other material resources' to be 'entirely given to the National Government as the central government of China'. Ironically, at the turn of 1949/50, after his victory in the Chinese civil war, Mao was to spend two months in Moscow negotiating the reversal of all the concessions made at Yalta at China's expense, with the exception of the independence of Outer Mongolia; and the final renegotiation was completed between Mao and Nikita Khrushchev in Peking (Beijing) in 1955.[21]

It was not until 15 June 1945 that the US Ambassador informed Chiang Kai-shek of the provisions of the Yalta Agreement on the Far East, by which time 'it was apparent that the Russians had already made the Yalta Agreement known to him'.[22] However, once its terms became publicly known in August, the Far Eastern part of Roosevelt's grand design received general applause on both sides of the Atlantic. The Sino-Soviet Treaty was seen as a victory for Chiang Kai-shek (who was probably less shocked by the Yalta Agreement than Mao was). Four years were to pass before the State Department publicly recorded that the failure to consult China in 1945 had been 'unfortunate'.[23]

Although Britain was one of the three signatories of the secret Yalta Agreement, Churchill was not consulted about its terms by the other two leaders either. He overruled Eden's and Alexander Cadogan's advice not to endorse the agreement, on the ground that his signature was necessary in order to preserve an active role for Britain in Far Eastern affairs. In his memoirs he later wrote that 'to us the problem was remote and secondary'.[24] Overwhelmed by the tide of the Pacific war, Britain had virtually abdicated its East Asian position to the United States.

Churchill's single positive success at Yalta related to Europe, with significant consequences after the war ended. Helped by Eden, he argued strenuously, against Stalin and Roosevelt, that France should both be given a zone of occupation in Germany and be made a member of the Allied Control Commission. At Roosevelt's first bilateral meeting with Stalin, he told him that the British wanted 'to have their cake and eat it' over France; he himself thought that a French occupation was 'not a bad idea ... but ... only out of kindness'.[25] On 10 February, however, Roosevelt told his two colleagues that he had changed his mind, thus paving the way for agreement on this issue.[26]

Churchill also achieved one negative success at Yalta: by his Fabian tactics he blocked the proposal for the 'dismemberment' of Germany, which it was agreed to refer to a committee under Eden's chairmanship. On the issue of German reparations, however, which was to exercise such a powerful influence at future allied meetings, Churchill could not prevent the protocol of the proceedings of the conference from recording the fact that the Soviet and US delegations had agreed that the Tripartite Reparations Commission, established in Moscow at the conference, should, 'in its initial studies' take 'as a basis for discussion' the Soviet suggestion that the total amount of reparations should be US$20 bn., of which half should go to the Soviet Union. But the Protocol did record the British view that no figure should be mentioned pending consideration by the Commission.[27] This debate was to be resumed at Potsdam.

The memory of Yalta has been immortalized by the photograph of the three leaders taken on 10 February 1945 in the patio of the Livadia Palace: Churchill smoking his cigar, Roosevelt haggard, a cigarette in his left hand, Stalin composed, with hands folded. Forty years on, the very word 'Yalta' has a sinister ring,* with the connotation either that the Polish agreement reached there was a dishonourable deal consciously concluded with Stalin by the two western leaders, or that Stalin somehow deceived them—either way, at Poland's expense. It is true that the language used in public by Roosevelt and Churchill immediately afterwards was coloured by the continuing strength, at that time, of pro-Russian feeling in both their countries. But neither

---

* A further tragic dimension was added thirty years later, when the full story of wartime and post-war repatriation to the Soviet Union was published. As Chapter 4 of Nikolai Tolstoy's book, *Victims of Yalta* (Hodder and Stoughton, London, 1977) makes clear, however, the Yalta agreement on prisoners of war was a 'ratification' of what the three governments had already agreed in the previous year. The first shiploads of Soviet citizens had left both the United States and Britain before the end of 1944.

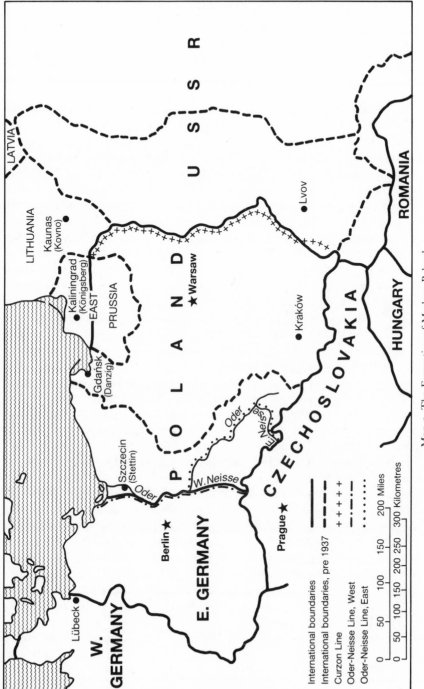

Map 2. The Formation of Modern Poland

Roosevelt nor Churchill deliberately sold Poland down the river in February 1945. (If any country was cheated at Yalta, it was surely China.) Nor was there any deception. Rather, their long negotiation with Stalin about Poland was conducted in earnest, but in a vacuum.[28]

After a week of argument, the agreement reached on Poland was this.[29] A Polish Provisional Government of National Unity should be established, 'which can be more broadly based than was possible before the recent liberation of Poland'. The provisional government functioning in Poland 'should therefore be reorganised on a broader democratic basis with the inclusion of democratic leaders from Poland itself and from Poles abroad'. The Soviet Foreign Minister and the US and British Ambassadors were 'authorised as a Commission to consult in the first instance in Moscow with members of the present Provisional Government and with other Polish democratic leaders from within Poland and from abroad, with a view to the reorganisation of the present government along the above lines'. 'Free and unfettered' elections were to follow as soon as possible 'on the basis of universal suffrage and secret ballot'; all 'democratic and anti-Nazi parties' were to have the right to take part and put forward candidates.

As for the frontiers of this new Poland, after Roosevelt had put in a muted plea for Polish retention of Lvov (the capital of Galicia), it was agreed that the eastern frontier should follow the Curzon line, with minor digressions. In the north and west, Poland 'must receive substantial accessions of territory'; the 'final delimitation' of the western frontier should await the peace conference. Although the contraction of eastern Poland to the Curzon Line had been inevitable from the Tehran Conference onwards, the Yalta phraseology about the western frontier papered over disagreement on the Soviet proposal that this frontier should follow the river Oder and the western river Neisse, thus incorporating in the new Polish state the whole of Pomerania and Silesia: a diplomatic battle that was to be fought for the second time at Potsdam. As we now know, however, over six months before the Yalta Conference Molotov had signed a Memorandum of Understanding with the Head of the 'Lublin Committee', Article 4[30] of which committed the Soviet Government to 'support the establishment of the Soviet–German frontier' along the line of the Oder–western Neisse;[31] to the east, both parties to this agreement accepted the Curzon Line.*

Whatever Churchill may have said and written subsequently about

* Although Polish sources offer some evidence for the belief that, at a still earlier stage, even Stalin was confused between the western and eastern Neisse, the territorial solution for Poland which he finally secured is a recognizable descendant of that proposed by the Tsarist Foreign Minister in 1914.[32]

Poland in particular and Eastern Europe generally, in private he does not seem to have had many illusions; and Milovan Djilas' recollection of what Stalin said to him two months after the Yalta conference has the ring of truth: 'this war is not as in the past; whoever occupies a territory also imposes on it his own social system. Everyone imposes his own system as far as his army has power to do so. It cannot be otherwise.'[33] Moreover, at Yalta itself Stalin's words on 6 February had been unmistakable. 'The Polish question is a matter of life and death for the Soviet Union' and 'No, better let the war with the Germans continue a little longer still, but we must be in a position to compensate Poland in the west at Germany's expense.'[34]

If it seems to us now that it might have been better if the hard facts of the Polish issue had been squarely faced by the Americans and the British at Yalta, it is also true that only if they had concerted firm Polish policies before they reached Yalta,[35] could they have modified Soviet policy over Poland. Failing that condition, the conclusion was foregone. And, as we have seen at the beginning of this chapter, the Anglo-American relationship was then in no state to fulfil that condition. Instead, at a time when the whole of Poland was occupied by the Red Army and when even the Polish Communist Party leadership in Poland was under no illusion about the real source of its authority,[36] the idea of Molotov sitting down with Harriman and Clark Kerr, as members of a triumvirate formed in order to decide which Polish leader was 'democratic' and which was not, belonged to Cloud-cuckoo-land from the outset. The agreement reached on Poland at Yalta lasted little more than a month. As will be seen in the next chapter, it was renegotiated—this time through a bilateral US-Soviet deal—at the end of May 1945.

Yet there remains one sonorous* document agreed at Yalta—the Joint Declaration on Liberated Europe[37]—which, in the light not only of the immediate sequence of events in Poland in 1945, but also of the horrors that were to follow in Eastern Europe during the final period of the Stalinist era, still provokes nausea across the years. This declaration, often confused with the Yalta Agreement on Poland, was not improved by the fact that its original drafters in the State Department almost certainly conceived of it as aimed as much at events in Greece as at those in Eastern Europe.[38] There were, however, those on the

* Under the terms of the declaration, in order to 'foster the conditions in which the liberated peoples may exercise' the rights laid down by the Atlantic Charter—'to choose the form of government under which they will live'—the three governments undertook jointly to 'assist any European liberated state or former Axis satellite state in Europe . . . to form interim governmental authorities broadly representative of all democratic elements in the population and pledged to the earliest possible establishment through free elections of governments responsive to the will of the people'.

American side who saw things clearly at the time. George Kennan, then Harriman's deputy at the US Embassy in Moscow, regarded the declaration as 'the shabbiest sort of equivocation'. And in a memorandum, which he sent to the Secretary of State on the eve of the Yalta Conference, the Secretary of War, Stimson (not a participant at Yalta himself) wrote these prophetic words:

Russia ... will claim that, in the light of her bitter experience with Germany, her own self-defense as a guarantor of the peace of the world will depend on relations with buffer countries like Poland, Bulgaria, and Romania, which will be quite different from complete independence on the part of those countries.[39]

A last word on Yalta may be left to de Gaulle, the leader of the French Provisional Government, who was not invited to take part in the conference, on the ground that 'such a debating society would confuse our essential issues', as Roosevelt put it to Churchill.[40] Instead, he was invited to associate his government with the Declaration on Liberated Europe and to co-sponsor the UN Conference at San Francisco. On 15 January 1945 he sent a memorandum to the three governments stating that French 'participation in these conferences seems vitally necessary', although the French Government 'suspects that this point of view is not shared by the other Allied powers'.[41] Twelve days later—on the eve of the Yalta Conference—he wrote in a memorandum to his Foreign Minister, Georges Bidault:

it is now too late to attend under the right auspices. Later on, we shall be in a far better position to discuss the European imbroglio from without, if we have not participated in the rigmarole to come, which may well end in rivalries between those present.[42]

1. Churchill, Roosevelt, and Stalin in the patio of the Livadia Palace, 10 February 1945.

2. Attlee, Truman, and Stalin at Potsdam, August 1945.

CHAPTER 5

Potsdam

From Yalta the three members of the Grand Alliance announced that the United Nations would be invited to send representatives to a founding conference of the world organization, which would open at San Francisco on 25 April. They also agreed that, after the UN Conference, their three foreign ministers should meet in London; 'permanent machinery should be set up for regular consultation between the Foreign Secretaries', who would meet 'probably about every three or four months'.[1] The UN charter was finally signed on 26 June at San Francisco. The London meeting began on 11 September, by which time the Heads of Government had held their last summit meeting, from 16 July to 2 August 1945, at Potsdam.

The cascade of events that filled the six months between the Yalta Conference and the end of the Second World War makes this period of history hard to distentangle, even today. At the time only one of the three leaders of the Grand Alliance held all the threads of the swiftly changing network of world power—military, economic, and atomic— in his hands: the President of the United States. How Roosevelt might have chosen to draw these threads together during these six months, and above all at Potsdam, remains one of the great 'ifs' of the history of our time. Truman can hardly be blamed for failing to draw them together effectively, because the political testament that Roosevelt left to him on 12 April, after what Churchill later called the 'costly weeks'[2] of his final decline, was at best fragmentary and at worst ambivalent.

Towards the end of Roosevelt's presidency the most striking example of the 'gay and apparently heedless abandon' with which he seemed to 'delight in pursuing two or more totally incompatible policies'[3] was the case of China, with ominous consequences for his successor. Moreover, there were major issues about which he was beginning to change his mind—for example, Indo-China.[4] No one can say whether, had he lived longer, he would have had even more significant second thoughts on other issues as well. What is clear today is that, at the moment when Roosevelt collapsed and died at Warm Springs, Georgia, barely three months before the atomic test at Alamogordo, New Mexico, by far the most critical decisions of policy required were those concerned with the atomic bomb. On 30 December 1944, the Director of the Manhattan Project, Major General Leslie Groves, had informed Marshall that he was 'reasonably certain' that the first bomb 'should be ready . . . about 1 August 1945'.[5] The man who in the event took these decisions, Truman, knew nothing about the atomic bomb until Stimson and Groves gave him his first briefing at the White House on 25 April (when, as a security precaution, they entered the building by separate doors).

Instead, the two international problems that immediately devolved on Truman on 12 April 1945 were quite different: they were—once again—Poland and, by now, the conduct of the final thrusts of the war against Germany. On both of these questions a frenetic triangular exchange of messages had developed during the last weeks of Roosevelt's life. Even at the time Churchill said in one of these messages to Roosevelt 'I hope that the rather numerous telegrams I have to send you . . . are not becoming a bore to you'; and in his memoirs he recorded that he had not realized how ill the President was.[6] Since the agreement on Poland reached at Yalta, not only had no elections been held in Poland,* but in Moscow the three-man commission, established by the Yalta agreement on Poland, had agreed on nothing whatever. Among the Polish leaders in London, even Mikołajczyk did not accept the Curzon Line as Poland's eastern frontier until he was obliged to do so in a statement that he issued in mid-April, under pressure from Churchill.[7] Meanwhile, in Poland itself, the composition of the Provisional Government had still not been 'reorganised'. Sixteen members of the Polish resistance who had gone to Marshal Zhukov's headquarters at the end of March, under a Soviet guarantee of immunity, then disappeared (in May the Soviet Government announced that they would be tried in Moscow for sabotage). Finally, the problem was compounded by the question of how Poland was to be represented at the San Francisco Conference. Since a Warsaw-led delegation was unacceptable to the US and British governments, Stalin retaliated by deciding that Molotov should not attend either—a decision which, after Roosevelt's death, he revoked on the eve of the conference, as a personal gesture towards the new President.

THE TWO PERSPECTIVES

In the controversy over Poland that developed between the three members of the Grand Alliance in the spring of 1945, the US

---

* Nor were they until January 1947.

Government took up a position somewhere between the Soviet position and the British. The Soviet position was expressed with the same brutal frankness as it had been seven years earlier. (When the French Ambassador called on Potemkin, in Litvinov's absence, at the Soviet Foreign Ministry, in an attempt to explain the 1938 Munich Agreement, Potemkin observed: 'Mon pauvre ami, qu'avez-vous fait? Pour nous, je n'aperçois plus d'autre issue qu'un quatrième partage de la Pologne').[8] For Stalin, in 1945, the Soviet Union simply had 'the right to demand for Poland a regime that would be friendly towards the USSR'. By 13 March the three Powers were, in Churchill's words, 'in the presence of a great failure and an utter breakdown of what was settled at Yalta'. At that stage Roosevelt disagreed; and it was not until 29 March that he sent Stalin a long message including a warning that any solution 'which would result in a thinly disguised continuance of the present Warsaw regime would be unacceptable'.[9] Even so, the last presidential message received by Churchill, on the day of Roosevelt's death, advised him (when speaking about Poland in the House of Commons) to 'minimise the Soviet problem as much as possible, because these problems, in one form or another, seem to arise every day, and most of them straighten out'.[10]

On a short-term view, this forecast proved accurate so far as Poland was concerned. On 30 May 1945, Hopkins, sent to Moscow by Truman, at last agreed with Stalin on the enlargement of the Polish Provisional Government, although when it came to the point, three of the seven 'non-Lublin' Poles invited to join the government refused portfolios. (Mikołajczyk became Deputy Prime Minister and Minister of Agriculture.) Most of the remaining sixteen ministers, in the State Department's words, 'may be Poles at heart but realise ... that their political strength comes from Moscow and not from the Polish people'.[11] Churchill grudgingly accepted what Hopkins had laboriously* negotiated in Moscow as 'no advance on Yalta', but 'an advance upon the deadlock'.[12]

For the reasons explained in the preceding chapter, this conflict over the composition of the Polish Government, taken together with the agreement subsequently reached on Polish frontiers at Potsdam, was a rearguard action. This was especially true of the US Government. Truman—by his own account—gave Molotov a drubbing when they first met in Washington in April 1945, but he was not prepared to set the other parts of the Yalta package at risk, however much—in Hopkins' words to Stalin—'the question of Poland had become a

---

* He had six meetings with Stalin, one of them alone (during which he sought to obtain the release of the *AK* Poles held in Moscow).

symbol of our ability to work out problems with the Soviet Union'. It is, therefore, in the context of the overall US-Soviet understanding, reached by Hopkins in his last mission to Moscow, that the agreement regarding the Polish Government has to be assessed. At a press conference held on 13 June 1945, Truman expressed his complete satisfaction and gratification with the results of the Hopkins mission—understandably, since they included the date of the Red Army's readiness to invade Manchuria (8 August); agreement that Chinese Nationalist troops should enter Manchuria simultaneously; a joint effort to help unify China under Chiang's leadership; final Soviet acceptance of the US position on UN Security Council voting (thus averting a breakdown at San Francisco), and agreement on the holding of the Berlin summit meeting in mid-July.[13] On the eve of this conference—on 5 July—both the US and the British governments simultaneously announced their recognition of the Warsaw Government as the government of Poland.[14] So ended a diplomatic battle that had lasted six months.

The second of the two immediate problems inherited by Truman from Roosevelt was military, although in Churchill's view, passionately expressed both at the time and after the war, it was also political. In the last fortnight of his life Roosevelt had found himself the target of a fresh barrage of messages from Stalin and Churchill. Although these messages, which were not directly related, arose from separate anxieties—Stalin's from news of the negotiations (secretly initiated early in March, in Switzerland, by the SS General Karl Wolff) for the surrender of the German forces in Italy, and Churchill's from General Eisenhower's telegram to Stalin informing him of the axis of his final advance in Germany—both touched on the same sensitive nerve: the question of whose forces should take Berlin and Prague (Vienna was entered by Soviet forces on 11 April 1945).[15]

Roosevelt received an insulting message from Stalin on 3 April, accusing the western allies of making a separate peace agreement 'on the basis of which Marshal Kesselring has agreed to open the front and permit the Anglo-American troops to advance to the east'. An eyewitness account of the President's reaction described 'his eyes flashing, his face flushed, outraged that he should be accused of dealing with the Germans behind Stalin's back'. The tone of Roosevelt's reply—'a feeling of bitter resentment . . . for such vile misrepresentations'—offers one of the few clues to the question of what a political duel between two leaders might have been like, had Roosevelt lived. For once, Stalin had overplayed his hand in his dealings with Roosevelt. But in the event he was lucky. In the words of a presidential message sent to Stalin on 12 April, the Berne incident had by then appeared 'to have faded into the past'. And Roosevelt died that afternoon.[16] This affair, like the Polish problem in the following month, was settled. In both Stalin seemed to have got what he wanted. Yet the question remains for future historians—and can be settled only when the Soviet archive is eventually opened—whether Soviet security interests in Central and Eastern Europe could not have been satisfied in 1945 by less crude methods and in a manner that did not lead, within a year, to Truman's conclusion that he was 'tired of babying the Soviets' and to Churchill's denunciation of the Iron Curtain at Fulton, Missouri.

At almost the same moment as Stalin was attacking Roosevelt about the Berne negotiations, Churchill took issue with him over Eisenhower's declared intention of 'driving eastward to join hands with the Russians or to attain general line of Elbe'; in his reply to Eisenhower's telegram announcing this intention, which he welcomed, Stalin described Berlin as having 'lost its former strategic importance'. To Roosevelt, Churchill described the matter of Eisenhower's telegram to Stalin as closed on 5 April 1945; but he did not give up his belief that 'we should shake hands with the Russian armies as far as to the east as possible and if circumstances allow, enter Berlin'.[17] A month later he was urging Eisenhower to capture Prague.[18] Thwarted on both counts, he none the less went on to urge upon Truman that, as a deliberate bargaining counter, there should be no western withdrawal from the areas held by their forces, at the moment of the German surrender, well within the occupation zone allocated (by previous tripartite agreement) to the Soviet Union. On 11 June, just before his press conference on the Moscow discussions, the President finally turned the Prime Minister down.[19] Only on the issue of Trieste did Truman end by agreeing with Churchill. What followed in Germany was described nine months later by an unrepentant Churchill, in Truman's presence at Fulton, Missouri, as an Anglo-American withdrawal 'to a depth at some points of 120 miles on a front of nearly 400 miles to allow the Russians to occupy this vast expanse of territory which the Western democracies had conquered'.[20]

In the light of the subsequent course taken by history in Central Europe, the decisions first, not to go for the capture of Berlin and Prague, and then, to withdraw westwards seemed extraordinary to western historians writing during the cold war; and the theme of the squandered victory has endured ever since. Today what seems no less remarkable is the Soviet decision not to go straight for Berlin in February 1945. Exactly why Zhukov's assault on Berlin was postponed in the first week of February is still not entirely clear; it is the subject of internal Soviet controversy.[21] For whatever reason, his directive of

26 January, which foresaw the capture of Berlin by 15–16 February, was not carried out; and one of the consequences of this change of plan by the Soviet High Command was that two months later Eisenhower was presented with the opportunity to thrust to Berlin instead. Had he done so, he wrote in his memoirs, it would have been 'more than unwise; it was stupid'.[22] Churchill, who took the opposite view, was not stupid, although he was sometimes unwise. In this instance what he proposed amounted in effect to adopting a united Anglo-American stance against the Soviet Union over Germany in 1945, three years ahead of the Berlin crisis and four years before the signature of the North Atlantic Treaty. If public opinion in the United States and in Britain was not yet ready to follow where he pointed in March 1946, in his Fulton address, would it have done so one year earlier?

More important for the purposes of a study of the Anglo-American relationship is the fact that, as 1945 unfolded, the essential difference of Anglo–American international perspective, already apparent at Yalta, became increasingly clear. This difference did not lie in a simple contrast between a naive American idealism and a mature British belief in what the Treaty of Utrecht had described, nearly two and a half centuries earlier, as 'a just balance of power, the best and most solid foundation of mutual friendship and a lasting general concord'. The real contrast lay rather between American perception of US national interests in Asia and their perception of US interests in Europe. In Asia, major military commitments were readily accepted; and, indeed, in establishing the post-war American line of defence in the Pacific (Hawaii–Micronesia–Philippines), US national interests were paramount, in American eyes, where 'the security of the United States and stability of the Pacific world' were at stake.[23] On the other hand, in Europe, as Roosevelt had already warned Churchill in 1944, Congress could not at that time be expected to support the deployment of American troops for more than two years after the end of the European war. At Yalta Roosevelt said exactly the same thing to Churchill and Stalin in the second plenary session of the conference, (even though he modified his remark, to some extent, in the next session).[24] Given the numerical superiority of Soviet ground forces in Europe, only a global US-Soviet deal could square this circle; what it might consist of was still unclear at the moment when Roosevelt died. He had stalled on the Soviet request for a US \$6 bn. reconstruction loan, first mooted by Molotov on 3 January 1945;[25] the outcome at Alamogordo could not be regarded as a complete certainty; and the Far Eastern war still had to be won.

In the words[26] used by Hopkins, talking to Stalin on 26 May, 'the interests of the United States were world-wide and not confined to

North and South America and the Pacific Ocean ... President Roosevelt had believed that the Soviet Union had likewise world-wide interests and that the two countries could work out together any political or economic considerations at issue between them'. The resemblance between this 'cardinal' concept, as Hopkins called it, and the later Kissingerian concept of the superpower relationship need not detain us at this point. In the summer of 1945 British interests were also world-wide. As the European war came to an end, the British Government planned to increase British involvement in the Pacific war substantially; Mountbatten was planning to retake Singapore by the end of 1945; and in June the Viceroy, Field-Marshal Wavell, began the first post-war attempt to grapple with the political problems of India. But the principal problem 'knocking at the door' for Britain was European, as it also was for the Soviet Union.[27] For reasons of European history, the British view of the post-war world was much less sanguine than that of the Americans; and the focus of their scepticism was Central Europe, where the Second World War had, for them, begun nearly six years before. It was against this background that Churchill called the last volume of his memoirs *Triumph and Tragedy*, because he believed that in the closing months of the war a great opportunity for a settlement with the Soviet Union in Europe had been missed.

It is this difference of Anglo-American perspective that explains Truman's refusal to visit Britain* on the way to Potsdam; his anxiety to avoid any suspicion of Anglo-American 'ganging-up' on the Soviet Union before the conference; and his decision (taken without consulting Churchill in advance) to send Hopkins to Moscow in May 1945.[28] Once he had decided to send Hopkins to Moscow, it was logical to send an equally trusted emissary to London, but Truman's choice could hardly have been more unfortunate: Joseph Davies, the author of a foolish book called *Mission to Moscow* (he had been US Ambassador to the Soviet Union before the war). He and Churchill argued, alone, for eight hours at Chequers. Davies' report to the President makes no mention of Churchill's 'Note by the Prime Minister on Mr Davies's Message', a long and withering memorandum textually reproduced in his memoirs; and in 1954 Davies was to dispute that he and Churchill had exchanged any documents at all. What is certain, however, is that Churchill interpreted the presidential message, which Davies conveyed to him orally, as meaning that there should first be a meeting between Truman and Stalin, which 'representatives of His Majesty's Government should be invited to join a few days later'. This hint of US-Soviet

---

* In the end he flew from Berlin to Britain, where he spent six hours in all, on 2 August 1945; he lunched with King George VI on board HMS *Renown*; later he embarked for home. This was the only visit that Truman paid to Britain during his presidency.

condominium led Churchill to inform Truman on 31 May that he would 'not be prepared to attend a meeting which was a continuation of a conference between yourself and Marshal Stalin ... we should meet simultaneously and on equal terms'.[29]

Churchill won this round. He had to accept the timing of the conference agreed on in Moscow by Hopkins, however; although, as it turned out, it was Stalin who arrived later than the other two leaders at the Potsdam Conference, whose first primary plenary session was not, therefore, held until 17 July. Nevertheless, neither in its date, nor in its venue, nor in its purpose did this conference resemble the meeting that Churchill had originally proposed* to Truman on 11 May: a show-down—'the grave discussion on which the immediate future of the world depends'. As late as 30 May Stalin and Molotov had to be assured by Hopkins that Truman really was proposing 15 July as the date for the conference—not 15 June, as Churchill had suggested to Stalin.[30]

THE ATOMIC DIMENSION (1)

In the run-up to the Potsdam Conference the records now publicly available reveal one Anglo-American lacuna, which, after full account has been taken of the constraints of secrecy, the slowness of communications, and the surge of problems that confronted both governments as the European war ended (including, in Britain, a general election), remains astounding to the modern reader. By the terms of the secret Quebec Agreement of 19 August 1943, the two governments were bound not to use the atomic bomb against third parties and not to communicate any information about it to third parties, except by mutual consent; and in the words of the secret Aide-Memoire of the conversation between Roosevelt and Churchill at Hyde Park on 19 September 1944, 'when a "bomb" is finally available, it might perhaps, after mature consideration, be used against the Japanese, who should be warned that this bombardment will be repeated until they surrender'.[31] On these great issues, therefore, there had to be not only Anglo-American consultation, but also the consent of both governments. There was; but the process was perfunctory.

Roosevelt and Churchill had both confined the atomic secret to a very small circle. Even Stimson had not seen the Hyde Park Agreement

---

* In his telegram of 11 May to Truman, Churchill said that he considered that 'the Polish deadlock can now probably only be resolved at a conference between the three heads of government in some unshattered town of Germany ... at the latest at the beginning of July'; and his 'iron curtain' telegram sent on the following day ended with these words: 'this issue of a settlement with Russia before our strength has gone seems to me to dwarf all others'.

before June 1945; Churchill had to send Truman a photocopy at Potsdam.[32] During the last four months of the war, however, in Washington the major atomic issues were repeatedly addressed by the President's advisers. In London ministerial discussion—as described when the relevant documents were examined, nearly twenty years after the war, by the official historian of the United Kingdom Atomic Energy Authority—was 'cursory'.[33] What happened was this. On 29 June 1945, the following minute was addressed to the Prime Minister by John Anderson:[34]

OPERATIONAL USE OF T.A.

Lord Cherwell will have told you of the private manuscript letter which I have had from Field-Marshal Wilson reporting some details which he has been given confidentially and unofficially of the outline plan for the first use of the weapon against the Japanese.

Field-Marshal Wilson has now reported that Mr Stimson proposed to raise this matter at a meeting of the Combined Policy Committee on 4th July, in order that the two Governments may record their decision on the use of the weapon, in accordance with the terms of the Quebec Agreement.

The Americans are now making their final plans and preparations for the use of the weapon. May I have your authority to instruct our representatives on the Combined Policy Committee to give their concurrence in the decision to use it?

Churchill simply initialled this minute. Field-Marshal Wilson, having been so instructed from London, gave British consent to the use of the atomic bomb at the meeting of the Combined Policy Committee* in Washington, adding that the Prime Minister 'might wish to discuss the matter with the President at the forthcoming meeting in Berlin'.[35] Such Anglo-American discussion as there was at Potsdam is considered later in the chapter; but, as Churchill recorded in his memoirs, 'there never was a moment's discussion as to whether the atomic bomb should be used or not'.[36]

Thus, in the summer of 1945, just when the two governments were at last reaching the three turning-points of the atomic drama—Alamogordo, Potsdam, and Hiroshima—their atomic dialogue, far from entering the phase of 'mature consideration' prescribed by the Hyde Park agreement, became muted. The dialogue did not cease, but instead of addressing the central issues, discussion focused on questions

---

* The (secret) Anglo-American atomic committee in Washington set up by the 1943 Quebec Agreement. Under the terms of the (equally secret) 1944 Declaration of Trust, one of this committee's most important functions was to direct the Anglo-American Combined Development Trust, established to gain control of supplies of uranium and thorium.

which today seem secondary: for example, the texts of the public statements to be made in Washington and London after the first atomic bomb had been dropped and that of the Smyth report. There is, moreover, no sign in Churchill's exchanges with Truman during the three months preceding the Potsdam Conference (the same is true of his memoirs) that the Prime Minister realized the link, clearly perceived in Washington, between the date of the conference and the date of the atomic test. It was Truman's aim, on Stimson's advice, that the test should if possible precede the conference,[37] and his last known instruction regarding the bomb—the manuscript note written at Potsdam in reply to Stimson's telegram of 30 July 1945, asking for stand-by authority to release the presidential announcement of the first atomic attack—speaks for itself.*

Seen from the London perspective of mid-1945, three main factors combine to account for the British detachment from this critical moment in history. Although the Anglo-American atomic partnership expressed in the Quebec Agreement and in the Declaration of Trust was the prize of one of Churchill's hardest transatlantic battles— perhaps the hardest of all—the Manhattan Project remained, operationally and financially, an American enterprise. Since the conduct of the Pacific war was predominantly in American hands, the use of the atomic bomb against Japan was seen as an American decision. True, the Quebec Agreement required the 'consent' of the British Government, but the primary responsibility was seen as resting with the President of the United States. Finally, there was Churchill's personal view of the Quebec Agreement, forcefully expressed in a minute to his scientific adviser, Frederick Cherwell, one year earlier:

I am absolutely sure we cannot get any better terms by ourselves than are set forth in my secret Agreement with the President. It may be that in after years this may be judged to have been too confiding on our part. Only those who know the circumstances and moods prevailing beneath the Presidential level will be able to understand why I have made this Agreement. There is nothing more to do now but to carry on with it, and give the utmost possible aid. Our associations with the United States must be permanent, and I have no fear that they will maltreat us or cheat us.[38]

In this particular minute Churchill was giving Cherwell guidance for a talk with the South African Prime Minister, Jan Christian Smuts, but it also helps to explain why in May 1945 he laid down that 'the question of machinery for Anglo-American consultations' was 'to emerge naturally' in the course of a talk that Wilson was then about to have with Marshall; and why he stressed his anxiety 'to avoid having to

---

* This note is reproduced opposite p. 65. The timing of the test made nonsense of Churchill's appeals for an earlier conference.

insist, at this stage, on any legalistic interpretation' of Clause 2 of the Quebec Agreement.[39] If such was the view from London, small wonder that in Washington no need was seen for anything more than a last-minute 'check with the British'.[40] In what exact proportions each of these factors contributed to the sum of the British rationale at the time, is debatable. The last was probably the most important of the three. What is certain is that it was not until after the two atomic bombs had been dropped on Japan that the British began to take an active part in the earnest debate about the bomb's implications that had begun in Washington at the beginning of May 1945.

Although Truman's first atomic briefing on 25 April lasted less than an hour, it had one immediate result: the formation, under Stimson's chairmanship, of an Interim Committee of advisers to the President. To the deliberations of this Committee the British made no contribution. If the advice that Stimson and his colleagues tendered in the course of the next two months seems confused to a modern reader, none the less Stimson himself from the outset was well aware of the fundamental choice that needed to be made (as he had put it to Roosevelt at their last meeting four weeks before the President's death) between 'a secret attempt at control of the object by the United States and Britain' and 'an international effort based on free exchange of scientific information and free access to the laboratories of the world'.[41] The main questions requiring decision were four, all of them predicated on a successful test at Alamogordo. Was the bomb to be dropped on Japan and when? Was the Japanese Government to be warned? Was the Soviet Government to be told about it in advance? Linked with this, there was the whole question of any future atomic relationship with the Soviet Union, which had been a controversial subject both in London and in Washington as early as the spring of 1944.[42]

On 6 June 1945 the Committee, which had held its first meeting on 9 May, recommended to the President that the atomic bomb should be used against a Japanese 'war plant surrounded by workers' homes', without warning, as soon as preparations could be made; the Soviet Government should not be informed about the weapon before it was used. In a memorandum which Stimson handed to the President, he added his own suggestion that the condition of a sharing of the secrets of the bomb with the Soviet Union should be either Soviet participation in an international control commission on atomic energy after the war, or a political accommodation in Eastern Europe. By 19 June, Stimson was describing the Interim Committee as 'thinking in a vacuum' until the Soviet Union had been dealt with. In late June the Committee changed their minds about warning the Soviet Union of the bomb's existence; they now recommended that the President should not wait until the bomb was used before informing the Soviet Government. Stimson conferred with the President on 2 and 3 July; and at the meeting of the Anglo-American Combined Policy Committee held on 4 July, he is recorded as saying:

... if nothing was said at this [the Berlin] meeting about the Tube Alloy weapon, its subsequent early use might have a serious effect on the relations of frankness between the three great Allies. He had therefore advised the President to watch the atmosphere at the meeting. If mutual frankness on other questions was found to be real and satisfactory, then the President might say that work was being done on the development of atomic fission for war purposes; that good progress had been made; and that an attempt to use a weapon would be made shortly, though it was not certain that it would succeed. If it did succeed, it would be necessary for a discussion to be held on the best method of handling the development in the interests of world peace and not for destruction. If Stalin pressed for immediate disclosure the President might say that he was not prepared to take the matter further at the present time ...[43]

A fortnight later, at Potsdam, Stimson was to have third thoughts, by which time Truman was receiving advice from other quarters, including the view of Byrnes, who (as well as being a member of the Interim Committee) had been Secretary of State since 3 July, that the bomb would now provide a way of ending the Japanese war before the Soviet Union could enter it.[44]

When Churchill met Truman at Potsdam, he added his voice to those who argued against insistence on the unconditional surrender of Japan.[45] But on the question of what was to be said to Stalin about the bomb, Churchill seems to have been only gradually persuaded that there was any real need to tell him anything at all. In a minute addressed to the Prime Minister on 12 July 1945, Cherwell described the arguments for and against talking to Stalin as 'fairly closely balanced'. As the Americans had contributed such an overwhelming proportion of the effort, however, it would not be 'easy to oppose Truman strongly, if he seems anxious to take the second course'—a course defined in the same minute as being 'to say, in broad general terms, that we have done a great deal of work on Tube Alloys and hope shortly to make an operational test, but refuse to say anything further'. These minutes show that Churchill had originally intended to write to Truman on this issue, but then decided to wait and talk to him at Potsdam. They also suggest a remarkable degree of unawareness in London, on the eve of the Alamogordo test, of the speed with which atomic events were then moving.[46]

For his part, Truman seems to have agreed with Stimson's conclusion, reached soon after the Interim Committee had been formed, that

US economic and atomic strength, taken together, constituted a royal flush, but the question remained: just how was the hand to be played? By the time that the President reached Potsdam this was still undecided. By an unhappy paradox, what happened at Potsdam reinforced the mistaken belief that the Soviet Union was either atomically ignorant or atomically incapable, or both.[47]

THE CONFERENCE

The Conference of Berlin, as it was officially called, was the last summit meeting of the Grand Alliance and the last East–West summit for ten years. It was also the longest. The protocol signed at Potsdam, south-west of the city, was the product of thirteen plenary sessions held in the neo-Tudor Schloss Cecilienhof, the former palace of the Crown Prince Wilhelm of Prussia. The communiqué of 2 August 1945 recorded that the three leaders left the conference, which had 'strengthened the ties between the three governments', with 'renewed confidence that their governments and peoples, together with the United Nations, will ensure the creation of a just and enduring peace'.[48] This expression of triumphant accord was not simply intended for public consumption. As Attlee put it in a letter that he wrote to Churchill from Potsdam, the three leaders parted 'in a good atmosphere'.[49] In fact the process over which they had presided was the first stage of a division of Europe more definitive than anything discussed at Yalta.

These two and a half weeks can be followed hour by hour and blow by blow, thanks to the abundant records (American, British, and Soviet) of the proceedings of the conference and (American and British) of the run-up to it. Interspersed with the series of tripartite plenaries, preceded by meetings of the three Foreign Ministers, there was from first to last a shifting pattern of other meetings—internal, bilateral, and on one day quadripartite—with ministers of the Polish Government; there was even one trilateral meeting of Chiefs of Staff.[50] Of all the American and Anglo-American meetings, those that mattered by far the most were concerned with the momentous atomic decision. This was reached during the weeks that the leaders of the three countries spent at Potsdam, but the issue was never discussed trilaterally at all.

At the trilateral level, discussion was twice interrupted, first by the British general election and then by Stalin's indisposition. On four days, therefore, there were no plenary sessions—on 26, 27, 29, and 30 July—nor could one be held on 16 July, because Stalin arrived a day late. Churchill's form at Potsdam was far from his best. Although Bevin picked up the threads with a firm hand on his arrival, the three days of British ministerial hiatus following Churchill's and Eden's departure from Berlin on 25 July did not help either the process of the conference or the British position in it. And although the principal questions discussed by the three leaders and their Foreign Ministers jointly at Potsdam were the same as at Yalta—Central Europe (in which Germany itself had now become the dominant issue), the Far East, and the establishment of peace-making and consultative machinery—a plethora of other, secondary questions was discussed in parallel with the main ones. In the tripartite plenary sessions, discussion ranged far and wide: Iran, Turkey, Italy, Romania, Bulgaria, Hungary, and Finland.[51] Bilaterally, there were several meetings of the (Anglo-American) Combined Chiefs of Staff, the British suggestion that they should meet in London before the conference having been turned down. With the planning date for the end of organized resistance by Japan fixed as 15 November 1946, they agreed in principle that a British Commonwealth land force, with air support, should participate in the Pacific war. As proposed by Marshall, they also redrew the boundary between the Asian commands; Mountbatten's South-East Asia Command now became responsible for Indo-China south of latitude 16;* so that, as an unexpected consequence, in September British troops were engaged not only in reoccupying British territories in South-East Asia, but also in landing in Saigon and Djakarta.[52]

On the main issues that the leaders of the three countries discussed together, the communiqué which they issued, like their smiling photograph taken on 1 August 1945, gave a misleading impression of allied unanimity. There were four cardinal differences between the reality and the appearance of Potsdam. First, the expression of tripartite solidarity obscured the extent to which the outcome of the conference had depended on a US-Soviet understanding, reached bilaterally between Byrnes and Molotov, regarding the critical questions of Poland and Germany—two questions which Byrnes deliberately and unequivocally linked. Secondly, the compromises agreed on these two questions at the very end of the conference (the last plenary session broke up just after midnight) formally left the future of Germany open, but in effect closed the options further. Thirdly, even the establishment of the five-power Council of Foreign Ministers, reassuringly sensible on paper, contained the seeds of future controversy. Finally, only four days after the conference a completely new factor was thrust into the equation of world power.

For the most part the decisions reached at Potsdam were worked out

* The first politico-military appearance of a line of demarcation that was later to become famous. Shortly afterwards the US and Soviet governments agreed on the 38th parallel as the line of demarcation between their forces occupying Korea.

by war-weary men—Truman was the exception in this respect—sitting in conference in the middle of the greatest demographic cataclysm in modern European history. After the guns had fallen silent, what was left was the noise of millions of tramping feet (those of what were euphemistically known as displaced persons). Moreover, the historian does well to remember the wise advice that events now in the past were once in the future. It is, therefore, not quite fair to blame the three Heads of Government now for their failure, in those difficult circumstances, to foresee all the consequences of what they jointly decided at Potsdam. On the other hand, it is harder to give the same benefit of the doubt to the simultaneous Anglo-American atomic dialogue, which was not much more impressive at Potsdam than it had been during the three previous months.

To begin with the first of these cardinal points—on the question of Central Europe, the three allies had no difficulty in settling the Soviet Union's claim to its share of East Prussia (the remainder became Polish).* But they resumed their argument over Poland's western frontier where they had left it at Yalta, with two important differences: there was now a Polish Government that they all recognized, whose leaders came to Berlin to join in pressing the case for the western river Neisse; and by now the German population of the western territories claimed by Poland, with Soviet support, had been largely replaced by Poles.[54] The argument in plenary began on 21 July; before leaving Berlin four days later Churchill declared, at his last meeting with Truman and Stalin, that if the conference ended with no agreement regarding the present state of affairs in Poland, with the Poles 'practically admitted as a fifth occupational power' of Germany, this 'undoubtedly would mark a breakdown of the Conference'.[55] This deadlock continued until Byrnes broke it on 30 July, at his bilateral meeting with Molotov, who at once expressed his 'gratification' in response. At this critical meeting, what the US Government conceded was Polish administration of German territory up to the western Neisse 'pending the final determination of Poland's western frontier', which 'should await the peace settlement'; Byrnes expressly linked this major US concession to Soviet acceptance of the US proposal regarding German reparations, put forward on the previous day at another bilateral US-Soviet meeting (this time Truman was present).[56] The essence of Byrnes' proposal on reparations was to get away from the

---

* On the future of East Prussia, in accordance with what had been agreed at Tehran, the Potsdam Conference 'agreed in principle to the proposal of the Soviet Union concerning the ultimate transfer to the Soviet Union of the City of Königsberg and the area adjacent to it' and 'the President of the United States and the British Prime Minister have declared that they will support the proposal of the conference at the forthcoming peace settlement'.[53]

Yalta concept of fixed sums by establishing the principle that each occupying power would meet its claims to reparations from its own zone of occupation, although, in addition, the Soviet Union would receive a percentage of industrial equipment either from the Ruhr or from the three western zones taken together.

These two bilateral US-Soviet meetings on 29 and 30 July were crucial to the outcome of the conference. Both of them stemmed from Byrnes' initiative. At the very end of the conference, Stalin had good reason to express his thanks to Byrnes, 'who seemed to work harder than anyone else', and he added 'these sentiments, Secretary Byrnes, come from my heart'.[57] By the time Bevin took over the British team, late on 28 July—Attlee hardly said a word—the most that he was able to do was to put forward a different, even more complicated, formula for the payment of reparations and to hold out until the afternoon of 31 July for the eastern river Neisse as Poland's western frontier. In the end he accepted the US-Soviet deal on both questions. The record of the British 'staff conference' held first thing on the morning of 31 July, at which Bevin decided to do this, stands as a model of resigned realism, although before finally agreeing to the western river Neisse, he succeeded in extracting assurances from the Polish delegation about elections in Poland.[58]

Churchill subsequently maintained[59] that, if he had returned to Berlin as Prime Minister, he and Eden would never have accepted the western Neisse as the Polish frontier, which not only ejected three million more German refugees into the western zones of occupation, but also deprived these zones of the benefits of Silesian coal. It is, however, hard to see what he could have done, beyond perhaps insisting that British dissatisfaction was recorded in the protocol in some form of words that did not wreck the whole conference, which had already gone on too long. Cadogan was not far out when he described the last plenary session in his diary as a meeting of the Big Two and a Half.[60] That, by the time of Attlee's and Bevin's arrival in Berlin, Byrnes 'believed he had the makings of a deal' is also fair comment, although the turning point of the conference came forty-eight hours after their arrival: it was not until 30 July that the vital territorial linchpin—US acceptance of the western Neisse as the Polish frontier—was inserted in the deal, unlike its financial component, regarding which Byrnes had already cast several flies over Molotov. In plenary session, American heat was beginning to be turned on even before Churchill and Eden left Berlin. On 24 July, Truman observed that he 'had much business in the City of Washington'. On 28 July Stalin told Attlee that 'the President was in a hurry', and earlier on the same day Cadogan recorded in his diary that he 'must try to sell . . . to

A. between the time of his arrival and this evening's meeting [Byrnes' outline of] how the major outstanding problems of the Conference might be solved'. He did so in a minute of limpid clarity, even though he found the subject of reparations 'almost unintelligible'.[61]

In the event, no German peace conference has ever been held. It was not until a quarter of a century later that a legal device was worked out whereby the problem of Poland's western frontier was resolved. (The formula used in the Treaty of Moscow in 1970 was repeated in the Final Act of the Helsinki Conference in 1975.)[62] Thus the political map of Central Europe was in effect redrawn at Potsdam; what was on paper provisional became permanent in practice. The Potsdam agreement on German reparations also went, unintentionally, a long way towards the territorial reshaping of Germany. Molotov at once countered Byrnes' proposal by asking: 'if reparations were not treated as a whole, what would happen to overall economic matters?' Byrnes insisted that his proposal did not affect the treatment of Germany as a single administrative and economic unit; and the 'Economic Principles' of the protocol of 2 August 1945 expressly stated that this was how the Allied Control Council would govern Germany.[63] But Molotov's question was to the point. And the British team were aware that the carefully worded formula finally agreed regarding reparations, after a long haggle, was not unifying but divisive. So too were some of the American team—as David Waley's memorandum of 2 August 1945 conclusively shows.[64]

Since this part of the Potsdam Agreement lay close to the heart of subsequent controversy that led ultimately to the division of Germany, the full text of Section III of the Potsdam Protocol has been added at Appendix 2. (Austria was expressly exempted from the payment of reparations, although the inclusion of a provision that 'German external assets' in Austria were to be a source of German reparations for the Soviet Union was to bedevil the Austrian question for years to come.)[65] Just how much trouble Section III of the Potsdam Protocol was storing up for the future becomes clear if three basic politico-economic facts of Europe in the immediate aftermath of the war are recalled. These were the Soviet determination to milk Germany dry—quite apart from the material havoc wrought by the German invasion of the Soviet Union, there was famine in the Ukraine as late as 1946–7—the French insistence that the amputation of eastern Germany should be matched with a similar amputation in the west as 'an essential condition for the security of Europe and the world', and the countervailing American and British refusal to allow the payment of reparations to be executed at the expense of feeding the German population of their zones of occupation: an expense which they foresaw would

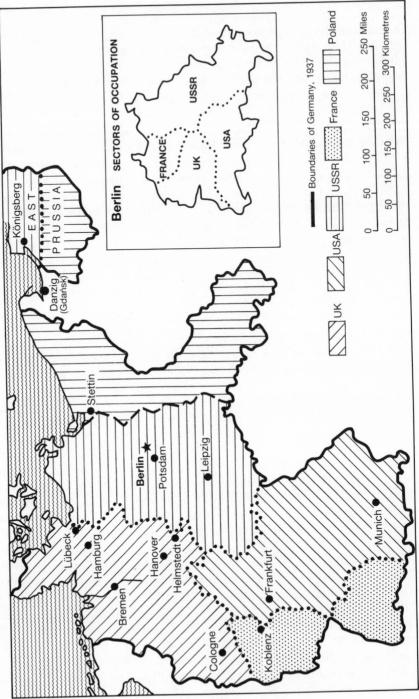

SECTORS OF OCCUPATION

Berlin

FRANCE

UK

USA

USSR

Boundaries of Germany, 1937

USA

UK

USSR

France

Poland

0   50   100   150   200   250 Miles
0   50   100   150   200   250   300 Kilometres

Königsberg

EAST PRUSSIA

Danzig
(Gdańsk)

Stettin

Berlin

Potsdam

Leipzig

Lübeck

Hamburg

Hanover

Helmstedt

Bremen

Cologne

Frankfurt

Koblenz

Munich

Map 3. Germany in July 1945

be their own, because both their zones had been historically dependent on food from the eastern part of Germany. French policy towards Germany, unrepresented at Potsdam, swiftly compounded the problem after the conference; and the plan subsequently put forward by the French Government provided for the separation from Germany (and the permanent military occupation) of the territory on the left bank of the Rhine, the separation also of the Ruhr industrial area (to be internationalized), the transfer to France of the Saar mines, and the incorporation of the Saarland in the French customs and currency system.[66]

What looked at the time like the most solid achievement of the Grand Alliance at Potsdam was the creation of the Council of Foreign Ministers, an expansion of the consultative machinery established at Yalta, which was described in the Potsdam Protocol as 'without prejudice to the agreement . . . that there should be periodical consultation between the Foreign Secretaries of the United States, the USSR, and the United Kingdom'. The concept of a Five-Power Council had originated at San Francisco. From there, on 19 June 1945, Stettinius had telegraphed an urgent memorandum suggesting it as an alternative to a fully fledged, formal peace conference, which would be 'slow and unwieldy', or a peace conference 'limited to a few States which would encounter much opposition on the part of States not invited to participate'. Just before the conference opened, Cadogan let the US delegation at Potsdam know that the British were in general agreement; and that, in view of the composition of the UN Security Council, it was reasonable to include China, as Stettinius had suggested.[67] When Truman put the US proposal forward at the beginning of the first session of the conference, Stalin at once questioned the inclusion of China in a body dealing with European problems, as Molotov already had done in Moscow before the conference began. Churchill supported Stalin.[68]

The compromise finally agreed regarding the Council of Foreign Ministers provided for a membership of five (therefore including China as well as France); London was to be 'the permanent seat of the joint Secretariat which the Council will form'; the first meeting was to be held there not later than 1 September 1945, although subsequent meetings might be held in other capitals; the Council's immediate task was to draw up peace treaties with Italy, Romania, Bulgaria, Hungary, and Finland; and the Council was to be 'utilised for the preparation of a peace settlement for Germany to be accepted by the Government of Germany when a government adequate for the purpose is established'. The protocol added this sentence: 'For the discharge of these tasks the

Council will be composed of members representing those States which were signatory to the terms of surrender imposed upon the enemy States concerned'; France was to be regarded as a signatory in the case of Italy.[69] It was the strict Soviet interpretation of this procedural sentence that was the formal reason for the breakdown of the London meeting of the Council of Foreign Ministers two months later. In the interval, however, there had been dramatic developments, unforeseen in the agreements reached by the three Heads of Government at Potsdam.

THE ATOMIC DIMENSION (II)

It will be recalled that the President's advisers had changed their minds at least once on the question of talking to Stalin about the bomb; and that Churchill had in the end decided to wait until he and Truman met at Potsdam before discussing this with the President. At this point chronology becomes extremely important. The critical dates were as follows. Stimson was already at Potsdam when Truman arrived there on 15 July; Churchill arrived on the same day; Stalin arrived a day later. On 16 July Stimson received, from George Harrison, Acting Chairman of the Interim Committee in Washington, the first, brief report* on the atomic explosion at Alamogordo, New Mexico which had taken place at 05.30 hours that morning. On 17 July, after lunching with Churchill, Stimson told him about the success of the test; Churchill was 'strongly inclined against any disclosure'; Stimson argued against Churchill's view 'at some length'.[71] (The first plenary session of the conference was held that afternoon.) At lunch on the following day, Truman asked Churchill, who had read the telegrams from Washington about the test, what he 'thought about telling the Russians. He [Truman] seemed determined to do this, but asked about the timing, and said that he thought that the end of the Conference would be best'. Churchill's reply was twofold: it 'might be better to hang it on the experiment, which was a new fact on which he and we had only just had knowledge', and 'on behalf of His Majesty's Government', he 'did not resist his [Truman's] proposed disclosure of the simple fact that we have this weapon'.[72] In his diary four days later[73] Stimson described Churchill as 'now not only ... not worried about giving the Russians information on the matter but ... rather inclined to use it as an argument in our favour in negotiations'.[74]

The final stage of Anglo-American agreement on the atomic bomb seems to have been reached at Truman's meeting with Churchill, who

* 'Operated on this morning. Diagnosis not yet complete but results seem satisfactory and already exceed expectations ...'[70]

3. Bevin and Attlee leaving Northolt aerodrome, London, to go to Potsdam, July 1945.

Sec War

Reply to your 41011
suggestions approved
Release when ready
but not sooner than
August 2.

HST

4. Truman's final instruction to Stimson about the press release regarding the atomic bomb (written at Potsdam).

had now read Groves' written report on the atomic test, on 22 July.[75] Two days later, in a directive approved by Stimson, the Commander of the US Strategic Air Forces, General Carl Spaatz, was instructed to deliver the first atomic bomb 'as soon as weather will permit visual bombing after about 3 August 1945'.[76] On the same day, after the plenary session had broken up, Truman 'casually mentioned to Stalin that we had a new weapon of unusual destructive force'. Stalin 'showed no special interest. All he said was that he was glad to hear of it and hoped we would make "good use of it against the Japanese"'. (24 July was also the day on which the Chiefs of Staff of all three countries exchanged detailed information about their plans for the war against Japan.) Churchill, who watched 'the talk between these two potentates' from about five yards away, recorded in his memoirs his own certainty that at 'that date Stalin had no special knowledge of the vast process' of Anglo-American research and of the 'heroic gamble' of the American production of the bomb.[77]

This erroneous conviction was by no means confined to Churchill. Moreover, long after Potsdam, it was widely believed on both sides of the Atlantic that it would be many years before the Soviet Union would be able to explode an atomic bomb—so much so, that nearly a month after the first Soviet atomic explosion had taken place the chief scientific adviser to the British Minister of Defence recorded, in a minute full of scepticism, his belief that it was 'quite possible' that the Russians had 'managed to be able to steal plutonium from the United States'.[78] Even Truman needed some convincing by his scientific committee before he released the news of the Soviet test to the public on 23 September 1949, over three weeks after the event.[79]

To conclude the atomic chronology of Potsdam, on 26 July 1945 the ultimatum[80] to Japan was published; on 30 July Stimson (by now back in Washington) informed the President that the 'time schedule for Groves' project' was making such rapid progress that it was essential that the presidential statement to be released after the first atomic attack should be 'available not later than Wednesday, 1 August'; and conscious that the conference was running late, Truman replied 'not sooner than August 2'.[81] From 31 July to 1 August there was a brief exchange of letters with Attlee, who—in the words of his biographer—'did not demur'; before Potsdam he knew extremely little about the bomb.[82] On 2 August the Potsdam Conference ended. Thereafter the momentum of bureaucracy and the vagaries of the weather seem to have taken over. After the first atomic bomb was dropped on Hiroshima, on 6 August, the timing of the second atomic attack, on Nagasaki, was purely military. In the seventy-two hours that separated the second attack from the first there was no review at political level, either in Washington or in London, or between the two capitals. In the words attributed to Churchill at Potsdam by an eyewitness, 'what was gunpowder? Trivial. What was electricity? Meaningless. This atomic bomb is the Second Coming in Wrath.'[93]

CHAPTER 6

The Advent of the Atomic Age: London and Moscow

IN the four weeks after Potsdam events moved with vertiginous speed: the two atomic attacks on Japan, in between the two the Soviet invasion of Manchuria,* and the Japanese surrender. Before the critical Anglo-American negotiations that at once followed this series of events (including their immediate consequence, the end of Lend-Lease) are discussed, in the next section of the book, this chapter will consider the impact on the trilateral allied framework of the advent of the atomic age.

In examining the impact on the Soviet Union of the launching of the atomic bomb, the modern historian has, in broad terms, a choice between three different approaches. He may look at the bomb by reaching back to the Anglo-American confidence, generally felt in 1945, in the prospect of an extended period of western atomic monopoly, which at that time found expression in the concept of the 'sacred trust'.[2] At the other extreme is the picture, drawn in later years by revisionist historians,[3] of the Soviet Union as the victim of atomic blackmail. No Soviet internal document is needed to establish that, to the Soviet Government in 1945, the sacred trust must have seemed poppycock. On the other hand, the blackmail looks rather less compelling when measured against the number of operable atomic bombs in the US atomic arsenal for the first two years after Hiroshima, perhaps no more than two in January 1947.[4] For this and for other reasons, the bomb was not quite the 'winning weapon' that it seemed to Bernard Baruch (who coined the phrase, in his speech to the UN General Assembly in June 1946) and many others at the time. In the existing state of the evidence—the Soviet sources fragmentary, the American abundant—the soundest approach is one based neither on

---

* The final Soviet decision to invade was taken on 7 August 1945. The troops who crossed the Manchurian border in the early hours of 9 August were the advance guard of a force of 1.5m. men, assembled over the previous six months. Given that at Potsdam Antonov had spoken of the second half of August, it is possible that the invasion was accelerated by the news of the atomic attack on Hiroshima on 6 August. If so, it was not brought forward by much; the planning of the operation had begun in April; and 8 August was the date that had been given to Hopkins by Stalin.[1]

what was believed in the west in 1945, nor on subsequent western projections of supposed Soviet thinking at that time, but on such Soviet sources relating to the early stages of the Soviet atomic project as have now become available, taken in conjunction with recorded Soviet behaviour at the first meeting of the Council of Foreign Ministers.

In September 1945 Stimson withdrew, exhausted, to his house in the Adirondack mountains in order to grapple with the problem once again. In the last memorandum that he wrote about it for the President before his retirement, he defined 'the problem of our satisfactory relations with Russia as not merely connected with but as virtually dominated by the problem of the bomb'. Abandoning as counterproductive his earlier idea of exerting pressure on the Soviet Union to undertake internal reform, he now argued that 'unless the Soviets are voluntarily invited into partnership upon a basis of cooperation and trust, we are going to maintain the Anglo-Saxon bloc over against the Soviet in possession of this weapon'. This would lead to 'a secret armaments race of a rather desperate character'. If the United States merely continued to negotiate with the Soviet Union, 'having this weapon rather ostentatiously on our hip, their suspicions and their distrust of our purpose and motives will increase'. Instead, what Stimson now recommended was an approach to the Soviet Union, which he described as:

a direct proposal after discussion with the British that we would be prepared in effect to enter an arrangement with the Russians, the general purpose of which would be to control and limit the use of the atomic bomb as an instrument of war and as far as possible to direct and encourage the development of atomic power for peaceful and humane purposes.

At the end of this memorandum he laid especial emphasis—'beyond all other considerations'—on the importance of this direct approach being made to the Soviet Union as 'a proposal of the United States, backed by Great Britain, but peculiarly the proposal of the United States'.[5] At the time Stimson believed that this memorandum had won the endorsement of the President, who set aside the whole of his Cabinet meeting on 21 September (the last day of Stimson's public service and also his seventy-eighth birthday) solely for discussion of this proposal. As Acheson later recorded in his memoirs, the wholly inconclusive discussion at the White House that morning was 'unworthy of the subject'.[6] Stimson's successive changes of mind may not have helped the President to make up his; but they were the result of an intellectually honest attempt by a distinguished American, who was one of the few men who realized all the dimensions of the problem, with which he had been closely concerned from the outset; and it is chiefly for that reason that his shifting opinions have been quoted at length in this and in preceding chapters, even though his final advice was not followed.

We shall probably never know what might have happened if Stimson's advice had been followed in September 1945, or if Truman had spoken differently to Stalin two months earlier. What we now know is that, if Stalin had not been kept informed of the progress of the Manhattan Project through espionage (notably by Klaus Fuchs from early 1944 onwards), he would certainly have been well aware of the state of the Soviet atomic project at the moment when Truman walked up to him at Potsdam on 24 July. It was on Stalin's personal instructions that responsibility for the problem of uranium was assigned to M. G. Pervukhin, Deputy Prime Minister and Minister for the Chemical Industry, in April 1942; later in the same year—an interesting sidelight—the codename given to the Stalingrad operation was *Uran;*[7] and it was to Stalin personally that G. N. Flerov, a young physicist, wrote from the Voronezh front appealing for the immediate establishment of a nuclear laboratory for the purpose of 'building the uranium bomb'. At the end of the year the State Defence Committee established the laboratory and appointed I. G. Kurchatov as scientific director.* Although work on the Soviet project did not begin in earnest until three months later and although its scale was small, by the spring of 1945 Kurchatov was able to begin work on the design of an industrial reactor for producing plutonium.[8]

Against this background, it is small wonder if, as recorded in Zhukov's memoirs, as soon as Stalin got back from the plenary session on 24 July, he said to Molotov: 'We shall have to have a discussion with Kurchatov about speeding up our work'. Another Soviet source, General S. M. Shtemenko, has been adduced as an indication that Stalin did not grasp what Truman had told him. In fact what Shtemenko's recollection (at second hand, of what General Alexei Antonov told him afterwards) does confirm is something different: that the Soviet General Staff knew nothing about the atomic project at that time; as in the United States and Britain, the early atomic policy decisions in the Soviet Union were political, not military.[9] Moreover, there is also the Soviet account of what Stalin said to

* In 1942 Igor Kurchatov, like Robert Oppenheimer, was not yet forty (he was not elected a full member of the Academy of Sciences until September 1943) and Flerov, his pupil, was under thirty. Early in 1942 Flerov was posted to the Voronezh front as an air force lieutenant. His letter to Stalin followed his deduction—from the silence about nuclear fission in foreign journals that he had been reading in the Voronezh University library—that the US atomic project was under way.

Kurchatov at a meeting held in the Kremlin after his return to Moscow '. . . the balance has been destroyed. Provide the bomb—it will remove a great danger from us': language that closely resembles that used in a British telegram sent from the Embassy in Moscow, assessing the Soviet reaction to the atomic bomb.[10] In August 1945 Kurchatov's estimate of the time that he needed was five years (he took exactly four). In any event, he and Pervukhin were appointed deputy chairmen of the Scientific–Technical Council which was at once formed, under B. L. Vannikov, Minister of Military Supplies. By November Molotov was saying publicly: 'we shall have atomic energy, and much else'.[11]

In 1945 Truman was not only new to the presidency and new to international relations, but—like almost everyone else—new to the great issues of the atomic bomb, about which he was assailed with conflicting advice. As Acheson remarked,[12] as late as January 1946, 'the Secretary of State and the President . . . did not have either the facts or an understanding of what was involved in the atomic energy issue'. Some of the conflicting advice was British; and, like the Americans, the British spoke with more than one voice (in his Fulton speech, Churchill argued against even the emasculated international-ism that underlay the Baruch Plan, which was presented to the United Nations three months later)[13] In all circumstances—and whatever the Soviet archive may one day reveal—Truman's initial decision is understandable: in effect, to expose the bare minimum of atomic surface to Stalin at Potsdam and then to pause for reflection in Washington. It is, however, now clear that, had he been aware of the true state of atomic affairs in the Soviet Union at that time, either he should have said more to Stalin than he did on 24 July, or he should have said nothing at all at Potsdam; and if he had said nothing, it would have been wise to follow up his silence with a personal message from Washington after his return. Instead, he fell between two stools. A half truth is risky in all diplomacy. At a minimum, this particular half truth, followed in less than a fortnight by the attack on Hiroshima, must have made Truman look silly in Stalin's eyes at the time. As for the subsequent maximalist Soviet interpretation of the events of 24 July to 9 August 1945, an example is offered by Zhukov's memoirs: 'the Government of the USA intended to exploit the atomic weapon for the attainment of its imperialist aims from a position of strength in the cold war'.[14] Whatever Stalin may have thought on 24 July, what we do know for certain is that, immediately after Hiroshima, he responded (to Harriman) with an atomic half truth of his own. And the fact that Stalin underestimated Truman is well attested. Their encounter on 24 July cannot have helped him to understand Truman's qualities. Certainly it is hard to conceive of a more inauspicious entry by the two future superpowers into the nuclear age.[15]

In the wake of what happened in the preceding month, the tactical course that Molotov followed at the meeting of the Council of Foreign Ministers, held in London at Lancaster House from 11 September to 2 October 1945, was predictable. As *The Economist* put it at the time, the Russians 'kicked at every door on the off-chance that some of them will be unlocked';[16] or as Byrnes described them after the conference collapsed, they 'are stubborn, obstinate, and they don't scare'.[17] If Potsdam was Byrnes' conference, this one was Molotov's. Whereas nothing had been done to establish a common western position in advance, Molotov had the advantage of having no one to consult but Stalin. As the Foreign Minister of the host government, Bevin proposed at the outset that the agreement reached at Potsdam should be interpreted as meaning that all five Foreign Ministers represented on the Council should take part in discussion, but that only those whose governments had signed the terms of surrender should participate in the Council's decisions. Molotov having—eventually—accepted this suggestion, all five ministers did take part in the first ten days of this meeting. On 23 September, however, Molotov asked for China and France to be excluded from discussion as well. Truman and Attlee sent personal messages to Stalin over his head in an attempt to get his instructions rescinded. They failed.* Bevin then went to considerable lengths—unsuccessfully—to break the deadlock, but Molotov made assurance double sure by demanding that the original decision to invite China and France should be deleted from the minutes of the conference. Tempers were lost; insults were exchanged; and after a final meeting that lasted into the small hours of 2 October, the conference ended without even a communiqué.

Byrnes[19] and Bevin shared the belief that it was Soviet ambitions in the Mediterranean, and particularly in Libya, that explained what happened at this conference. This belief was almost certainly mistaken.[20] True, Molotov did make a strong bid for Tripolitania, but Anglo-American recognition of the regimes in Romania and Bulgaria was an objective far closer to Soviet interests. Moreover, it was Stalin who later said to the Yugoslav Foreign Minister: 'Do you think that Great Britain and the United States will permit you to break their line of communication in the Mediterranean? Nonsense. And we have no

---

* The wartime illusion that Stalin was the dove and Molotov the hawk persisted in the West. Djilas observed the important point that they addressed each other in the second person singular.[18]

navy'.[21] In any case, on a broader view from Moscow, the logical sequel to the events of August 1945 was surely for Molotov to be as difficult as possible at the conference in September. Indeed, Bevin himself told his Cabinet colleagues on 11 October 1945 that he 'thought that many of the difficulties met with in the recent meeting of the Council of Foreign Ministers could be attributed to Russian resentment at having been excluded from this [sc. atomic] development' Moreover, at least one senior British official in London took the view that, in acting as he did at the conference '... in substance, though not in form, M. Molotov was right'.[22]

At their London meeting the Council of Foreign Ministers made a bad beginning. This was only their first meeting; fourteen months later peace treaties with all European countries except Germany and Austria were signed. Some of the political stereotypes that were later to become familiar features of the international stage, however, first took shape at this conference: Molotov's *nyet* and Bevin's doggedness. Only the American part remained uncertain. Although the logical outcome of Molotov's tactics in London might appear to be a greater measure of Anglo-American understanding, in the event there was little trust between Bevin and Byrnes. As will be seen in the next section of this book, by the end of 1945 the Anglo-American relationship in general had reached a low ebb.

On 24 November 1945 it was neither from Byrnes himself nor from the British Embassy in Washington that Bevin learned of Byrnes' suggestion to Molotov that the three foreign Ministers should meet on 11 December in Moscow. Instead, this news first reached London in a telegram from the British Ambassador in Moscow, to whom his US colleague there had shown the text of Byrne's message to Molotov of the previous day. Whereas Molotov 'beamed with pleasure', it took Bevin almost a fortnight to decide to join his two colleagues in Moscow.[23] In an effort to repair the damage caused in London, Winant, who had reported to Byrnes that what he had done had been 'deeply resented by both Bevin and the Cabinet', went to the remarkable length of arranging a teletype conference between Bevin and Byrnes at the US Embassy in London on 27 November. It was a frosty conversation; two days later Winant, claiming that he 'did eliminate anger from the interchange', reported that Bevin was 'desperately anxious' to talk to Byrnes before the proposed meeting in Moscow; but on 29 November Byrnes laid it on the line: 'Molotov has agreed December 15 ... Please let me know whether you agree'. It was not until 6 December that Bevin's agreement was communicated to the State Department—'in deference to your strong views'.[24]

As was to be expected from its genesis and from Bevin's hesitations—

and as the British press was quick to point out—this was essentially a US-Soviet conference.[25] And, specifically, Bevin failed to secure agreement to his proposal[26] for a tripartite commission on Iran.* In general, however, the Moscow conference of the three Foreign Ministers proved to be a marked contrast to the London conference of the five. In the communiqué[28] issued there on 28 December 1945 they agreed on procedure paving the way for a peace conference to draw up treaties with Italy, Romania, Bulgaria, Hungary, and Finland; on the establishment of a Far Eastern Commission and an Allied Council for Japan, in Washington and Tokyo respectively; on the establishment of a Joint US-Soviet Commission in Korea; on the desirability of withdrawal of Soviet and American forces from China 'at the earliest practicable moment'; and on (minor) measures to broaden the Romanian and Bulgarian governments. Potentially most important of all— they also agreed to sponsor a resolution, at the first General Assembly of the UN in January 1946, to set up a UN Atomic Energy Commission.[29]

In the process of winning these successes, it was not only Bevin whom Byrnes snubbed; he also offended first Vandenberg and then—fatally for his own secretaryship of state—Truman himself.[30] Publicly, the President pronounced himself satisfied with what the Foreign Ministers had accomplished in Moscow, which would, he thought, have 'constructive results'.[31] Privately, Byrnes received a reprimand from Truman when he joined the President on the *Williamsburg* immediately after his return from Moscow. Accounts of what passed between the two men on the yacht on New Year's Eve vary. Six years afterwards, however, the text of a manuscript letter from Truman to Byrnes was published, dated 5 January 1946 (after the *Williamsburg* altercation, but before the President's statement about the Moscow meeting), which Truman later included in the first volume of his memoirs.[32] Today this letter is remembered for its vivid concluding sentence: 'I'm tired of babying the Soviets'. Although the letter was almost certainly never sent and in all probability Truman wrote it out in longhand simply as a means of clearing his own mind, it remains a significant piece of evidence of the moment at which the President's personal view began to change, in step with the change in Congressional opinion,

* The Azerbaijan crisis did indeed erupt two months later: the first occasion on which the United States publicly adopted a robust line against Soviet policy in a third country—and moreover a country where the principal western economic interest was not American, but British (the Anglo-Iranian Oil Company). During this episode the two governments, in marked contrast to Byrnes' failure to support Bevin over Iran in Moscow, successfully directed their actions in parallel, if not in concert, towards an identical aim. Although this crisis is often cited as a post-war turning-point in Washington, most accounts over-simplify what was then happening on the ground in Northern Iran.[27]

which was already under way as 1945 drew to a close and which would gradually gather momentum through the following year. Perhaps too it goes some way towards explaining, if not the President's unusual decision to invite [33] the Leader of the British Opposition to deliver an address to Westminster College, at Fulton in his home state, at least what was going on in his mind when he accompanied Churchill there two months later.

5. Churchill and Truman at Fulton, Missouri, 4 March 1946—the occasion of the speech in which Churchill proposed a 'special relationship' between the United States and Britain.

6. Bevin, Marshall, Molotov, and Bidault at the meeting of the Council of Foreign Ministers, Moscow, April 1947.

PART III

ANGLO-AMERICAN DISCORDS, 1945–49

CHAPTER 7

The Invisible Keystone of the Arch

IN the preceding account of the way in which the Anglo-American relationship operated within the trilateral framework of the Grand Alliance in 1945, the main emphasis has been on US policy, because at Yalta, and still more at the Potsdam and Moscow meetings, the initiative rested largely in American hands. Within the trilateral framework, the initiative remained there through 1946, which was a transitional year. This section of the book will assess the principal bilateral negotiations between the United States and Britain that followed the end of the Second World War. Each of these three sets of negotiations stemmed from a British initiative;* all of them lay close to the heart of the Anglo-American relationship; and all of them ended in disappointment, in varying degrees, for Britain.

On 10 November 1945, the Prime Minister arrived in Washington in order to conduct personally with the President (and with the Canadian Prime Minister) the first post-war round of atomic discussions. Attlee's visit coincided with the beginning of the crunch in the Anglo-American financial negotiations, in which he took no part; these were being conducted in Washington, in the wake of the US Government's abrupt termination of Lend-Lease.† Three days before Attlee's arrival in Washington, Bevin had announced in the House of Commons the formation of an Anglo-American Committee of Enquiry, which was to spend the first four months of 1946 deliberating over the problem of Palestine. These nuclear, financial, and Middle Eastern issues form the subject of the chapters that follow, to which there has been added, as a coda, an account of a further issue of Anglo-American discord, whose global consequences were in time to prove the gravest of all: the rift that by the end of 1949 divided the two countries over policy towards the new China.

---

* Historically, it was Truman who first raised the Palestine issue, in the memorandum that he sent (to Churchill) at Potsdam on 24 July 1945 (see Chapter 10). But it was Bevin who proposed the Anglo-American Committee of Enquiry.

† The abolition of the Combined Boards followed, in September 1945. The Combined Chiefs of Staff Committee retained a somewhat shadowy existence, which continued until the formation of NATO.

THE FIRST PHASE

An understanding of the four and a half post-war years of atomic discord between the United States and Britain first requires a summary of the main points of the three wartime agreements: the Quebec Agreement of 19 August 1943; the Declaration of Trust of 13 June 1944; and the Hyde Park Aide-Memoire of 19 September 1944.[1] Chapter 5 has already described how, in the final weeks of the war against Japan, the two governments complied with their obligation under Clause 2 of the Quebec Agreement, to seek each other's consent before the atomic bomb was used. Under Clause 4 of this agreement, the Prime Minister had 'expressly' disclaimed any British interest in the 'industrial and commercial aspects' of atomic energy 'beyond what may be considered by the President of the United States to be fair and just and in harmony with the economic welfare of the world'. As has also been noted in the same chapter, this agreement established the atomic Combined Policy Committee (CPC) in Washington, between whose members and their technical advisers there would be 'complete interchange of information and ideas on all sections of the project'. Clause 5 ended with two provisions:

(c) In the field of scientific research and development there shall be full and effective interchange of information and ideas between those in the two countries engaged in the same sections of the field.

(d) In the field of design, construction and operation of large-scale plants, interchange of information shall be regulated by such *ad hoc* arrangements as may, in each section of the field, appear to be necessary or desirable if the project is brought to fruition at the earliest moment. Such *ad hoc* arrangements shall be subject to the approval of the Policy Committee.

In the following year the Declaration of Trust established the Combined Development Trust, whose purpose was, under the direction of the CPC, 'to control to the fullest extent practicable the supplies of uranium and thorium ores', the supplies required by the Trust being 'held by it in trust for the Two Governments jointly'. Finally, Clause 2 of the Hyde Park Aide-Memoire said: 'Full collaboration between the United States and the British Government in developing tube alloys for military and commercial purposes should continue after the defeat of Japan, unless and until terminated by joint agreement'.

These agreements, known at the war's end only to a very few in the three* countries concerned, have been described by the British official

* Though bilateral in form, the wartime agreements were trilateral in practice. Canadian involvement was the result partly of the importance of Canadian uranium resources and partly of the Anglo-Canadian heavy water project undertaken in Canada at Chalk River, during the war. After the war the atomic relationship remained triangular—an important factor in the evolution of modern Canada.

most closely connected with the post-war atomic negotiations as 'among the most bizarre diplomatic documents ever concluded' and, therefore, providing 'a weak basis' for the British claim that the governments should continue their atomic collaboration and exchange of information in time of peace.[2] Bizarre they certainly were, but so was the subject; and although it could be argued, on a strict interpretation, that their validity was confined to the war, it was not only the hapless Aide-Memoire (the American copy of which was lost for several years), but also the final clause of the Declaration of Trust that expressed the intention of the two Heads of Government that Anglo-American atomic co-operation should not end with the war. As Churchill himself put it in a telegram sent to his colleagues on 21 September 1944, his agreement with Roosevelt had been based on 'indefinite collaboration in the post-war period subject to termination by joint agreement'.[3]

The initial phase of the post-war Anglo-American atomic dialogue, which lasted ten months, was the first major test of the quality of the relationship between the two countries after the Second World War, and, in particular, of the accuracy of Churchill's forecast of indefinite collaboration in the atomic field. In a special message to Congress of 3 October 1945, recommending the establishment of a US Atomic Energy Commission, Truman also announced that he would 'initiate discussions, first with our associates in this discovery, Great Britain and Canada, and then with other nations ... in the field of atomic power'. Five days later, at an impromptu, but on the record, press conference given at Linda Lodge, near Tiptonville, Tennessee, he told reporters that only the United States had the combination of capacity and resources necessary to produce the atomic bomb; if other countries were to catch up with the United States, they would 'have to do it on their own hook, just as we did'.[4] On 23 October the Senate established a special committee on atomic energy; three days later Brien McMahon, the senior Democrat on the committee, became its chairman.

Meanwhile, things had moved very slowly in Britain, partly because there was not a single member of the new government who had been one of the small circle of atomically initiated ministers authorized by Churchill when he was Prime Minister; so much so that the chairman of the Advisory Committee on atomic questions appointed by Attlee on 21 August 1945 was, by virtue of his wartime atomic experience, John Anderson, who sat on the Opposition front bench, even though he was not formally a member of the Conservative Party. Thus the records show Attlee (who until December 1946 was also Minister of Defence) and Bevin following an atomic learning curve similar to the one along which Truman had moved in the early months of his presidency. Attlee communicated his first reaction, originally written in manuscript, to his colleagues at the first meeting of the GEN 75 Committee on 28 August: 'The most that we may have is a few years start. The question is what use we are to make of that few years start'. After ruling out an 'Anglo-American Hegemony in the world', he went on to say that joint action by the three great powers 'based upon stark reality' and a declaration by them 'that this invention has made it essential to end wars' seemed to him the only feasible course and the only one 'to offer a reasonable hope of staving off imminent disaster for the world'. He concluded his memorandum:

I can see no other course than that I should on behalf of the Government put the whole case to President Truman and propose that he and I and Stalin should forthwith take counsel together ... I believe that only a bold course can save ... civilisation.[5]

It was Bevin who then proposed that as a first step the Prime Minister should send Truman a message 'setting out the whole case and suggesting that a review of world policy was needed' in the light of the atomic bomb.[6] His colleagues agreed. The letter that finally went to the President four weeks later had been shown in draft to Churchill, who—as well as some trenchant comments about Atttlee's proposed 'act of faith' in the Soviet Union—had this to say about his own atomic agreements with Roosevelt:

This almost amounts to a military understanding between us and the mightiest power in the world. I should greatly regret if we seemed not (not) to value this and pressed them to melt our dual agreement into a general agreement consisting, I fear, of pious empty phrases and undertakings which will not (not) be carried out ...[7]

It was Attlee's letter of 25 September, followed by a telegram three weeks later offering to come to Washington, that led to the tripartite talks there in early November. In the intervening six weeks the ebb and flow of British views resembled earlier American fluctuations. As late as 11 October Bevin took the view that

there seemed ... to be everything to gain and little to lose by making them [sc. the Soviet Union] parties to the knowledge of the process and trusting to their good faith in observing an international agreement for the control of the weapon ... we should take the risk of giving this information to the Russians in the interests of our foreign policy.

A week afterwards he said that 'further study had convinced him' that 'recent difficulties with Russia' could not be attributed to the issue of the atomic bomb. He then added a vital proviso: 'it must be clear what should be asked for in exchange for this knowledge'.[8] By the time that

Attlee reached Washington his views had also hardened; nor could he ignore the force of Churchill's reminder quoted above.[9]

Five days of deliberation in Washington, partly in the presidential yacht, *Sequoia*, resulted in three documents: two tripartite agreements at Head of Government level and, at official level, an Anglo-American memorandum of intention. Of these three documents, only the first was public: the Tripartite Washington Declaration.[10] This declaration, which was read out verbatim in the White House on the morning of 15 November by the President 'in his flat, Missouri accent', while 'Attlee and King sat slumped in their chairs ... blinking eyes that struck reporters as bloodshot', was the product of much drafting and counter-drafting.[11] It called for the establishment by the United Nations of a commission to make specific proposals on atomic energy. The second agreement consisted of three secret sentences, hurriedly signed by Truman, Attlee, and King on the same morning, which 'reproduced the essence of the 1943 understanding on [atomic] interchange'.[12] The third document was equally secret: a memorandum of intention, jointly drawn up by Major General Groves and Anderson and dated the following day, which listed six points to be considered by the Combined Policy Committee in preparing a further document that would supersede the Quebec Agreement '*in toto* together with all other understandings with the exception of the Combined Development Trust Agreement, which should be revised in conformity with the new arrangements'.[13] Ironically, in the event, the two secret documents were to lead nowhere, whereas the public declaration had a swifter result than its drafters intended at the time.

Truman and Byrnes refused Vannevar Bush's advice that influential senators should be brought into these discussions. Instead, Tom Connally, the Chairman of the Senate Foreign Relations Committee, and Vandenberg were invited only to the concluding White House ceremony, which they reached 'seething with resentment' and left the moment that Truman had finished reading the tripartite declaration (without waiting, that is to say, for the customary photographs). As Bush recorded, this mistake was dangerous.[14] On the British side, a three-power agreement, based on the hope of a virtual monopoly of atomic raw materials, contrasted with a foreign affairs debate held on the eve of the Washington conference in the House of Commons, in the course of which Bevin had noted Parliament's agreement 'on the imperative necessity of Britain retaining her moral lead in the world'.[15] At least one American was aware of this contrast—Groves, who later described the British attitude at this conference as 'there [*sic*] usual saintly position'.[16]

The kindest explanation of this confusion in British logic in

November 1945 is also the simplest: the ministers responsible had not yet made up their minds—a view that is supported by the minutes of the Cabinet meeting held the day before Attlee left for Washington (the only occasion on which the Labour Cabinet, as opposed to a Cabinet Committee, ever discussed the issues of the atomic bomb until December 1950, when once again the occasion was a Prime Ministerial visit to Washington). In 1945 Attlee secured the Cabinet's endorsement of the general line taken in the memorandum that he had submitted to his colleagues on 5 November. In effect, this backed the horse both ways: 'no attempt should be made to restrict the development of atomic energy by any country, in view of the impossibility of effective control'; it was 'essential that atomic weapons should be available to restrain aggression'; but those who developed atomic energy should be determined 'to live up to the principles and purpose of the [UN] Charter', and to back up its authority by using their atomic weapons against an aggressor if the occasion arises'; and at the same time 'a real attempt must be made' over the next few years 'to build a world organisation upon the abandonment of power politics'.[17]

As it turned out, British minds had to be concentrated very soon afterwards. The Tripartite Declaration proposed that the UN Commission should be set up 'at the earliest practicable date' and 'instructed to proceed with the utmost despatch'. Nevertheless, a final line—'there should be consultation at the earliest possible date with the other permanent members of the Security Council'—appears to have been deleted from the final draft.[18] Although, in the words of the historians of the US Atomic Energy Commission, the next logical step was to give the Soviet Union 'a special invitation to share the initiative',[19] for the first week after the conference both Truman and Byrnes maintained the opposite position. Then, on 23 November, Byrnes sent Molotov the telegram proposing a meeting of Foreign Ministers, which caused Bevin the anguish described in Chapter 6.

As has been noted in the same chapter, potentially the most important single item among the agreements reached by Byrnes, Molotov, and Bevin in Moscow at the end of December 1945 was the decision to co-sponsor a resolution establishing the UN Atomic Energy Commission. In January this commission was duly set up. Its subsequent history, and the negotiations with the Soviet representatives in New York regarding international control of atomic energy, which took up the second half of 1946, lie outside the scope of this book, not because of the fact that they ended in deadlock, but because the British role in these negotiations was minimal from start to finish. The American proposals for the control of atomic energy by an International Atomic Development Authority, known to history as the

Baruch Plan, were doomed to failure, primarily because they jettisoned the one assurance on which Stalin had insisted in his talks with Byrnes in December 1945: that the Soviet right of veto in the Security Council would be unimpaired.[20] Yet it was only with reluctance that the British Government accepted Baruch's elaborately hedged proposals, from the day on which he presented them to the United Nations in June up to and including the final vote* in New York on 31 December 1946. Thus, in Bevin's words, 'there was always a risk that they might put forward proposals which we would regard as dangerously weak, in order to secure agreement'.[21] The remark ascribed to Bevin,—'let's forget the Baroosh and get on with making the fissle'[22]—summed up British thinking as it developed through 1946.

The British Government's formal decision to make the bomb was finally taken on 8 January 1947—'formal', because it had long been preceded by other ministerial decisions that unmistakably pointed the way, notably those taken on 18 December 1945 and 25 October 1946.[23] This contrast between the American and the British attitudes towards the issue of international control of atomic energy in 1946 illustrates not just a difference of governmental policy, but a difference between the nature of the two societies at that period. In Britain it was not only the existence, functions, and deliberations of the small *ad hoc* atomic committee of senior Cabinet Ministers that were shrouded in secrecy; so was the whole complex of issues surrounding atomic energy. Parliament never debated these issues, nor was there any equivalent of McMahon's senatorial committee. Whereas the US Atomic Energy Commission published semi-annual reports, the public apathy in Britain about atomic issues—once the first shock of August 1945 had passed—continued until almost the end of the post-war quinquennium.[24]

The reasons underlying both the apathy of public opinion and the reticence of the small inner circle of 'atomic' ministers and officials in post-war Britain have been the subject of searching analysis.† Taken together they had one specifically Anglo-American consequence. By 1946 the mutual ignorance that had been a feature of the transatlantic relationship in the inter-war years, though lessened by the wartime experience, had not yet been succeeded by the detailed knowledge, both academic and official, of each other's political system that is taken

---

* The final vote in the UN Atomic Energy Commission was a formal victory for the Baruch Plan; the Soviet Union abstained. The Commission's life dragged on until May 1948, when it was wound up.

† Looking back after a quarter of a century, Roger Sherfield took the view that secrecy 'probably was overdone, but it did not matter much', whereas in her official history, Margaret Gowing ascribed the British secrecy primarily to 'awe and fear'.[25]

for granted in both the United States and Britain today. Moreover, British experience of Washington decision-making in time of war, centralized in the White House, was not a good guide to the much broader process that was, by the end of 1945, resumed in the United States. Because the few who took atomic or atom-related decisions in Britain could do so—as they themselves saw it—on the merits of the case and with little or no heed paid to parliamentary, still less public, opinion in their own country, they did not, with few exceptions, find it easy to understand the different parameters within which corresponding decisions were reached in the United States.

The US Administration had to take account not only of conflicting views within the bureaucracy—notably between the War Department and the State Department—but also in Congress and among the lobbies of varying opinion, nationalist and internationalist, across the country. In order to secure his primary objective—legislation guaranteeing civilian control of atomic energy in the United States—Truman had to steer a middle course between these views. This course involved much presidential tacking: for example, Truman's approval of the contrasting appointments—from opposite political poles—of Baruch as US Representative to the UN Atomic Energy Commission (soon afterwards described by Byrnes, who made the appointment as Secretary of State, as 'the worst mistake I have ever made') and of David Lilienthal, the former Chairman of the Tennessee Valley Authority,[26] as Chairman of the US Atomic Energy Commission.* Already at the turn of the year 1945/6 the pendulum of American opinion on atomic issues was swinging to an extent that alarmed those on the British side in Washington, particularly James Chadwick[27] and Roger Makins,[28] who were aware of the need for haste if the British Governments were to get what they wanted from the two secret documents agreed in Washington on 15 and 16 November 1945.

The first of those documents simply said this:

We desire that there should be full and effective cooperation in the field of atomic energy between the United States, the United Kingdom, and Canada.

We agree that the Combined Policy Committee and the Combined Development should be continued in a suitable form.

We request the Combined Policy Committee to consider and recommend to us appropriate arrangements for this purpose.[29]

---

* The nationalist and internationalist views in the United States were epitomized by Baruch and Lilienthal respectively. It is, however, important to realize that many Americans—for example, Bush—of the internationalist persuasion agreed with the nationalists in opposing atomic co-operation with Britain, but for a different reason: they believed that it would reduce the chances of reaching agreement with the Soviet Union. Byrnes agreed with Bush.

The six points which the Groves–Anderson memorandum listed for the Combined Policy Committee's consultation were specific. They represented a precise calculation by two men, each of whom knew the subject inside out, of where American and British national interests lay. For the United States, the essential objective was the extension, if possible, of Anglo-American control of supplies of uranium and thorium world-wide, to the point of becoming a virtual monopoly. (This was an illusion, because there were uranium resources in Eastern Europe.) Since the United States' residual moral obligation to Britain—the legacy of the wartime atomic relationship—was a constantly diminishing factor, Britain's rights in the Trust were by far the most valuable card in the British hand, although it was a card that could be played only with the greatest care; without joint action with the United States, Britain might well have been unable to secure uranium for its own atomic project; and moreover it was paid for in sterling.[30] For Britain, the crux was the resumption of the atomic interchange, virtually in accordance with the formula agreed in 1943. For the rest, the memorandum consisted of a compromise, whereby the United States would no longer be obliged to seek British *consent* for the use of atomic weapons—as required by the Quebec Agreement—but need only *consult* the British and Canadian governments; there would also be an obligation to consult together before any atomic information was disclosed to other governments or negotiations about atomic energy were entered into with other governments. In return—though this was not explicitly stated—Britain recovered the freedom, in effect renounced in 1943 at Quebec, to exploit atomic energy for its own 'industrial and commercial' purposes. Finally, the new agreement that was to replace the Quebec Agreement would be tripartite, if the Canadian Government so wished.

Within six months of their signature in Washington the agreement recorded in both these documents had sunk, though not without trace; it was salvaged, to some extent, in January 1948. In 1946 the formal death blow was delivered by Truman in his telegram of 20 April. In it he told Attlee that, had anyone informed him five months earlier that the United States should 'obligate' itself to furnish the engineering and operational assistance for another atomic energy plant, he 'would not have signed the memorandum'. He went on:

I would not want to have it said that on the morning following the issuance of our declaration to bring about international control we entered into a new agreement, the purpose of which was to have the United States furnish the information as to constitution and operation of plans which would enable the United Kingdom to construct another atomic energy plant. No such purpose was suggested by you or thought by me.[31]

Attlee responded on 6 June with a long telegram going over the whole history of the Anglo-American atomic relationship of the previous five years.[32] There was no reply.

These were the last formal exchanges, but in reality the critical dates came earlier in the year. On 15 February 1946 the Combined Policy Committee was informed of the British decision, taken two months earlier, to build a large-scale reactor to produce plutonium.[33] The moment of communicating this information, although it was in no way novel, was unfortunately timed. It followed hard on the heels of the public disclosure of the spy ring in Canada, which included a British physicist, Alan Nunn May, who had worked on the Montreal atomic project (the news of his arrest first appeared on the same day as Churchill's Fulton speech). On 1 April the Senate Atomic Energy Committee considering McMahon's bill began a radical amendment of the section of the bill dealing with 'Dissemination of Information'; and by 10 April this section, renamed 'Control of Information', provided for the full range of penalties for criminal actions involving unauthorized dissemination of 'restricted data' which was defined in terms that allowed very little latitude. The McMahon bill was passed by the Senate on 1 June and, further strengthened by the House of Representatives, it received presidential approval two months later.[34]

Neither Truman nor Attlee comes well out of this affair. In Acheson's words, 'a Government, having made an agreement from which it had gained immeasurably, was not keeping its word and performing its obligations'.[35] The British Government, on the other hand, relied on the belief that its own interpretation of American commitments to Britain in the atomic field would be accepted in Washington, without seeking to ensure that those responsible for atomic legislation in Congress were made—directly or indirectly— aware of these commitments. (It is a measure of the strength of the nationalist view of atomic issues in the United States in 1946 that even Baruch was assailed by the Hearst press for proposing to give 'foreign masters the American secret of the atomic bomb'—attacks which continued right up to the end of the year.)[36] The wisdom of the few on the British side—Chadwick wanted to talk privately to McMahon, but was overruled—who realized that the British Government was going about the matter in the wrong way, was justified years later when McMahon told British Ministers that, had he been properly briefed about the history of US atomic co-operation with Britain and Canada, his bill would probably not have needed to have been passed on such restrictive lines.[37] Be that as it may, after 12 May 1947, when Acheson at last gave the Joint Congressional Committee, in secret, an oral summary of the Quebec and Hyde Park Agreements, the terms of reference of the Combined Development Trust, the membership of the

Combined Policy Committee, and the outline of Anglo-American agreements on raw materials, the Committee's initial expression of shock at the Administration's conduct of such a controversial policy behind Congress' back for so many years was not followed by leaks to the press or by an outcry against the Administration.[38] Whether or not one State Department participant in the post-war atomic negotiations between the United States and Britain was right in the view which he expressed years later, that the McMahon bill was seen by the Administration as 'an easy, tacit way to end the relationship begun by Roosevelt and Churchill',[39] the atomic relationship between the two governments was certainly something that the Administration did not strive officiously to keep alive.

This first phase of Anglo-American post-war diplomacy left the British licking wounds that took many years to heal. But the voices on the British side that carry most conviction forty years afterwards are those of the leading members of the British atomic team: John Cockcroft, lamenting Britain's 'masterly inactivity' in atomic affairs in Western Europe* into which the British Government was forced by the US Government; Christopher Hinton, expressing the heretical view that the McMahon Act was the best thing that could possibly happen because it would make the British think for themselves; and Chadwick, though well aware of the overall importance of the Anglo-American relationship to British foreign policy, asking 'are we so helpless that we can do nothing without the United States?'[41]

THE SECOND PHASE

Despite the failure of the first post-war attempt, the Combined Policy Committee remained in existence and the Combined Development Trust continued to function in Washington. In the second attempt, the initiative was not British, but American. So also was the pace. After less than a month of negotiation, a new tripartite—for the Canadian Government was again a party—atomic agreement† was concluded on 7 January 1948.

---

* In the immediate post-war years the Anglo-American atomic relationship was paramount for Britain. Although the British wartime atomic debt to France was very considerable, French requests for atomic collaboration were resisted without too much difficulty, given that Jean-Frédéric Joliot Curie, the head of the *Commission à l'Energie Atomique*, was, if not a member of the French Communist Party, a Communist sympathizer. The claims of the Belgian Government were harder to refuse because of the help that they had given over the supply of uranium from the Congo during the war. There were 'even a good many misunderstandings between London and Ottawa ... not handled in the British side with much consideration for Canadian sensitivities'.[40]

† The text of this agreement, the so-called *Modus Vivendi*, is public, but the principal British documents relating to it are in the 'retained' category and therefore not available in the Public Record Office. Unascribed references and quotations relating to this in pp. 89–91 are thus indebted to Chapter 8 of Margaret Gowing's official history (written with full access to the British records).

In the intervening two years both the international climate as a whole and the relative political strengths of the United States and Britain had undergone a change. 1947—most significantly, perhaps, what happened in Paris on 2 July 1947—marked the onset of what Walter Lippmann had in that year christened the 'cold war'. In this new political environment the ease of communication between officials of the two governments across the negotiating table was greater than it had been two years earlier. The intellectual firepower assembled on both sides was also impressive. To take two examples, one on each side of the Atlantic (and both born in the same month of 1904): in London, Bevin brought back Roger Makins from Washington in 1947 to become, first, Assistant Under-Secretary and then, in the following year, Deputy Under-Secretary at the Foreign Office; he was also directly responsible to the Prime Minister as chairman of the atomic energy official committee in Whitehall.[42] In Washington, after Marshall had taken over the State Department in January 1947, one of his first appointments was that of George Kennan,* who headed the newly formed Policy Planning Staff for its first two and a half years, with the preparation of a new policy for Europe as his first task. Marshall's initial instruction to Kennan—'avoid trivia'—remains a model of ministerial brevity.

Yet, for the British, the summer of 1947 also brought the first of a series of economic crises. In July 1947 the same volume of *Foreign Affairs* that carried Kennan's 'X Article' also published an article by the Soviet economist Evgenii Varga on 'Anglo-American rivalry and partnership'. Varga's article included the statement that Anglo-American 'relations, viewed in perspective, appear a peculiar combination of antagonism and cooperation, as a result of which the United States is constantly gaining ascendancy over Britain, reducing her more and more to the status of a second-rate power in both economic and political respects'. The subtitle of his article was 'A Marxist view'.[44] Varga's observation, quoted above, however, was in keeping with a sentence in an assessment written with a fresh eye by Oliver Franks, not long after he had taken over the Washington Embassy in 1948:

the whole question of our relations with the Americans on atomic energy questions seems to me to be increasingly bound up with the larger issue of the extent to which the Americans are prepared to treat us on more or less equal terms as a first-class power.[45]

* He was well known to members of the British Foreign Service from his service in Moscow. It was from there that in February 1946, he sent his memorable 'long telegram' on Soviet intentions and how the West should handle the Soviet Union. This was followed by his anonymous 'X Article' published by *Foreign Affairs* in 1947, which launched the concept of 'containment'.[43]

It is in the light of Franks' astringent observation that this second phase of Anglo-American nuclear diplomacy has to be considered today.

In the hasty negotiations, conducted almost entirely at official level, that preceded the atomic agreement of January 1948, American and British motives were fundamentally different. Uppermost in American minds was their fear that, by the end of 1949, US stocks of uranium would be almost exhausted, coupled with the wish to acquire uranium held in Britain as cheaply as possible. (For reasons that are still not clear, the British did not at first realize that it was this that had impelled the urgent US approach in December 1947.) And in the process the US Administration was determined to be rid of Clause 2 of the Quebec Agreement, by which 'members of the Joint Committee on Atomic Energy, particularly Senator Vandenberg and Senator Hickenlooper, had been very disturbed'; they had 'urged most strongly' that the British right of veto should be abrogated.[46] The basic British aim was to 'wipe out the misunderstanding and bitterness of the past and put our relationship on a solid basis', and to do this before 'the rats' could get at it (a particular British anxiety was that Marshall Aid might be used as an American bargaining counter: a suggestion of the Defense Secretary, James Forrestal, which was in the event overruled by the State Department).[47] The form of the 1948 agreement was designed to avoid both the requirement of Article 102 of the UN Charter—that every treaty and international agreement entered into by any member of the UN must be registered as soon as possible with the UN Secretariat and be published—which had caused the first misgivings, both in London and in Washington, about the secret 1945 agreements, and the need for reference to Congress. Unlike in December 1945, the US Government was now bound by the strict provisions of the McMahon Act, which severely circumscribed both the form and the substance of the new agreement. Instead of signing an agreement, the procedure adopted this time was simply to hold a meeting of the Combined Policy Committee, at which—on 7 January 1948—each of the representatives of the three countries in this committee formally declared his government's intention to proceed on the basis of a *Modus Vivendi*: a document consisting of eight articles* and three annexes.[48]

Under the terms of the first of these articles, all earlier atomic agreements were regarded as null and void, with certain exceptions, one of which was the (public) tripartite Washington Declaration of 15 November 1945. Of the three annexes to the new agreement, the first

* The text of these eight articles will be found in Appendix 3.

virtually guaranteed to the United States the allocation of all supplies of uranium available from mines in the Belgian Congo in 1948 and 1949, together with a formula that provided for these supplies to be topped up, if required, with two-thirds of the unallocated stocks already held in Britain. The third listed nine 'suitable subjects in which cooperation and the exchange of information, at the present time, would be mutually advantageous' to the three signatories.

Other than its provisions with regard to raw materials, this agreement had no set term. It is not the first temporary accommodation in diplomatic history to have had definitive results: Clauses 2 and 4 of the Quebec Agreement lapsed. Thus, the British Government quietly dropped the secret right of veto on American use of atomic weapons, which it had held, under the Quebec Agreement, for four and a half years. Nor did the right of consultation, envisaged in the abortive memorandum of 16 November 1945, survive. With only one exception,[49] no British minister or official appears to have expressed any concern about this issue at the time. It was this concern, belatedly felt, that formed the background to Attlee's decision to make his second visit to Washington, nearly three years later. However, because the crisis of December 1950 was part of the nadir of the Korean War—and this Anglo-American problem was not resolved even then—it has seemed best to reserve it for consideration in Chapter 18.

As for the flow of secret atomic interchange released by the meeting of the CPC on 7 January 1948, this varied greatly from topic to topic. Although it certainly produced some very useful information at a critical moment for the British atomic project, so far as 'Topic 6'—entitled 'Fundamental properties of reactor materials (i.e. solid state physics, basic metallurgy)'—was concerned, American permission for discussion of this, which was high on the British list of priorities was withheld.[50] The US Atomic Energy Commission's rules governing the programme of technical co-operation embodied in the agreement were such that within a year the Chairman of the Commission, Lilienthal, recorded his view that 'the kind of in-between world in which we are living is a source of irritation to our friends and to us and worsens rather than improves relations with the United Kingdom and Canada'.[51]

1948 was the year of the Berlin blockade and also of the negotiations that led up to the North Atlantic Treaty. The closeness of Anglo-American co-operation in these fields of defence in no way lessened their dissension in atomic matters. On the contrary, as this year went forward, the Anglo-American atomic argument broadened out into a further issue: the vulnerability of Britain in the event of war, and hence the possible loss of atomic information and material. From here it was not a long step in Washington's logic, on the eve of the signing of the North Atlantic Treaty in 1949, to the idea that the right context in which to set any fresh approach to the Anglo-American atomic relationship was the integration of defence policy, conceived as a whole. Fresh tripartite negotiations in this new context did not begin until a meeting of the CPC held on 20 September 1949.[42] In preparation for this meeting, the Administration had at last cleared the ground with Congressional leaders, at a memorable meeting held at Blair House on 14 July 1949, when the entire Anglo-American atomic background was explained to them by Truman, Acheson, and Eisenhower.[53] In consequence, the constraint that had made it impossible even to mention the words 'atomic weapon' in the *Modus Vivendi*, now no longer applied.

The essence of the 1949 US proposals was to maximize the conservation of atomic materials, research, production, and the storage of atomic weapons in North America; to minimize them in Britain; and, in return, to supply Britain with American atomic weapons. On such a proposal the chances of a meeting of Anglo-American minds were minimal. Nevertheless, the ball was still, in a formal sense, in play between the two governments when just over a month later—in Acheson's words—'a bomb exploded in London':[54] the arrest of Klaus Fuchs. This arrest would certainly have brought the negotiations to a halt, but in fact the new US Defense Secretary, Louis Johnson, had already rejected the most recent British counter-proposals. Following Fuchs' arrest, Johnson went on to end US participation in the Combined Development Agency; and he also tried, unsuccessfully, to abrogate what remained of the *Modus Vivendi*.[55] The *Modus Vivendi* survived through the early 1950s; but in any case by then the second phase of the nuclear age—that of the hydrogen bomb—was beginning.

Thus the last attempt to remodel the Anglo-American atomic relationship undertaken in the 1940s ended in a miasma of espionage; a year later Donald Maclean (then Head of the Foreign Office North American Department) fled to Moscow. By comparison with Fuchs', Maclean's direct contribution to the Soviet atomic programme can hardly have been large, but during his time at the Washington Embassy he had been British secretary of the Combined Policy Committee and British representative in the Combined Development Trust/Agency. By a final twist of Anglo-American fate, in September 1948 (when Maclean left Washington) the Kremlin must be assumed to have had full knowledge of the previous five years of the Anglo-American atomic relationship—nearly a year before the Blair House meeting and at a time when in the British Parliament this knowledge was confined to a handful of members on the two front benches.*

The cause for the deepest regret arising from over four years of
Anglo-American atomic negotiations is that no joint attempt was made
to solve what was surely the crucial atomic problem: the issue of
international control. Such an attempt might well have failed. But the
fact that those years coincided with the final outbreak of Stalin's
political paranoia (which lasted until his death, in March 1953) is not,
I suggest, a sufficient excuse. Nor did the issue take either the
Americans or the British by surprise after Hiroshima, although the new
British Government did have to do some rapid homework. As early as
May 1944, Cherwell had expressed to Churchill his belief that 'plans
and preparations for the post-war world or even the peace conference
are utterly illusory so long as this crucial factor is left out of account';
reporting in the same month from Washington to Anderson, Campbell
had included in a long letter (which was submitted to the Prime
Minister) the view that 'it is more and more evident that opinion about
the nature of long range control of this special military technique,
including the ultimate question of entry of other nations into the
control, is becoming a fundamental point'; and in a manuscript letter
which Smuts wrote to Churchill on 15 June 1944, he said: 'If ever there
was a matter for international control, this is one'. Instead, the atomic
issue that preoccupied Churchill at his final meeting with Roosevelt—
on board the U.S.S. *Quincy* in Alexandria harbour on 15 February
1945—was not international control of atomic energy, but the American reaction to the prospect of Britain 'going ahead with [its] own
work on Tube Alloys'. As we now know, he took this opportunity to tell
the President, in Hopkins' presence, that the British would be doing
this. According to Churchill's own record, 'The President raised no
objection of any kind. He ... mentioned September for the first
important trials.'[57]

What would have happened, had Roosevelt survived and Churchill
been re-elected in 1945, no one can say. Certainly, to have had a
chance of achieving any measure of success at all, an Anglo-American
effort to solve the problem of control would have had to have begun, at
the latest, immediately after the attacks on Hiroshima and Nagasaki.
In the event, six months later the Acheson–Lilienthal Report, published on 28 March 1946, proved to be the 'last best hope' in the early
post-war period.[58] In this, as in the deliberations of the Interim
Committee in the previous year, the British played no part.

* The only parliamentary statement in four years was the Defence Minister's reply to a
question put on 12 May 1948: '... all types of weapons, including atomic weapons, are being
developed'. Asked to elaborate his reply, he refused.[56]

In the narrower, bilateral atomic context, a judgement on these years depends largely on what view one takes of the *Modus Vivendi* of 7 January 1948. This agreement has loomed larger for British than American historians. 'In the long history of strange atomic energy agreements, the *Modus Vivendi* emerges as the strangest of them all' is the verdict of the official historian of the United Kingdom Atomic Energy Authority, who doubted whether the crumby half loaf with which the British contented themselves in 1948 was ever worth what they had given up for it.[59] On the other hand, for the principal British negotiator of the *Modus Vivendi*, the British cancellation of the (veto) Clause 2 of the Quebec Agreement was balanced by the American cancellation of Clause 4: 'it was a Roland for an Oliver ... they were well out of the way'[60]—precisely the American view. That this was also the view taken in Whitehall at the time is clear. What is less clear is why the 'consultation' formula, agreed two years earlier between Groves and Anderson—a hard-headed American and a hard-headed Scot— was allowed to go by the board. And would the British Government not have done better, at the turn of 1947/8, simply 'to drop the whole thing and ... go it alone'? This was indeed what Lilienthal believed to be the right course for both governments, and what he said, according to his journal, at the Blair House meeting in July 1949.[61] Moreover, in the end, this is virtually what did happen.[62]

The reasons why, for over four years, neither government was prepared to 'drop the whole thing' were radically different on each side of the Atlantic. And it is this difference that largely explains the failure of each attempt to renegotiate the Anglo-American atomic relationship made in those years. For the Americans, until the Soviet atomic explosion of 29 August 1949 finally dispelled the illusion of a western monopoly of atomic raw materials, the attainment of this monopoly was the paramount consideration. That the British were willing to return repeatedly to the negotiating table and to concede what they did, in particular in January 1948, is convincing evidence of the great store set by the few atomically initiated ministers and officials in London on crowning the overall Anglo-American relationship, in secret, with a privileged atomic understanding: the invisible keystone of the Anglo-American arch. Another ten years were to go by before this stone was at last put in place, in very different circumstances.[63]

CHAPTER 8

The American Loan to Britain

By common consent, the Marshall Plan is the greatest act of American statesmanship since the Second World War. If the Anglo-American financial relationship during the years 1945-9 is considered in isolation from what went before, the Marshall Plan then becomes its highlight. But this simple approach leaves unanswered the question why the brilliant success of 1948 was followed by the sterling crisis of 1949. It also fails to take full account of the consequences in the longer term. It is, therefore, preferable to reserve the Marshall Plan, which was in any case a US-European enterprise, for later treatment, and to set the sequence of bilateral Anglo-American financial developments during the post-war quinquennium in a broader context. This more complex approach to the two countries' financial relationship with each other during these years will oblige us to look outside the time-frame both of Part III of this book and indeed of the book itself, in particular at the second half of 1944.

Just as there is general agreement that the swift British response to Marshall's speech coincided with the beginning of a new phase in the overall Anglo-American relationship after the war, so too there can be no doubt about the date of which this relationship hit rock bottom: 13 December 1945. Unlike the disagreements between the two governments discussed in previous chapters, which were going on at the end of that year behind the scenes, this was a public event—the vote on that day in the House of Commons on the Anglo-American Financial Agreement: 345 votes in favour against 98, but the large number of abstentions made it an unimpressive victory for the Government. How did it come about that so many British Members of Parliament, on both sides of the House, including Churchill himself and a future Labour Prime Minister (James Callaghan), reacted as they did to an American line of credit of US$3.75bn., at an interest rate of 2 per cent, repayable over a period of fifty years, with a grace period of six years and a provision for waiver of interest? The sum* of the US loan was greater than the entire potential lending power of the International

---

* The figure of $3.75bn. was reached by Truman's splitting the difference between Vinson's figure of $3.5bn. and the State Department's figure of $4bn. The subsequent Canadian loan brought the total sum available to Britain up to US$5bn.

Bank at that time. Moreover, the same agreement wiped out all but US$650 m. of Britain's Lend-Lease debt, which was to be repaid on the same terms as the loan. (The Soviet Union received nothing at all; discussion of a possible US loan to the Soviet Union petered out in 1946; and the issue of Soviet Lend-Lease debt was not resolved until Nixon's visit to Moscow in 1972). Viewed in themselves—without the accompanying conditions—the size and the terms of the US loan to Britain would seem generous today. Comparisons are difficult, because—at the time of writing—commercial interest rates are about five times higher than those prevailing in 1945; and in order to qualify for a loan of such long maturity today, the per capita income of the borrowing country would have to be less than US$795. Nevertheless, the softest possible terms available now to the poorest category of *developing* country from the International Development Agency, whose loans rarely exceed US$150 m., would be a maturity of fifty years, a grace period of ten, and no interest.[1]

In order to understand the fierce British reaction at the time, we need to consider not only the starkly contrasting state in which the Second World War had left the western world's two largest economies, but also the philosphical difference between the American and British concepts of international economic management at the war's end. In the United States of 1945 there was a political determination to bring both American power and American ideas to bear in the creation of a new international economic order.[2] This new order was confidently expected to benefit the world as a whole as well as to advance the political and commercial interests of the United States. The thought that the interests of the United States and those of other countries might not be identical was scarcely entertained; the only limiting factor then in sight was Congressional opinion.

The massive expansion of American industrial output and the simultaneous impoverishment of Europe during the Second World War had left the United States in an extraordinary economic position in 1945; but the impact of the American conceptual approach to the problems of post-war international economic management, whatever flaws the subsequent course of events may have revealed, was hardly less formidable. The basic dogma of economic faith, passionately believed in both by the State Department under Cordell Hull and by the Treasury under Henry Morgenthau, was simply that 'if goods can't cross borders, soldiers will'. Nor was the United States' principal ally in the war exempt from its share, in American eyes, of responsibility for the economic disasters of the inter-war years. In the worlds of Hull's celebrated indictment[3] of the Ottawa Agreements (which had established the network of Imperial Preferences in 1932), these were 'the greatest injury, in a commercial way, that has been inflicted on this country since I have been in public life'. Britain had, so to speak, been warned; and part of the warning had been put on record from the outset of Lend-Lease* over three years before the war ended.

The American remedy for the economic evils of the 1930s was to provide the post-war world with a multilateral trading system, under which all barriers to international trade and international payments would be both reduced and rendered non-discriminatory. The instruments devised by the United States, jointly with Britain, for the introduction of this system were to be three entirely new multilateral institutions: the International Monetary Fund, the International Bank,† and the International Trade Organization. Of these, the first two were established pursuant to the Final Act of the UN Monetary and Financial Conference, signed at Bretton Woods, New Hampshire, on 22 July 1944; the third was still-born, being replaced in effect by the General Agreement on Tariffs and Trade (GATT), negotiated at Geneva three years later. In the United States, support for the concept of a multilateral international trading system was virtually unanimous; the question for debate was the extent to which the American taxpayer should contribute to its establishment and its support—an issue finally resolved on 18 July 1945, when the Senate passed the bill authorizing US participation in the International Monetary Fund and the International Bank. So strong was American belief in the rightness of the proposed multilateral trading system and so great was the Administration's anxiety to overcome conservative Congressional opposition to the Bretton Woods agreements that their spokesmen—including men like Dean Acheson and William Clayton, who should have known better— made extravagant claims about what the two new institutions could be expected to achieve; and they flatly denied the need for any other measure to finance the post-war reconstruction of Europe and Asia.[4]

The new institutions were predominantly an Anglo-American creation. Nevertheless, it was symptomatic of the difference between the American and the British attitudes towards them that, at the moment when the long Congressional debate on the Bretton Woods agreements ended, Parliament had not even begun to consider them; the deadline, under the terms of the Final Act, was 31 December 1945.[5] The British attitude was generally cautious and—by 1945—increasingly fragmented by political considerations. On the left of the political spec-

---

* 'The elimination of all forms of discriminatory treatment in international commerce' was enshrined in Article 7 of the Mutual Aid Agreement (establishing Lend-Lease) signed by the two governments in February 1942.

† Its full title was, and remains, the International Bank for Reconstruction and Development, although in recent years it has been generally known as the World Bank.

trum, the crucial question was whether the new system would conflict with the prior domestic claims of full employment; for the right, the Imperial Preferences were sacred; and the American case against these preferences was not helped by the US Administration's abandonment, at the end of 1944, of the earlier proposals for a single-stroke, across-the-board, reduction of tariffs. In almost all quarters in Britain there was doubt whether the United States would, when it came to the point, be willing to accept the full responsibilities of the world's greatest creditor.[6]

Given such wide differences of economic philosophy, it is remarkable that agreement was reached at Bretton Woods at all. Would it have been, if death had removed Keynes' intellectual authority from the transatlantic economic debate two years before the end of the war, instead of in April 1946? However that may be, in both the United States and Britain the memory of the economic consequences of the Peace of Versailles was vivid. Underlying the Bretton Woods agreements there was also an Anglo-American consensus: the fundamental fear of post-war planners on both sides of the Atlantic that the all-out industrial performance of the war years would, sooner or later, be followed by a depression,* and their determination not to repeat the self-defeating mercantilism of the inter-war period. The initial economic failures after the Second World War notwithstanding, it remains a signal Anglo-American achievement to have convened the Bretton Woods conference over a year before the war ended and almost a year before the United Nations Organization itself was established. By comparison with the economic illiteracy of Versailles, the Bretton Woods Final Act was a monument of informed far-sightedness.

Intellectually, both these institutions were the fruit of months of Anglo-American discussion, in which Keynes led the British delegation, while Henry Morgenthau's assistant at the Treasury, Harry Dexter White, was effectively his American opposite number. Financially, the United States and Britain were to become the two largest contributors to both the Fund and the Bank. The Bretton Woods agreements might have been capable of avoiding much post-war controversy and confusion in the international financial field but for one fatal defect: the total volume of the Fund's quotas and of the Bank's subscriptions, agreed in the Final Act, was far too small for either institution to make more than a marginal contribution to the immense problems that, in the event, the task of post-war reconstruc-

---

* This was the fate for the western economic system predicted in the Kremlin. Evgenii Varga, Head of the Soviet Institute of World Economy and Politics, was reprimanded for taking the opposite view, which was condemned as 'bourgeois objectivism'.[7]

tion presented. In the form that the International Monetary Fund assumed in the Final Act, its resources were only a third of what Keynes' earlier plan for an international Clearing Union had originally envisaged.

To explain how this came about would lead us even further back in history.[8] For present purposes it is enough to recall that at its outset (the Final Act) the Fund's capital was agreed at US$8.8bn., and that, following the Soviet Government's failure to ratify the Bretton Woods agreements, its capital was less than US$8bn. As for the Bank, the first of its objectives listed in its Articles of Agreement was 'to assist in the reconstruction and development of territories of members . . . including the restoration of economies destroyed or disrupted by war'; and the last, stated in the same article, was 'to assist in bringing about a smooth transition from a wartime to a peacetime economy'. But of the Bank's authorized capital of just under US$10 bn., only 20 per cent was subject to immediate call; of that percentage, only 2 per cent was payable in gold or US dollars; and the Bank's loans and guarantees were limited to 100 per cent of its capital.[9] Thus, although the institutions established by the Bretton Woods agreements were to flourish later on,* they were not geared for the immediate post-war years. Nor was the modern World Bank's vital role—the provision of project finance for developing countries—a major part of the original Bretton Woods design, although the first of the International Bank's articles did include a reference to 'less developed countries'.

The contrast between the economic condition of the United States and that of Britain in the summer of 1945 can be summed up in two remarks made at the time. The new Secretary of the Treasury, Fred Vinson, observed that 'the American people are in the pleasant position of having to learn to live 50 per cent better than they have ever lived before'[10]—a statement that reflected the fact that US national output had more than doubled in real terms during the war.† In London, Keynes' memorandum on 'Our Overseas Financial Prospects', written the day before VJ Day (only a month after Orme Sargent's 'stocktaking' memorandum), warned the government of the prospect of a 'financial Dunkirk'.[11] A week later Truman‡ announced the end of Lend-Lease:[12] in Churchill's words in Parliament, a 'rough

---

* And without the Parkinsonian proliferation characteristics of so many post-war bureaucracies. In 1980, for example, the staff of the IMF numbered 1,530: a statistic for which the author is indebted to the *UN Yearbook*, 1983, cited by Robert O. Keohane and Joseph S. Nye, *Foreign Policy*, No. 60 (Fall 1985), p. 156.

† American GNP rose from US$91bn. in 1939 to 210bn. in 1945; manufacturing volume nearly trebled.

‡ Both Acheson and Clayton were abroad at the time. The termination applied equally to all allies of the United States, although shipments to China were later authorized.

and harsh' decision.[13] It was not only that, at a single stroke, Britain had lost the vital prop of Lend-Lease which, together with Canadian Mutual Aid, had sustained its economy for over four years of war. That was blow enough in itself. At the Quebec Conference a year earlier Churchill had sought to pave the way for a smooth transition; he had temporarily succeeded;[14] and in 1945 even Keynes had assumed that Lend-Lease would be maintained for at least a few weeks, if not months. In reality, the prospect was even starker than Truman's stroke of the pen implied. One of the conditions of Lend-Lease had been that it should not be used to build up British financial reserves; and until the very end of the war* US pressure to keep British reserves down was rigorously exerted.

By the end of the war total British reserves amounted to little more than £1bn. On the eve of the war Britain's gold reserve alone had stood at £650m.—in very different circumstances. Over £7bn. (a quarter of Britain's national wealth) had been sold to pay for the war. British invisible income from overseas had been halved. The so-called sterling balances, which were soon to become the source of a major Anglo-American misunderstanding, amounted to over £3bn.[16] Because they had been swollen by British debts incurred during the war to Commonwealth countries and to Egypt,† in Washington they were seen as comparable with Lend-Lease debt, incurred by Britain during the same period for similar purposes. On this view, the term 'balances' was euphemistic. In London, on the other hand, these sums were primarily regarded as an integral part of the sacrosanct obligations of the Bank of England, the central manager of an international currency—sterling—in which the world's confidence must be maintained: not debts, therefore, but balances. (This was the paramount consideration, but the balances were also seen in Whitehall as a reserve of future purchasing power for British exports and in certain cases— notably India and Egypt—as a potential aspect of post-war foreign policy.) Semantics apart, the British post-war financial dilemma was, at the least, bound to be complicated by the paradox that, in order to finance its war effort, as well as the vast sums borrowed from the United States and Canada, Britain had also become heavily indebted to the countries of the sterling area, for whom the City of London was the traditional supplier of capital.

Against this sombre background, the earliest date by which Keynes' memorandum could project a balance in Britain's overseas payments,

---

* Even as late as Potsdam Truman sent the Prime Minister a memorandum alleging that British gold and foreign exchange holdings were higher than they should have been—a charge vigorously repudiated by the Chancellor of the Exchequer.[15]

† India and Egypt together accounted for almost half the sterling balances in 1945.

with extreme good fortune, was 1949. Meanwhile, as Keynes put it, 'Ministers would do well to assume that ... we are ... virtually bankrupt', facing over the next three years a 'deficit of the order of $5 billion which can be met from no other source but the United States'. In the event, it was not until 1950 that Britain achieved its first post-war balance of payments surplus, after two major financial crises, each of which did indeed bring the country to the brink of bankruptcy.

Whereas seven British Cabinet Ministers had crossed the Atlantic to attend the Ottawa Economic Conference only thirteen years earlier, neither in the 1945 nor in the 1947 financial crisis does the Chancellor of the Exchequer or any other minister seem to have contemplated a journey to Washington (Attlee's visit was concerned with quite other matters). Instead, in 1945 Keynes was once again appointed Britain's negotiator. Among the papers relating to Anglo-American wartime economic discussions this jingle was found:[17]

> In Washington Lord Halifax
> Once whispered to Lord Keynes:
> 'It's true *they* have the money bags
> But *we* have all the brains'.

A glance at the quotation from the Foreign Office memorandum of March 1944, mentioned in Chapter 3, is enough to confirm that, as these lines suggest, the attitude then adopted by the British official mind towards the United States was not characterized by intellectual modesty. In 1945, in the marathon negotiations that began in Washington on 13 September, Keynes was initially over-sanguine—he at first hoped for US$6bn., either as a grant or as an interest-free loan—and hampered by unrealistic instructions from London; he was unable (because of his health) to fly home for consultations; and he was the last man likely to get on with Vinson, of whom he made merciless fun.[18]

In the agreement finally signed in Washington on 6 December 1945, the pride and the prejudice of both sides were given full rein. Among the many Anglo-American compromises that had been embodied in the Bretton Woods Final Act, one was now renegotiated. The very first of the Fund's Articles plainly set out the signatories' commitment 'to promote exchange stability, to maintain orderly exchange arrangements among members, and to avoid exchange depreciation' and 'to assist in the establishment of a multilateral system of payments in respect of current transactions between members and in the elimination of foreign exchange restrictions which hamper the growth of world trade'. But Article XIV (entitled 'Transitional Period') made it equally clear that, for the first five years after the Fund began operations, there would be no penalties for a country retaining restrictive practices on payments and transfers for current international transactions. In the negotiations in Washington just over a year later, one of the US Government's essential conditions for the granting of the loan to Britain was British agreement to renounce four of the transitional years allowed by Article XIV of the International Monetary Fund. Faced with American pressure to make currently earned sterling freely convertible by 31 December 1946, the best that Keynes could get was convertibility twelve months after the effective date of the Loan Agreement; hence the significance that the date 15 July 1947 subsequently assumed.[19]

The issue of the sterling balances was fudged. Although, in the words of Keynes' first biographer, the Americans were 'assured . . . that she [Britain] would have dealt with the matter in her own way and have it completely under control before the date on which sterling should become convertible',[20] all that the loan agreement itself contained was a vague 'every endeavour' formula in Section 10, which committed the British Government to 'make arrangements . . . for an early settlement' of those balances on the basis of a division into three categories: those to be released at once and to become convertible, those to be released by instalments from 1951 onwards, and those to be 'adjusted as a contribution to the settlement of war and post-war indebtedness'.[21] Whereas, in American eyes, it should have been possible to wipe out these balances, just as the US Government had, simultaneously with the Financial Agreement, cancelled the US$20 bn. of debt outstanding under Lend-Lease,[22] in Whitehall no one dreamed of doing any such thing. Instead as a British Treasury memorandum put it, nearly four years later, 'We have failed to reach agreement with anybody . . . While, therefore, we cannot be accused of being in default under Section 10, we certainly cannot claim that we have achieved the objects set out in it'. This memorandum, circulated by the President of the Board of Trade to his colleagues on the eve of Bevin's and Cripps' departure for talks in Washington to resolve the second British financial crisis, in 1949, reached the candid conclusion: 'we are somewhat vulnerable'.[23] Thus the financial paradox at the heart of the sterling area remained; and in the first financial crisis, which followed convertibility in 1947, it was to prove a significant contributory factor. By then the sterling balances had risen to £3,559m.[24] And when, over twenty years later, the nettle of the sterling balances was grasped at last, they still amounted to about £4bn., although by that time the principal holders were different from those of 1945.[25]

Congress debated the loan to Britain at leisure. Paradoxically, in the end this proved to be an advantage for Britain. Whereas the Administration had rested its case with Congress largely on the speed with which the Loan Agreement, by comparison with the articles of the International Monetary Fund, would oblige Britain to lift controls over its international payments,* by the time Congress at last reached the crunch, in mid-1946, other arguments were beginning to tell far more on both sides of Congress, as the trend of American opinion towards the Soviet Union began to change. In the early months of the debate it was touch and go. On 22 April, however, Vandenberg, earlier a sceptic, swung his weight behind the loan, for foreign policy reasons.[27]

In Britain, on the other hand, speed was of the essence, with 31 December 1945 set as an additional deadline in Parliament for membership of the International Monetary Fund and the International Bank.† The record of the debate in the House of Commons makes dismal reading. Logic was clogged by self-pity. The Government paid the penalty for its failure to educate public opinion to the realities of the country's post-war predicament. The wiping out of Lend-Lease debt was largely ignored and the Loan Agreement, coupled with Bretton Woods, was seen as a betrayal of the wartime principle of equality of Anglo-American sacrifice. For Churchill, the convertible article of the Loan Agreement was 'so doubtful and perilous that the best hope is that in practice it will defeat itself and that it is in fact too bad to be true'.[28] The voice of reason was heard five days later, in the House of Lords. Ridiculing the concept of building up

a separate economic *bloc* which excludes Canada and consists of countries to which we already owe more than we can pay, on the basis of their agreeing to lend us money they have not got and buy only from us and one another goods we are unable to supply,

Keynes in his peroration described the 'determination to make trade truly international' as an 'essential condition of the world's best hope, an Anglo-American understanding, which brings us and others together in international institutions which may be in the long run the first step towards something more comprehensive'.[29]

Keynes' appeal was powerful, but lonely. Although Parliament passed the legislation before it, the national mood was that of *The Economist*, in whose words: 'our reward for losing a quarter of our total national wealth in the common cause is to pay tribute for half a

---

* Moreover, Clayton, who was Vinson's deputy in these negotiations, again made unrealistic claims for the liberalizing effect that the agreements would have, particularly in relation to the sterling area dollar pool.[26]

† Parliament also had before it the 'Proposals for consideration by an International Conference on Trade and Employment', agreed by the two governments and approved in November 1945.

century* to those who have been enriched by the war'. Beggars, the same article concluded, could not be choosers, 'but they can, by long tradition, put a curse on the ambitions of the rich'.[31] As Keynes had himself warned the Cabinet, in his memorandum four months earlier, 'in practice we shall in the end accept the best terms that we can get. And that may be the beginning of later trouble and bitter feelings'.[32] In virtually every field, the first Christmas after the war marked the nadir of Anglo-American relations in the 1940s.

* At the time of writing, the annual servicing of the US loan amounts to approximately US$132m. The last payment of principal is due to be made in 2005 and the final interest payment in 2006.[30]

CHAPTER 9

After the Loan

In the event, 1946 went better for Britain, in economic terms, than the negotiators of the US loan had expected. British exports, which at the war's end were running at one third of their pre-war level, reached 111 per cent of that level in the final quarter of 1946. Only US$600m. of the loan were drawn. But the following year was a different story. As late as December 1946 the US Federal Reserve Board had believed that 'the world's unsatisfied needs for dollars in 1947 would amount only to US$3.5 billion—needs which could be financed without difficulty out of foreign gold and dollar holdings, the British credit, and Export–Import Bank loans. The remainder ... could be satisfied without excessive strains by the new international institutions'.[1] This forecast was swiftly disproved. The US export surplus, which had been running at US$8.2bn. in 1946, rose to 11.3bn. in 1947. The combined dollar deficit of Britain and the sterling area in 1947 exceeded US$4bn. Under this strain, by the middle of August 1947 only US$850m. of the US loan remained undrawn; and the outflow of capital from London had reached a rate that would have exhausted the rest of the loan in another month.[2] Yet not until 18 August was a team of officials sent to Washington to patch up a fig-leaf agreement* suspending the convertibility of the pound sterling.

It remains a matter of debate which was the primary cause of the 1947 disaster—not the weather, although the European winter was certainly the worst in living memory. Today, there is general agreement that there was more than one cause and that the fundamental financial imbalance of 1947 was deep-rooted. The convertibility deadline of 15 July—afterwards used by the British as the scapegoat—was not in itself the cause, but rather the trigger that set off the mechanism of a disaster already far advanced. Instead of waiting until five minutes to midnight, why did the two governments not decide earlier to postpone convertibility? Even today, the records do not throw much light on this question. Admittedly, Snyder would have been even less sympathetic to such an idea than Vinson, whom he had by then

---

* By terms of notes exchanged in Washington on 20 August, the convertibility of sterling was suspended: an action described as 'emergency and temporary'.[3] In the event, the convertibility of sterling, like that of other western European currencies, was not restored until December 1958.

succeeded at the Treasury, but in the summer of 1947 there were others in Washington, such as Marshall, whose views were different. In London, Keynes was dead; Bevin's energies were largely taken up with the Marshall Plan; and the Cabinet as a whole wasted precious weeks agonizing, not about the fate of the pound, but about whether or not to nationalize the iron and steel industry.[4] Perhaps the nearest approach to an answer to this question is that, until the very last moment, the British pride and the American prejudice that had combined to produce the terms of the Financial Agreement of 1945, were too powerful to allow either side to face reality.

However that may be, there was not even the elementary exchange of statistical information across the Atlantic which might at least have set the signals to amber. It was only a bare three weeks before the convertibility date that the British Treasury (pleading 'a most complex web of transactions to be analysed' and that it had been 'loaded with work') provided 'the facts of the British financial position', which the US Ambassador claimed to have been seeking for the previous three months; and even then Dalton is recorded as telling Clayton only that the 'burden of convertibility' would be 'difficult'.[5] In fact, the two Treasuries' behaviour towards each other resembled that of two British generals in an earlier British fiasco, described in the anonymous doggerel of the time:

> Great Chatham, with his sabre drawn
> stood waiting for Sir Richard Strachan;
> Sir Richard, longing to be at 'em,
> stood waiting for the Earl of Chatham.[6]

At the time this disaster had three dimensions: domestic British, transatlantic, and the effects on the international economic community as a whole. The British Cabinet's reaction to the financial crisis was such that *The Economist* headed a leading article 'Moral Crisis in Downing Street'; and the US Ambassador even reported to the State Department 'growing suspicions among members of the Cabinet that we are . . . quietly intriguing to force a coalition'.[7] In the internal party political crisis that followed, Attlee survived as Prime Minister largely thanks to Bevin's personal loyalty. The outcome in Britain was three years of Crippsian austerity. In the Opposition parties of both countries there was much initial *Schadenfreude*. Taft's description of the loan on 23 July as 'nearly all spent on food, films, and tobacco' was echoed a fortnight later by Churchill.[8] The press of each country—and not just the popular newspapers—hurled thunderbolts of blame across the Atlantic. For the Americans, the socialist government of Britain, perceived as profligate and incompetent, was the villain of the piece;

for the British, their country was, in *The Economist*'s words, 'being driven into a corner by a complex of American actions and insistencies [*sic*], which, in combination, are quite intolerable'. The next sentences of this article went so far as to say:

not many people in this country believe the Communist thesis that it is the deliberate and conscious aim of American policy to ruin Britain and everything that Britain stands for in the world. But the evidence can certainly be read that way. And if every time that aid is extended, conditions are attached which make it impossible for Britain ever to escape the necessity of going back for still more aid, obtained with still more self-abasement and on still more crippling terms, then the results will certainly be what the Communists predict, whether or not it is what the Americans intend.[9]

It says much for the revived strength of the Anglo-American relationship across the board in 1947, by comparison with two years earlier, that it was able to absorb a financial shock of such severity. What followed the crisis over the next eight months, in the financial field, can be briefly recounted. After the Anglo-American exchange of notes in Washington on 20 August 1947 the financial chaos in London was halted; in September Section 9 of the Loan Agreement, forbidding discriminatory quantitative restriction of British imports, was effectively set aside;[10] in October the GATT was signed in Geneva (the conclusion of a multilateral conference at which the two principal negotiators, the US and British governments, in the end, agreed on what amounted to a limited trade-off between discriminatory tariffs on the one hand and Imperial Preferences on the other), in December the US Government agreed to Britain's drawing the last US$400m. of the 1945 loan;[11] and four months later Truman signed the Economic Cooperation Act (the Marshall Plan) into law.

If that were the whole story, the crisis of 1947 might be regarded as catharsis rather than trauma. In one way the crisis was indeed cathartic. The idea of Britain as a major economic rival, and of the sterling area system, centred in London, as a threat to American world trade, which had underlain both Congressional attitudes and those of men like Baruch at the end of the war, was largely dispelled. Although half the world's trade was still carried on in sterling, by September 1947 Britain was visibly helpless without American financial support. Moreover, American opinion could not be indifferent to the fact that Britain still maintained over a million men under arms. By then American fears were no longer of the snares of a 'Santa Claus philosophy', but rather of the menace of Hannibal at the gates. To the new phase in the overall relationship between the two countries that was then taking shape, the crisis of that summer gave a powerful impetus. Yet the crisis was also traumatic, not just because of its immediate effects, for which the two governments succeeded in finding remedies, but because of the scars, left well beyond the boundaries of the Anglo-American relationship by their abortive attempt at multi-lateralism, which took many years to heal.

The brave Anglo-American vision embodied in the Bretton Woods agreements was followed in only three years by an economic crisis on a scale that made it once again respectable, intellectually and politically—especially in Britain—to retreat into less courageous and less enlightened ideas of international economic management, far removed from both the letter and the spirit of Bretton Woods. The United States and Britain each bears a share of the blame for this false start. At the time this was obscured, first by an economic achievement and then by yet another setback. It was an irony of the financial crisis of 1947 that just over a month later, on 22 September, British foreign policy secured one of its greatest successes in the entire post-war period: the signature in Paris by the sixteen western European Foreign Ministers, with Bevin in the chair, of the preliminary report on European Economic Cooperation: the European response to the Marshall Plan. Britain's allocation of Marshall Aid under the European Economic Recovery Programme for 1948–9 amounted to US$1,263m. 1948 became a boom year for Britain,[12] whose exports in that year reached a level 150 per cent above that of 1938. Nevertheless, in the summer of 1949, Britain faced its second post-war economic crisis.

The dimensions of this crisis may be judged by the fact that by September 1949 British reserves had fallen to US$1,410m. As in 1947, more than one adverse factor came into play, but the truth—of which the market was keenly aware and, therefore, took full advantage—was that British reserves had not yet risen to a point at which they could cope with what was described to the Cabinet Economic Committee in mid-June as 'a lot of items in the dollar balance all going wrong at once', in a recessionary phase in the American economy (although, as it turned out later, this was a temporary phenomenon), which reduced dollar earnings from the second quarter of the year onwards. To this Cripps' initial reaction, at the next meeting of the Committee, was that the United Kingdom's own dollar expenditure should be cut immediately by not less than 25 per cent.[13] The British dilemma was not made any easier by the fact that, as Franks (recalled from the Washington Embassy for consultations) reminded the Committee, by the end of June 'neither the President nor the Secretary of the Treasury had admitted the fact of a recession in the United States'; and it had been borne in on him before he left Washington that the British Government must 'avoid giving the impression that, in times of crisis, the United Kingdom always adopted a restrictive policy'.

Again, as in 1947, the Cabinet hesitated, but this time they spent many hours addressing the problem. On 1 July the Economic Policy Committee met twice. Both Cripps and Bevin opposed devaluation. Cripps, a determined man of strong views, was supported in this by the heads both of the Treasury and the Bank of England; he had, moreover, told journalists in May that devaluation was 'neither necessary, nor would it take place',[14] thus reaffirming the commitment to maintain the pound's value given to Britain's trading partners in and outside the Commonwealth. At the first meeting on 1 July Bevin drew a parallel with 'the situation which had existed before the negotiation of the Atlantic Pact.* At that time, our choice lay between trying to continue as an intermediate power or aligning ourselves with the United States'. The present problem 'could not be settled on a purely financial level and must be discussed on the political plane'. What was needed was 'a wholly new approach to the question, to bring the sterling and dollar areas into equilibrium'.

There was another important difference from 1947, in that this time the transatlantic channel of communication was open; American official opinion was running—discreetly—in favour of devaluation; but at a meeting with Snyder on 9 July this advice was resisted by Cripps, who then left for a five weeks' stay in a Swiss sanatorium, to recover from a sudden illness. (Bevin's physical decline also became clearly visible from the autumn of 1949 onwards.) When the Cabinet finally bit the bullet, on 29 August 1949, their secret decision was the result of what can perhaps be described as a revolt of the second tier both of ministers and of the bureaucracy.[15] Cripps and Bevin then left for talks in Washington. Three weeks later the pound was devalued, from US$4 (or, precisely 4.04) to $2.80.

The British records now reveal the striking extent to which the crisis of 1949, unlike that of two years earlier, was then believed by the British Government to have turned out for the best. From London a telegram sent to Bevin (by then in New York) jointly by the Prime Minister and the Chancellor on the day after the devaluation—which was announced in Washington on 20 September—informed him that it had 'enjoyed . . . a far better press than could . . . have been expected'.[16] True, the outcome could have been worse. As late as 3 September Acheson had stressed to Franks the reality of the anxious doubts 'on the American side whether the British wished to work towards a free world in which exchanges between [the] dollar and the sterling areas were at a high level, or whether they wished to build up a soft currency world centred round the sterling area and cut themselves off from North

* The North Atlantic Treaty had been signed on 4 April 1949.

America': a process which today might be called 'decoupling'. From Washington, on the day of the announcement of devaluation, Franks reported that in his view the 'two great gains' were the 'creation of an entirely new spring of goodwill and confidence ... on the American and Canadian sides and a corresponding enhanced degree of readiness of help', and 'a real conviction that the three countries ... have it in their power to remake the economic and trade pattern of the Western world if they agree together'.[17] In Whitehall, by the end of the year, Robert Hall recorded the 'legitimate satisfaction' that 'we are now on the right path'.[18]

The British decision to devalue the pound was indeed preceded by tripartite (US-Anglo-Canadian) economic talks at ministerial level, at which it was agreed, among other things, to make informal 'arrangements for continuing consultation'; and the conclusion of the long communiqué issued in Washington on 12 September 1949 was that, once Marshall Aid had come to an end, 'a satisfactory equilibrium between sterling and dollar areas' was in sight.[19] By the end of 1950, Hugh Gaitskell, who had just succeeded Cripps as Chancellor, was able to announce in Parliament that Britain would make no further call on Marshall Aid. And, looking still further on into history, in the 1950s the western economic community embarked on its longest period of growth since the beginning of the century. Between 1950 and 1962 aggregate industrial output of advanced western countries was to double.* Can this eventual outcome fairly be regarded as the fulfilment of the hopes pinned by Bevin on his 'wholly new' political approach in the dark days of July 1949? The fact that at that time Bevin was opposed to devaluation, to which he saw his 'political' solution as the alternative, makes this view hard to sustain. Nor can this 30 per cent devaluation of the pound be brushed into the margin of history. Although the tripartite communiqué of 12 September said nothing about devaluation, its verbiage masked a decision, already taken by the British Government and desired by the United States Government, which formed part of the beginning of a major geopolitical shift.

Historically, it was not inappropriate that, as we have seen in Chapter 7, on the same day as sterling was devalued, American representatives finally unveiled to their British colleagues, at a meeting of the atomic Combined Policy Committee in Washington, the most nationalistic in the series of atomic proposals secretly discussed between the two governments during these four years. On 23 September 1949 Truman released the news of the first Soviet atomic explosion, over

* On the eve of the First World War their production stood at a level about 50 per cent above that of 1900.

three years earlier than the British.* Almost simultaneously the Chinese People's Republic was proclaimed in Peking. And October was the month in which the last step in the political division of Germany was taken. A new era was about to begin in the pivotal year, 1950.

Franks' personal assessment on 20 September 1949, quoted above, not only stressed what he considered to be the positive aspects of the outcome of the economic crisis; it also expressed his belief that the Americans had 'decided to regard us once more as their principal partner in world affairs, and not just a member of the European queue'. Whatever may have been said to the British, American records show that this was not quite how the US Administration, among themselves, saw Britain at the time. In his report to the American members of the Combined Policy Committee on 13 September, the day after the tripartite economic talks ended, the Under Secretary of State, James Webb (who had previously been Director of the Budget), had indeed said that 'one of the most important conclusions' of the talks was that 'the three countries were partners in the economic crisis that had developed and that they would have to work their way out of it as partners'. But this conclusion had more to do with the fact that the United States and Canada were Britain's chief creditors than with a bilateral Anglo-American partnership in the direction of world affairs. Webb's introductory remarks at this meeting were followed by a long account, given by Kennan, of a detailed study of Britain's role conducted over the preceding weeks by the Policy Planning Staff in the State Department. After deploying four† arguments that show sympathy and disenchantment with current British policy in roughly equal measure, Kennan went on to tell his colleagues that the conclusion reached in the Department was:

that an attempt should be made to link the United Kingdom more closely to the United States and Canada and to get the United Kingdom to disengage itself as much as possible from Continental European problems. While the United Kingdom would continue to be a staunch ally of Western Europe, it should assume more nearly the role of adviser to Western Europe on its problems, and less the role of active participant. It was hoped that the United Kingdom could be persuaded to disengage itself quietly from the Council of Europe.[21]

---

* The first British atomic explosion took place at Montebello, Australia, on 3 October 1952.[20]

† First, the United Kingdom 'tended to exert a retarding influence on Western European plans for closer political and economic integration'. Secondly, it would 'act as a deterrent' to bringing Eastern European countries 'eventually ... into the orbit of Western Europe'. Thirdly, it was 'rather unlikely that the British would be willing to go along with' the development of a firm Franco-German relationship, if they were deeply involved in a programme of Western European unification. Finally, 'in trying to play a role of leadership on the Continent, the United Kingdom was continually finding itself entering into commitments on which it could not deliver'.

A planning staff seldom makes policy. Too much should not, therefore, be read into the record of a single meeting of this kind. Nevertheless, the record of what was said there recaptures the atmosphere in Washington at the critical juncture at which this meeting was held. So too does Webb's report[22] of remarks made by the Secretary of Defense, Louis Johnson, at a private lunch a month earlier. Johnson then said:

the United Kingdom was finished, there was no sense in trying to bolster it up through the ECA, MAP, NAP [sc. the Economic Co-operation Act, the Mutual Defense Assistance Program, the North Atlantic Pact], or assistance in the field of atomic energy. Even the Canadians ... were disturbed with the prospect that we might give atomic secrets to the British ... As the Empire disintegrated we should write off the UK and continue cooperation with those parts of the Empire that remained useful to us.

Acheson did not share Johnson's jaundiced view of Britain.* Yet he too had reached a pessimistic conclusion about Britain's role in Europe, whose accuracy was to be demonstrated six months afterwards, by which time 'for Britain, the aftermath of the 1949 sterling crisis saw defeat emerge from the jaws of victory'.[23]

Had there been a British fly on the wall at the meeting in Washington on 13 September 1949, however, he might have derived some encouragement from some of the remarks made by the participants. Indeed, the reflections of the Policy Planning Staff during the British financial crisis had even found their way into the *Wall Street Journal*:

An economic union of the United States and Britain. That startling idea is being seriously discussed privately by a growing group of key State Department officials as 'the only way out' of Britain's permanent dollar shortage. These officials think Britain could but won't solve her own problems. For political-military reasons, they figure, the United States must solve them for her permanently. Whether it matures or not, this sensational proposal is significant, for it shows how little faith the administration now has that England can be rescued by any mild or temporary aid ... This all adds up to the proposition that the two countries should merge their economies almost as completely as those of the 48 United States. Yet the two nations would keep their political independence. How this could work out is a question the State Department men have not quite cleared up.[24]

They certainly had not. There was no Congressional support for such a concept, nor for that of an Anglo-North American (or North Atlantic) free trade area—a mirage which grew out of it and was to beguile some people in Britain in the years that followed.

---

* There was, moreover, no love lost between the two men.

What is common to all these American records is the absence of any recognition of the concept of a shared Anglo-American world leadership. The furthest that the State Department was prepared to go was to 'recognize and support the British in their role as leader of the Commonwealth and their attempts to strengthen it ... in close association' with both the United States and Western Europe.[25] As later chapters will show, British anxiety about American willingness to recognize them as their partners was already evident in the preliminary discussion between the two governments of the Marshall Plan, and it was to become a recurring theme. To assess this question now would be to anticipate discussion that properly belongs to later chapters of this book, as does the related question of the attitudes of the two governments towards the question of European integration. Here, two major consequences of the crisis of 1949 must be recorded. First, the effect in London: the massive devaluation was accompanied by what the most recent history of the post-war Labour Government has described as a 'partial isolationism' in British foreign policy and 'almost insular implications in terms of foreign trade and finance'.[26] The siege economy was maintained and even intensified, so much so that—to cite only one example—when the British balance of payments at last achieved a substantial surplus, in 1950, a system of food rationing that had lasted over ten years was left largely intact by the British Government.[27]

Secondly, the British decision to devalue affected other countries. It was not just that after the devaluation of September 1949 Britain's trading partners were obliged to follow suit. The British decision to devalue was taken in intimate consultation with the US Government, but with no warning given to the French, even though the French and British Governments had just spent a week in confidential talks in Washington, without a word about devaluation. The first that the French Finance Minister learned, in Washington, of the British decision to devalue was on the eve of the public announcement—and then not from Bevin or Cripps.[28] This slight was to be repaid, with interest, by the French Government in May 1950.

CHAPTER 10

The Middle East

The blow-by-blow treatment of the sequence of events described in the three preceding chapters, warranted by aspects of the Anglo-American relationship that are arcane or long since forgotten, is inappropriate to a discussion of parts of the world that are familiar from our television screens today. In the next two chapters, therefore, a more thematic approach is adopted, supplemented by chronological appendices (4 and 5) covering the period 1945–9.

During these years there were two critical areas of the world where the United States and Britain each took pride in its own position: respectively, China and what would today be called the Middle East.* Through its particular historical experience, each country saw itself as the repository of a special knowledge of, and hence a special responsibility for, one of these areas, about which neither government was inclined to listen to the advice of the other at the war's end. Each country had a sense of national mission—the one in China and the other in the Middle East—sometimes amounting to a belief in its own infallibility, which led it to cling to geo-political objectives long after the march of events had proved them to be unattainable. For each of them, its special region turned out to be the Achilles' heel of its foreign policy after the Second World War. And what happened in these four years, in relation both to China and to the Middle East, was not a good advertisement for Anglo-American co-operation.

Both in the United States and in Britain domestic political considerations were an important dimension in the formulation of foreign policy towards these areas. This applied particularly to the United States, where the power of the Zionist movement (repeatedly under-estimated by the British Government) critically affected US policy towards the Palestine issue; while that of the China Lobby ended by virtually paralysing US policy towards China altogether. So far as the Middle East is concerned, however, it is also clear now that

* The term Middle East began to drive out the others during the Second World War, when the 'Grey Pillars' building in Cairo housed the enormous GHQ, Middle East Forces. But the older term, Near East, remained in use on both sides of the Atlantic in the years 1945–9; and it was on the countries traditionally composing the area once defined as the Levant that US and British policy in this region was primarily focused during this period.

this was an area that demanded a bipartisan approach in Westminster if ultimate disaster was to be avoided. A fusion of Churchill's view of Palestine with Bevin's vision of a new Anglo-Arab partnership in the Middle East would not, I suggest, have been wholly impracticable if it had been attempted at the outset. As it was, the cut and thrust of party politics was left to take its toll.

THE IMPERIAL BACKGROUND

The American delusion about China was matched by the obsession of British policy-makers with the Middle East after the war. As will be seen in later chapters, it was not until nearly five years after the end of the war in Europe that the British Cabinet Defence Committee decided that, in the event of war, Europe would take priority over the Middle East. In order to understand what happened in the focal part of the Middle East, above all Egypt and Palestine, the peculiar characteristics of the British position in the region as a whole must first briefly be considered. If the North American colonies formed the core of the first British Empire, for a century and a half after the American War of Independence India was the core of the second. In 1945, however, India was—for the Empire—a wasting asset. In so far as the British Empire still had a jewel, by then it was—for economic and strategic reasons—the Middle East, which was, paradoxically, neither British nor an empire.

Aden and the Gulf protectorates apart (where British influence had penetrated much earlier, not from Britain itself, but from India), British paramountcy in the Middle East was a comparatively recent phenomenon—the outcome of the dissolution of the Ottoman Empire, which Britain had replaced as the dominant power in the region. On the map, impressive patches of red were splashed across this vast area, its western approaches guarded by two solidly British possessions—Malta and Gibraltar. In reality, partly because the dissolution of the Ottoman Empire had taken place in stages over the half century culminating in 1918, these territories formed a political patchwork of extraordinary complexity. In 1945, the only sovereign British territory in the whole region consisted of two crown colonies—Cyprus and Aden—lying on its northern and southern extremities. With these two exceptions, the dependent countries in the region consisted of an assortment of (League of Nations) mandated territories, protectorates, 'protected states', and—occupied temporarily—the colonies of the former Italian Empire, which had been conquered by the British during the war. All the other countries were independent, although in practice the degree of independence varied. In particular, Egypt, which owed its independence to Britain after the First World War, was bound to Britain by a treaty of military alliance which had another eleven years of legal validity. The Anglo-Egyptian Sudan, although constitutionally a condominium, had in practice been administered by British officials for over twenty years.

Thanks to its Ottoman inheritance, Britain had acquired in the Middle East a unique status, which may perhaps be called an imperial position. For an understanding of the political intricacies of this motley collection, the British Government relied mainly—Palestine, Cyprus, Aden, and the Sudan were the principal exceptions*— on a small corps of specialists, the Levant Consular Service. This service (which was merged in the new, post-war British Foreign Service, under the Eden–Bevin reform) long antedated the final breakup of the Ottoman Empire, as its name implies. The network of relationships established with local leaders and the knowledge accumulated over the years by British oriental secretaries and oriental counsellors (in Constantinople, their title was dragoman), scattered first across the territories of the Ottoman Empire, and later in the successor territories, was such that in 1945 the British Government was accustomed to being kept better informed about what was going on behind the scenes in Cairo than it was about happenings in Moscow—sometimes perhaps even than in Washington. Although the Levant Consular Service included its full share of eccentrics, it remains more memorable for the qualities of some very distinguished men. Reader Bullard, for example, who was a Russian-speaker as well as an oriental linguist, was remembered with admiration in Moscow by his wartime Soviet opposite number a quarter of a century afterwards.[1] And after Walter Smart's[2] death, one of his many friends rightly recalled that although no one could have looked less like a pro-consul, 'no Englishman, perhaps since Cromer, did so much to uphold and broaden Britain's position in the Levant'.

THE BRITISH POST-WAR APPROACH

In 1945, the question central to British policy towards this region was twofold: how to modernise Britain's relationship with the individual countries and with the forces of Arab nationalism as a whole; and how simultaneously to retain its own position, military and economic, as the region's paramount power. To both parts of this central British

---

* The first three of these territories were administered by the Colonial Service; the last had its own Sudan Political Service. The Indian Political Service staffed posts in the Gulf and in parts of Iran.

preoccupation Egypt held the key. The Second World War had transformed the Middle East. Before the war its importance had been seen mainly—though not exclusively—in terms of imperial communications with India and beyond. Through the accidents of war, a massive military base had been built up in Egypt, on a scale and covering an area wholly unforeseen by the drafters of the Anglo-Egyptian Treaty in 1936. As late as July 1945, long after the tide of war had receded from Africa, about a quarter of a million troops were stationed in Egypt, which had become the nodal point of what the British Chiefs of Staff still described as the Imperial Strategic Reserve.[3] Imperial eastward communications remained important; but two other factors were, from the end of the war onwards, common to every British military assessment of the Middle East: the need for a strategic base to defend the region in the event of another war and the need for airfields within bombing range of 'important industrial centres of Russia'.[4] The growth in the region's economic importance can be measured from the fact that, whereas 19 per cent of Western Europe's oil supplies came from the Middle East in 1938, ten years after the war this figure had risen to 80 per cent. And in 1945 Britain paid for its oil imports from the region in sterling.

Politically, the war had given a powerful fillip to the forces of nationalism, both within each Middle Eastern country and across the Arab world. The British had themselves contributed to this surge of Arab nationalism during the war years in a number of ways; most important of all perhaps, by their promises of independence to Syria and the Lebanon, which they had honoured. Soon after the end of the war, Bevin effectively committed the new British Government to eventual Sudanese self-determination. There was moreover a fundamental contradiction implicit in two earlier British commitments: the Balfour Declaration of 1917 and the Palestine White Paper of 1939— the former promising the British Government's 'best endeavours to facilitate the achievement' of 'the establishment in Palestine of a national home for the Jewish people', and the latter stating that after 1944 no further Jewish immigration into Palestine would be allowed 'unless the Arabs of Palestine are prepared to acquiesce in it'. Against this ambivalent background, Palestine at first formed only a small part[5] in the British grand design for the Middle East.

Because, only two years after the Second World War ended, the British Government announced its intention of relinquishing the Palestine mandate (which it did, nine months after handing over power in the Indian sub-continent), it has sometimes been supposed that at this juncture British economic difficulties at home (which were indeed immense) brought about a failure of British political will in the Middle East. This is not so. The one emotion absent from British records relating to the Middle East in the post-war quinquennium is self-doubt. How, then, in British eyes, did the Anglo-American relationship affect the British post-war approach to the central question, as defined above? To begin with, not very much.

It was indeed already evident during the war that British policy would be in trouble if two powerful strands of American opinion—the pro-Zionist and the anti-colonial—were ever to coalesce in relation to this region. But as the war developed, even though Roosevelt was both pro-Zionist and deeply distrustful of the British Empire, his foreign policy in the Middle East seemed to be taking a different turn. The reason was simple: oil. The history of the US involvement in Saudi Arabia from 1943 onwards lies outside the time-frame of this book. However, if in that year anyone on either side of the Atlantic had been asked to forecast which, if any, country of the Middle East was most likely to become the focus of Anglo-American disagreement after the war, the answer would probably have been not Palestine, but Saudi Arabia. In the event, in spite of the suspicions harboured on both sides, this was averted; in effect, Britain recognized the special US position in Saudi Arabia. And, on his way home from Yalta, Roosevelt took matters a stage further by assuring King Ibn Saud on 14 February 1945 that 'he would do nothing to assist the Jews against the Arabs and would make no move hostile to the Arab people'.[6] In the last weeks of his life Roosevelt was beginning to change his mind on several major issues of foreign policy. How he would have reacted to post-war developments in this field, including the impact of the public discovery of the full horror of the Holocaust, is a matter for speculation. What the British Prime Minister, also pro-Zionist, was thinking of, however, not long after Roosevelt's meeting with Ibn Saud, was 'the idea that we should ask them [the Americans] to take it [Palestine] over'. Three weeks before he left office, in a minute to the Colonial Secretary, Churchill wrote: 'I believe we should be the stronger the more they are drawn into the Mediterranean. At any rate the fact that we show no desire to keep the mandate will be a great help. I am not aware of the slightest advantage which has ever accrued to Great Britain from this painful and thankless task.' The Chiefs of Staff, instructed to comment on Churchill's idea, gave it a dusty answer on 12 July 1945. The 'grave disadvantages' that they saw in his suggestion were based on three main considerations, which are quoted in full here because they were profoundly to influence the thinking of the Government which took office a fortnight later, in spite of the pro-Zionist sympathies of several members of the Labour Cabinet:

We should be adjudged to have abandoned our predominant position in the Middle East. This abrogation of responsibility would have evil consequences, not only in Middle East countries but in India and beyond.

The safeguarding of our strategic interests in the Middle East would virtually depend upon the policy pursued by the United States in Palestine. If that policy, over which we should have little if any control, were inimical to our interest, we should be faced with an increased internal security commitment in the Middle East and an embarrassing conflict of policy with the United States.

The effect upon Russia of American control in Palestine can only be guessed. The Foreign Office advise us that Russia might see in this change of responsibility some covert threat to herself. Moreover, the Russians might be ready to set themselves up as champions of the Arabs if in the event the Americans pursued a Zionist policy in Palestine and we felt obliged to acquiesce in it.[7]

This exchange confirms that Churchill was not being wise after the event when, a year later, he told the House of Commons that 'we should ... as soon as the war stopped, have made it clear to the United States that, unless they came in and bore their share, we would lay the whole care and burden at the foot of the United Nations Organisation'.[8] Instead, what developed into a three-year Anglo-American diplomatic battle over Palestine was joined at Potsdam. Truman's memorandum of 24 July (sent to Churchill, acknowledged by Attlee) expressed 'the hope that the British government may find it possible without delay to take stages to lift the restrictions of the [1939] White Paper on Jewish immigration into Palestine' and asked for 'ideas on the settlement of the Palestine problem, so that we can ... discuss the problem in concrete terms'. A month later he wrote to Attlee asking for 100,000 Jews to be admitted into Palestine.[9]

The British Government was unwise not to have accepted the figure of 100,000[10] in 1945, when there were about 50,000 Jewish refugees in western-occupied zones of Europe. Instead, the Foreign Office reaction to Truman's memorandum of 24 July was that his request had been made 'largely for the record'. Whatever Bevin's later mistakes over Palestine may have been, he did not accept this initial official advice. While still at Potsdam he minuted to Attlee: 'I consider the Palestine question urgent and when I return to London I propose to examine the whole question, bearing in mind the repercussions on the whole Middle East and the USA'.[11] Three and a half months later, when announcing the formation of the Anglo-American Committee of Enquiry in the House of Commons, he uttered the prediction, with which Churchill was later to taunt him: 'I will stake my political future on solving this problem'.[12] The press conference that Bevin gave after his statement on 13 November 1945, the first of his series of public gaffes over Palestine, reverberated round the world. The question whether, and if so, to what extent, Bevin was anti-Semitic has been much discussed. Rather than add to the answers already suggested to this question, I prefer to quote the considered conclusion of an historian who has gone to great lengths in looking into this matter, with the help of Israeli and Zionist archives as well as other sources:

Bevin's personalization of British policy by staking his own political future on its success, his egocentric and emotional temperament, his disposition to fight when faced with opposition and not to guard his tongue when angered all combined to make him vulnerable to ... charges [of anti-Semitism] ... Bevin paid a high price for these failings. The time and effort he devoted to trying to find a compromise solution to the problem of Palestine were ignored. Henceforward he became the principal target of Zionist propaganda which progressively identified him with the Nazi image the Jews had learned to hate. And since he naturally resented being compared with Hitler and the organizers of the death camps, this in turn had its effect on Bevin's own attitude. Prejudice was cumulative, on both sides, making it more difficult for Bevin to form a cool judgement and to disengage from a problem which had defeated and marred the reputation of every British minister who touched it ...[13]

Because of its potential impact on Anglo-American relations across the board, the Palestine question was treated as a special case almost from the outset. On the other hand, no serious effort to concert British and United States policy in the Middle East as a whole was made until more than two years after the Second World War. In Cairo, for example, little attention was then paid by members of the British Embassy to the views of their US colleagues.[14] Indeed, in a letter written to the Foreign Secretary in December 1945, Halifax took issue with Sargent's description of British policy in the Middle East as being 'in fact, to build up a kind of Monroe system in that area'.[15]

Having secured a breathing space over Palestine, it was vital for the British Government to lose no time in solving the second part of its twofold central problem: how to secure the position of Britain as the paramount power of a region which had changed dramatically during the Second World War, but which was linked to Britain by military treaties mainly negotiated in the very different conditions of the inter-war years. Instead, a whole year was wasted (as has already been noted, it was a practical impossibility for the weight of Bevin's personal authority to be deployed in all directions at once). Egypt was the crux of the matter. Was a Middle East strategic base still needed;

and if so, did it have to remain in Egypt? If the base could be relocated elsewhere, a great burden would be lifted from the long-troubled Anglo-Egyptian relationship, which could then be recast in a joint effort for mutual defence. In Bevin's mind, the concepts of military and economic partnership went hand in hand. In the economic field, his enthusiasm far exceeded the confines of a regional programme of technical assistance, based in Cairo; it extended to a grandiose vision of industrial development, such as would be made possible by the final utilization of the waters of the Nile.[16] Politically, 'as Foreign Secretary of a Labour Government', he looked 'beyond the present Egyptian leaders'[17] to the new class that would sooner or later take their place: the natural allies, so it seemed, of the leaders of European social democracy. Whatever could be achieved on these lines in Egypt would serve as a model for a new British relationship with the other countries of the region as a whole; and if this could be achieved, the problem of Palestine might prove to be peripheral.

For Bevin to succeed over Egypt, he had to overcome opposition on several different fronts in Britain—sceptical advisers, who were aware that for the time being business could be done only with leaders of the semi-feudal regimes that held power in such countries; British public opinion, with which Egyptians (as opposed to the peoples of other countries of the region) were unpopular;[18] 10 Downing Street, where Attlee (who was then also Minister of Defence) was reluctant to agree to the allocation of scarce resources to the defence of what he called 'deficit areas'; the Chiefs of Staff, who, although they bent with the wind, never gave up their fundamental belief that 'if we have to abandon Egypt, we must abandon our status in the Middle East altogether';[19] and finally Parliament, where Churchill's support of this strategic view was coloured by the memory of a subaltern who had taken part in the Battle of Omdurman.

When the bulldozer moved forward at last, the Foreign Secretary produced for his colleagues a cogent memorandum contesting simultaneously the objections both of the Prime Minister—and behind him, the Chancellor—and those of the Chiefs of Staff. Although the Prime Minister's objections had been directed primarily against the Chiefs of Staffs' extravagant aspirations in relation to the former Italian colonies, his conclusion, dated 2 March 1946, had ranged far wider: the necessity to 'review with an open mind strategic conceptions which we have held for many years' and to avoid giving 'for sentimental reasons based on the past ... hostages to fortune'. Its penultimate sentence ran:

It may well be that we shall have to consider the British Isles as an easterly extension of a strategic area the centre of which is the American Continent rather than as a Power looking eastwards through the Mediterranean to India and the East.[20]

In parenthesis, as we now know, although the US Joint Chiefs of Staff's first post-war 'tentative over-all strategic concept', produced in Washington three months later, shared two of Attlee's pessimistic assumptions, it took the view that Egypt and, 'if feasible', India were territories which, like the British Isles, were areas where air bases must be established and defended, in the American interest.[21]

In his memorandum of 15 March 1946, Bevin deployed against Attlee not only the ideological argument already quoted in Chapter 3, but also a geo-political concept: the British position in the Mediterranean was an essential prerequisite for exercising British influence and for protecting British interests in both southern Europe and the Middle East. Militarily, he came out strongly in favour of relocating the Middle East base, regardless of the 'great capital outlay', in Mombasa, so that 'the whole heart and centre of command shall be on British territory'. What—rightly—caused him 'very great concern' was the fact that this centre was still 'in another country's territory [Egypt]'.[22]

The Chiefs of Staff did not accept the Mombasa proposal; nor did they take advantage of the loophole left in Bevin's memorandum by examining the one and only alternative sovereign territory that might have served as a site for the Middle East headquarters and base: Cyprus (in the end a subsequent British Government was to do just this, far too late in the day). The bulldozer was not to be stopped, however. By the middle of April the Chiefs were reporting to the Cabinet, though in carefully hedged terms, on the possibility of the withdrawal of 'all combatant troops' from Egypt;[23] in the Cairo Embassy the uncompromising bulk of Miles Killearn (the architect of the 1936 Treaty) was replaced[24] by Ronald Campbell's keen intellect; and on 7 May the British Government took a calculated risk by announcing in the House of Commons, after the report of the Anglo-American Committee of Enquiry into Palestine had been published, but before Anglo-Egyptian defence negotiations had begun in earnest, its conditional willingness to withdraw all forces* from Egypt.[25] Whether or not this risk was justified (as Churchill instantly maintained), three blunders had been made. The negotiations had been left dangerously late. The Secretary of State for Air, William Stansgate,† was not at all the right man to conduct them. And—perhaps worst of all—the price of the Chiefs of Staffs' acquiescence in withdrawal from

---

* They were withdrawn from Cairo on 4 July 1946.
† He was obliged to retire from the government in October 1946.

Egypt was that Palestine should be substituted for Egypt; Palestine was indeed by then the practicable alternative as a site for the Middle Eastern base; the belief that the facilities which the British sought in Cyrenaica—and eventually obtained, through the Anglo-Libyan treaty of 1953—could be regarded as an adequate substitute for the massive installations built up over years in Egypt, though often expressed by the Foreign Office, was never really accepted by the Ministry of Defence; and in the end the Foreign Office came round to the view that the Egyptian base was irreplaceable.[26]

At the end of October 1946, Bevin, who had by then taken over the Anglo-Egyptian negotiations himself, finally succeeded in reaching agreement with the Egyptian Prime Minister, Ismail Sidky, in London. On Sidky's return to Cairo, however, the agreement collapsed under the weight of the contradictory interpretations that each government placed, not on the provisions relating to Egypt itself, but on its Sudan Protocol. Of all the political complexities in the whole region, none equalled those of the Anglo-Egyptian Sudan, a condominium established in 1899. There was perhaps only one man (the Foreign Office Legal Adviser, Eric Beckett) in Whitehall who fully understood all the ins and outs of its complex constitution, which was a legal minefield. The attempt to cross it was none the less made by these two experienced negotiators: Sidky, old, ill, but astute; and Bevin, three months after his first major heart attack,* but in complete command and determined to make his Egyptian policy work. The compromise formula for the Sudan on which they finally agreed—nominal sovereignty for Egypt, but no practical change within the Sudan, whose ultimate goal would still be independence—meant different things in London and in Cairo. The Bevin–Sidky agreement, therefore, fell apart in a matter of weeks; and Sidky resigned on 9 December 1946.[27]

Not that Bevin and his advisers conceded defeat. On the contrary, he was to prove again and again, both in his Middle Eastern and in his other policies, his capacity for picking up the fragments of a shattered plan and beginning afresh. British forces remained in the Canal Zone; the 1936 Treaty (on which, with US support, the British Government rested its case) survived the challenge mounted against it at the UN by the Egyptian Government in 1947; and the Treaty was not formally denounced by Egypt until October 1951, after yet another British attempt to negotiate its revision—this time with a Wafdist government[28]—had failed. Nevertheless, the heart of the British post-war design for the Middle East had, in fact, been mortally damaged. The Bevin–Sidky failure was followed a year later by the fiasco of the

---

* He collapsed on the voyage to New York immediately afterwards.

Anglo-Iraqi Treaty of Portsmouth, which the Regent of Iraq repudiated after riots in Baghdad. And Palestine had now moved from the wings, to a position uncomfortably near the centre of the Middle Eastern stage, where it was to become the subject of bitter Anglo-American controversy.

THE AMERICAN POST-WAR APPROACH

Unlike the single-minded British post-war approach to this region, the American approach cannot be summed up in a sentence. In very broad terms, there were in the initial phase three strands of opinion among those concerned with making policy in official Washington—the belief that, although military power in the region was British, this power was based on out-moded colonial principles, and that the moral leadership in the Middle East was, therefore, American; the view of those for whom the combination of the region's oil resources and its air bases represented a major western interest; and the conviction of others, whether for idealistic reasons or for reasons of domestic policy or both, that the problem of Palestine could not, as the British hoped, be kept in the margin of wider issues of foreign policy, and that it must somehow be solved. Common to all three was the determination not to commit US troops to the Middle East. Towards Palestine, in particular, the twists and turns of US governmental policy repeatedly bewildered the British Government at the time—and may even now confuse the historian seeking to pick an impartial course through this diplomatic maze—especially as these different strands of opinion in Washington both overlapped with each other and evolved considerably during the four years that spanned Truman's first move over Palestine, made at Potsdam, and the conclusion, by the newly independent Israel, of the last of the series of armistice agreements with its Arab neighbours, whom it had just defeated in the first Arab–Israeli War.

The first school of thought had underlain the Rooseveltian attitude towards the British Empire as a whole. In the second school, Loy Henderson, Director of the Near Eastern and African Affairs Division of the State Department, and James Forrestal, the first US Defense Secretary, were among those who came nearest to concurrence in the British view of Arab friendship and co-operation as the *sine qua non* of both oil and bases. With the onset of the cold war, the second view strengthened and the first view weakened in force. Listening to official Americans of the second school of thought, British officials heard what they wanted to hear, much of it based on an assumption that was shared in Whitehall: namely—the words are Acheson's—that the movement to end the Diaspora 'obscured the totality of American interests'.[29] Nevertheless, what was decisive in the United States was not these first two schools of Washington thought, but—as the British were slow to realize—the third, which included the White House. Although Truman's mind was by no means made up about the Palestine issue when he took office, in the following eighteen months he became increasingly convinced that the solution of this problem must be the one passionately urged by the Zionists and their supporters: partition.

The White House's channels to Zionist opinion were kept open in a most effective way, mainly by three men: Clark Clifford, Edward Jacobson, and David Niles.[30] In 1945 the President of the Jewish Agency for Palestine and of the World Zionist Organization was Chaim Weizmann—a moderate, a British national, and a life-long angophile—who three years later was to become the first President of Israel. When Bevin met this 'tragic, formidable and politically embarrassing'[31] statesman in London, just over a month before publicly staking his political future on a negotiation that could succeed only if handled with extreme delicacy of touch, he dealt him a gratuitous rebuff.[32] By contrast, from the end of 1945 onwards one of the most potent influences on US presidential thinking, repeatedly exerted at critical moments in the Palestine drama, was that of Weizmann.

PALESTINE: THE ANGLO-AMERICAN ATTEMPT

In March 1946 Weizmann told the Anglo-American Committee of Equiry that injustice in Palestine was unavoidable; in his view (which sums up the tragic dilemma of Palestine after the Second World War), the decision to be taken was 'whether it is better to be unjust to the Arabs of Palestine or to the Jews'.[33] Yet—remarkably—in the first twelve months, Anglo-American agreement on a solution for Palestine was twice reached on paper, although on neither occasion did the two governments ratify what had been agreed. The unanimous report of the Anglo-American Committee, released on 30 April 1946, recommended not partition, but a bi-national state;* 100,000 Jewish victims of Nazi persecution were to be admitted into Palestine at once; and the Haganah, the Jewish defence force, was to be disarmed. This compromise commended itself to neither government. On the very day on which the report was released, Truman fired one of the shots from the hip for which his presidency was to become famous: without consulting

---

* The report did not in fact use the word 'bi-national', however. At that time the Jewish population of Palestine numbered 560,000; the Arab population was nearly double. By 1950 the Jewish population of Israel numbered over one million; 700,000 Palestinian Arabs had fled.

the British Government, he publicly welcomed the Committee's recommendation that 100,000 entry certificates should be issued at once, leaving the rest of the recommendations in the air.[34] Attlee responded the next day by telling the House of Commons that the British Government would not proceed until the 'private armies' of both sides in Palestine were disbanded.[35]

When Attlee spoke, seven British soldiers had already been killed by Zionist terrorists; three months later the British Headquarters at the King David Hotel in Jerusalem was blown up by the (Jewish extremist) Irgun Zvai Leumi; the British GOC in Palestine used unvarnished anti-Semitic language; and Bevin added fuel to the flame with a rash incursion into American domestic politics in the speech that he delivered to the Labour Party Conference at Bournemouth.[36] Nevertheless, on 24 July 1946 Anglo-American agreement over Palestine was again reached—this time by teams of officials of both sides meeting in London. This, the so-called Morrison–Grady plan, dropped the bi-national approach of the Anglo-American Committee, substituting the concept of provincial autonomy, with the British, as the trustee government, retaining certain central administrative powers, the way being left open either for partition or for a bi-national state.[37] The admission of 100,000 Jews was—belatedly—accepted by the British Government, provided that the plan's constitutional proposals were also accepted.

In Washington, after a favourable first reaction, Truman bowed before the Zionist storm that followed the leaking of the Morrison-–Grady proposals; on 12 August he telegraphed Attlee that 'the opposition in this country to the plan has become so intense that it is now clear it would be impossible to rally in favour of it sufficient public opinion to enable this Government to give it effective support'.[38] The British Government then resorted to a tactic often adopted in later years as a substitute for an effective colonial policy: representatives of both sides were invited to confer in London. While this conference, which convened on 9 September 1946, was in session, Truman again cut the ground from under British feet, this time with his celebrated Yom Kippur statement, which in effect endorsed partition as the solution of the Palestine problem. Looking back in his memoirs, Acheson described the statement (although he did not think so at the time) as 'of doubtful wisdom'.[39]

Attlee's response on the same day—an acid telegram to Truman drafted in his own handwriting, without a word crossed out[40]—could not alter the fact that on 4 October 1946 this phase of Anglo-American co-operation had effectively been closed. Nevertheless, through the year that passed between Truman's decisive statement and the UN

General Assembly's vote in favour of the partition of Palestine, close contact was maintained over Palestine between the US and British governments. This contact was, however, largely sterile and at times— even though the year included the Truman Doctrine, the Marshall Plan, and the first British financial crisis—publicly acrimonious.* By now it was not only the US Government that was at sixes and sevens where Palestine was concerned. As has been observed in the previous chapter, the crisis of the Palestine mandate coincided with the crisis in Anglo-Egyptian relations. Moreover, confusion in Whitehall, where Cabinet responsibility for Palestine policy was shared by two Departments of State, received an added stimulus when—ironically, on 5 October 1946 (the day after Truman's Yom Kippur statement)— Attlee replaced as Colonial Secretary the ineffective George Hall by the co-founder of the Fabian Colonial Bureau and a pro-Zionist, Arthur Creech Jones. As such, he was in favour of partition, as indeed by this time were some senior British officials in the Middle East.[42] On the other hand, Bevin was increasingly inclined towards surrendering the mandate, rather than accept any responsibility whatever for a solution that would humiliate the Arabs. Between the two Secretaries of State stood the Chiefs of Staff, whose recurring theme in the Whitehall minuet that now developed was the strategic necessity of retaining military facilities in Palestine, which they considered 'of special importance in this [Middle Eastern] general scheme of defence' and as 'a screen for the defence of Egypt'. Given the commitment to withdraw from Egypt, 'we must be able to use Palestine as a base for the mobile reserve of troops which must be kept ready to meet any emergency throughout the Middle East'.[43]

THE REFERRAL TO THE UNITED NATIONS

In Palestine, where a division had been moved from Egypt at the end of 1946, the situation became increasingly ugly† in the final year of the mandate, in spite of the presence of 100,000 British troops. Creech Jones gave up his advocacy of partition, in the face of the Chiefs of Staffs' objections; a final compromise, which he cobbled together with Bevin (known briefly to history as 'the Bevin plan'), was rejected by both the Arabs and the Jews; and the Palestine Conference collapsed. On 18 February 1947—the day on which the Cabinet fixed an

---

* For example, Truman was 'outraged' by Bevin's reference, in his speech in the House of Commons on 25 Feb. 1947, to the Palestine problem being 'made the subject of local elections' in New York.[41]

† The author drove through Palestine in July 1947, the month in which the Irgun hanged two British sergeants and the *Exodus*, carrying 4,500 Jews, was seized off Palestine by the Royal Navy. They were returned to Germany.

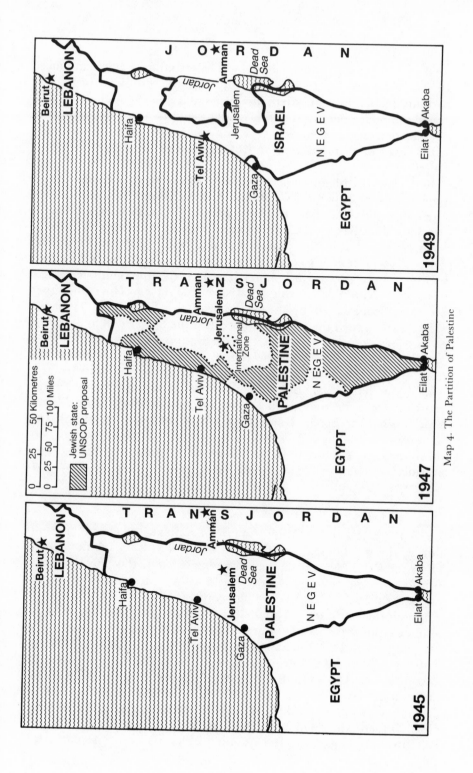

Map 4. The Partition of Palestine

unalterable date for British withdrawal from India—Bevin announced the Government's decision to refer the Palestine problem to the UN; and this decision was followed, just over seven months later, by Creech Jones' announcement to the UN 'in all solemnity' that, in the absence of a settlement, the British Government 'must plan for an early withdrawal of British forces and of the British administration from Palestine'.

Both these British announcements were interpreted by Zionists at the time as machiavellian.[44] We do not, however, need to accept all the intentions then ascribed to these British declarations in order to conclude that they were made more palatable for the British Government by two major miscalculations: neither the extent of Truman's personal commitment to partition nor the final alignment of UN votes was accurately assessed in London. The record of the skulduggery that went on in New York during the last, hectic days of November 1947 has been documented by the historian of the Truman presidency.[45] What really tipped the scale, however, was the Soviet Union's decision to cast its vote on this issue with that of the United States: a development that the British were not alone in failing to foresee, but from which they—unlike the Americans—drew the unwarranted conclusion that Israel, run by communist immigrants from Eastern Europe, would become a 'spearhead of communism' in the Middle East.[46] What the Soviet Union's support of the partition of Palestine did ensure was that it received rather more than the required two-thirds majority of votes in the UN General Assembly (33–13, with ten abstentions) on 29 November 1947.

THE PENTAGON TALKS

Looked at solely in the Anglo-American context, the course of events during the six months that followed the UN vote in relation to the future of Palestine may appear paradoxical. On the one hand, the British Government simply announced a date on which British civil administration in Palestine would come to an end: 15 May 1948. Through thick and thin Whitehall stuck to this date, but that was all; although the British Government would 'do nothing to obstruct the carrying out of the UN decision in favour of the partition of Palestine, British troops and the British administration [in Palestine] should in no circumstances become involved in enforcing that decision or in maintaining law and order while the UN Commission enforced it'.[47] On the other hand, Washington (the White House always excepted) now adopted what had been the British view: that the partition of Palestine was dangerous and impracticable. The State Department, the Department of Defense, and the Central Intelligence Agency now joined forces in opposing it; and they proposed, as an alternative, an international trusteeship temporarily to replace the British mandate until an Arab-Jewish agreement could be worked out.

If the distance between the views of US and British officials at this juncture was, on the face of it, so small—and the political stakes in 1948 were so high—was the logical outcome a final Anglo-American compromise designed to avert the regional war that became increasingly threatening? At the beginning of January 1948 the Arab League announced that the armies of its member states would occupy the whole of Palestine once the British withdrew. Moreover, only nine days after the UN vote in favour of partition, Marshall and Bevin, meeting in London, recorded their governments' approval of the conclusions that had been reached at Anglo-American talks held over a period of three weeks, in secret, by senior officials of the two governments in Washington; the so-called Pentagon Talks. By agreement between the two teams taking part in Washington, however, Palestine had been 'regarded as a thing apart and not to be debated in these discussions'.[48]

These talks, which covered the economic and social, as well as political and military, aspects of policy towards the Middle East, had originated in a personal proposal put forward by Bevin. Prompted by the Anglo-American dissension that had arisen in relation to another British proposed withdrawal—from Greece—he had suggested, in the words reported by the US Ambassador, Lewis Douglas:

a joint review of the whole position in the Middle East . . . for the purpose of aiming at a gentleman's understanding in regard to a common policy and joint responsibility throughout the area, with Britain acting as the front and ourselves [the Americans] supplying the moral support.

The only specific result of the Pentagon Talks was the assurance of US support for Britain in securing 'the necessary strategic facilities in Cyrenaica'. More generally, however, the two teams recorded their agreed view in October 1947 that if the conclusions of the talks, which were set out at great length, were accepted at governmental level, 'it would mean that our two governments would be following in the Eastern Mediterranean and the Middle East parallel policies based on our main objective of maintaining world peace'. Both the President and the British Cabinet did accept all that their officials had agreed in these talks. The senior Foreign Office participant, Michael Wright, claimed on his return to London that 'it is not the Americans who have altered our policy, but we who have secured American support for our position'.[49] But Wright's euphoria, perhaps understandable if it is measured against the reassuringly harmonious language of the

'General Statement by the American Group' in the (US) record of the Pentagon Talks, also has to be assessed against the omission from the talks of the single issue that mattered most at the time: Palestine.

THE END OF THE MANDATE

There was more than one reason why there was no Anglo-American meeting of minds over Palestine in the final months of the British mandate. On the ground, the authority of the mandatory power was disintegrating in the early months of 1948. The financial cost of maintaining the British presence in Palestine, moreover, was higher than the sum which, a year earlier, the British Government claimed to be no longer able to afford for the defence of Greece. On the higher political plane, in London there was not only a feeling of once bitten, twice shy, but also doubt whether the combined weight of the Washington bureaucracy would this time be capable of prevailing against the White House. This doubt proved well founded. What happened in Washington and New York between 19 March—the day of which Truman wrote 'this morning I find that the State Department has reversed my Palestine policy' (by announcing in the UN that the US Government would support trusteeship as a temporary solution for Palestine)—and the evening of 14 May 1948, when Truman disowned the US delegation in New York and recognized the State of Israel, is a matter largely of American, not of Anglo-American, history.[50] It, therefore, does not call for analysis here. In the course of these eight weeks, Truman kept partly in the dark and then openly overruled the American whom he most respected, Marshall. That the President was able to do this without having to look for another Secretary of State was due in large measure to Marshall's personal loyalty, summed up in the remark ascribed to him in a story related by Dean Rusk, twenty-six years later:

No, gentleman, you do not accept a post of this sort and then resign when the man who has the Constitutional authority to make a decision makes one. You may resign at any time for any other reason, but not that one.[51]

THE SEQUEL, 1948–9

Bevin received the news of Truman's recognition of the State of Israel, with what was recorded in the Foreign Office as 'extreme displeasure'.[52] Douglas telegraphed from the US Embassy in London in these terms:

I am convinced that crevasse widening between US and Britain over Palestine cannot be confined to Palestine or even to the Middle East. It is already seriously jeopardising foundation-stone of US policy in Europe—partnership with a friendly and well-disposed Britain. Irrespective of rights and wrongs of question, I believe worst shock so far to general Anglo-American concept of policy since I have been here was sudden US *de facto* recognition Jewish state without previous notice of our intentions to British Government.[53]

This recognition, moreoever, came hard on the heels of the collapse of Bevin's second attempt to revise a British treaty of alliance with a major Arab country (Iraq) where, as in Egypt a year earlier, he was obliged to fall back on reliance on the terms of the pre-war treaty. Only with Transjordan* (a mandated territory until 1946) did he succeed in negotiating a new treaty of alliance, two months before the expiry of the Palestine mandate. Once war had broken out between the Arabs and the Israelis over the division of the Palestine spoils, the terms of this new treaty, signed by Bevin against the advice of Orme Sargent, potentially ranged the United States and Britain, in the words used by Douglas at the end of the telegram quoted above, 'indirectly ... on opposite sides of a battle line scarcely three years after May 8, 1945'.

In assessing why Bevin was willing to expose Britain's relationship with the United States to such a risk in the very months when the two governments were working together, hand in hand and hour by hour, to combat the Berlin blockade, it must be borne in mind that 15 May 1948 was not then universally accepted as the beginning of a new era in the Middle East. On the contrary, the emergence of the state of Israel was seen in Whitehall as a challenge for British policy in the region, where—now that the Anglo-American abscess had been lanced—the two countries could work together again; and, provided that the boundaries of Israel could be 'radically altered', the new state 'might be the least of many evils'.[54]

That the British should have regarded this juncture, after three years of frustration, as an opportunity to regain the initiative in the Middle East[55] might seem far-fetched, were it not borne out by the records of both governments. At the beginning of 1949 British policy in the Middle East was described by Robert Lovett, Under Secretary in the State Department, as 'one of containing the Israelis even at the risk of permanently estranging them'. This policy, in the US view, was 'unrealistic, and they [the Americans] were not prepared to support it'.[56] Six months later a review of policy was conducted by the Foreign Secretary, in conference with British ambassadors summoned home from the Middle East for consultation. By then the British Government

---

* The (Transjordanian) Arab Legion, which by the end of the Second World War numbered 8,000 men, was commanded by General John Glubb. Transjordania received a British subsidy of about £2m. in 1945.

had recognized Israel *de facto* (on 29 January 1949, only two days before the US accorded *de jure* recognition). Yet in the summer of 1949—at a moment of high international drama in other fields—the British still felt their position as the dominant power in the Middle East so little impaired that, without any trace of irony, Bevin's memorandum circulated to his Cabinet colleagues after the Foreign Office conference included the statement that the British Government 'would see no objection to the development of normal trading relations between Israel and her Arab neighbours on a basis of complete independence'. The same memorandum described British influence in the region as still 'greater than that of any other foreign power'; alignment of policy within the United States was essential, but 'need not involve the loss of our special position'; and thought was 'being given to the desirability of a general Middle East Pact complementary to the Atlantic Pact'.[57]

Much had happened in the eighteen months that preceded this conference. The 1948 model of a new Anglo-Arab relationship had fallen at its first fence, in Baghdad. But it was not abandoned. And the fresh stimulus for British policy in the Middle East during the first six months of Israel's existence was an attempt to contain the new state of Israel within boundaries that would be reduced in accordance with the general Arab interest and with the specific interest of Britain's ally, Transjordan, for whom the British Government for several months sought to secure a land-link running across the Negev. For this attempt US support was enlisted and was at first forthcoming. Immediately after the UN mediator, Folke Bernadotte, was assassinated by the Stern gang, on 18 September, both the Foreign Office and the State Department publicly endorsed his second plan, whose main elements were international control of Jerusalem; Galilee to be Israeli; and the Negev to be Arab. Just over a month later—and one week before the US presidential election—Truman issued a statement which in effect supported the Israeli claim to the Negev. After his election he confirmed his statement in a letter to Weizmann.[58] Thus Marshall was once again overruled and the British Government dismayed. British records of the time abound in analogies with Munich.[59] The laboriously documented agreement reached at the Pentagon Talks counted for little or nothing. Not only did Bevin tell Marshall 'categorically that if the Israel forces should attack Transjordan proper at any time, the treaty of assistance with Great Britain would be immediately operative', but when Israeli forces crossed the Egyptian frontier after defeating the Egyptian army at the turn of the year, the British Government came close to invoking its 1936 treaty with Egypt.[60] The winter of 1948/9 was the lowest of the recurring low points touched by the Anglo-American relationship in relation to the Middle East during the second half of the 1940s.

In the end it fell to Hector McNeil, Minister of State at the Foreign Office, returning to his desk after reflecting for some days during the New Year break, to address to Bevin a blunt minute of nine pages, which offered him the 'not very palatable advice' that 'we should cut our losses and try and get the best settlement now possible as quickly as we can'. Two sentences of this minute hit the nail on the head. In the first McNeil wrote: 'As long as America is a major power, and as long as she is free of major war, anyone taking on the Jews will indirectly be taking on America. I consider that this is one of the essentials of which our Middle East policy must take account in the near future'. In an earlier passage in the same minute, he expressed the view that 'development in the Middle East in the near future depends primarily, if not exclusively, upon the United States, and of the Middle East states Israel will have a pre-emptive claim upon such funds'.[61] In the words of Bevin's biographer, the Foreign Secretary 'did not come round easily'.[62]

Nevertheless, by the time that the decade drew towards its close, the Anglo-American bitterness aroused by the Palestine issue had largely subsided. Over the Middle East as a whole—and over Anglo-American attitudes towards it—an uneasy calm settled, characterized by the Tripartite Declaration of May 1950. Acheson afterwards wrote of this self-denying ordinance issued by the governments of Britain, France, and the United States: 'rarely has so large an undertaking had so short-lived an effect'.[63] The calm was indeed shattered as the new decade went forward. Meanwhile, however, it remained the US view that the defence of the Middle East was a British responsibility.[64] And the British themselves, even after at last deciding that the defence of Europe must be their first priority, had not altered their view of the vital significance of the region: 'whoever controls the Middle East controls the access to three continents'.[65]

Since Britain was then the paramount power in the Middle East— Saudi Arabia apart—the primary British responsibility for what happened there during these early post-war years cannot be shrugged off. But the fact that no sustained attempt was made by the two governments to secure a less fragile Middle Eastern settlement after the Second World War, with less blood spilled, must be counted an Anglo-American failure. Arguably, even if such an attempt had been made, it would still have been thwarted by the strength of the nationalistic feelings of both sides in the region. Like China during the same years, the Middle East was the focus of passionate beliefs and the scene of great events.[66] Moreover, the clash of powerful personalities involved

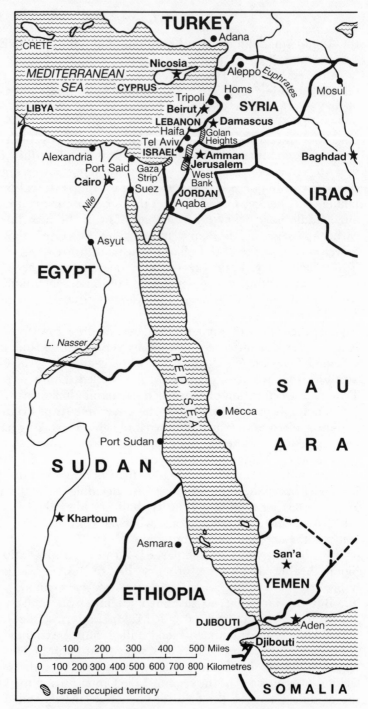

Map 5. The Middle East Today

in the unfolding of these events—Bevin, Churchill, Truman, Weizmann—counted for much. In London the force of Bevin's personal
imprint on the conduct of British policy in the Middle East was for five
and a half years undiminished either by the series of reverses that his
policy suffered or by his health. In Washington, whatever the alignment of the bureaucracy at any given moment, it was Truman who
remained the ultimate arbiter of US policy towards Palestine. It was
not only in Moscow that Truman was underestimated in the first phase
of his presidency. And, whether or not Bevin really reminded Truman
of John L. Lewis, he did little—so far as the Palestine issue was
concerned—to soften the general impression that he made on the
President and Byrnes when he first met them at Potsdam: 'graceless
and rough'.[67]

In the grim perspective of the Holocaust, no President of the United
States could have acted very differently from Truman in May 1948.[68]
And, however clouded Bevin's judgement became over the Palestine
issue, it would have needed all the combined authority of the Prime
Minister and the Foreign Secretary to have overridden the belief of
senior British officials of the period in the essential beneficence of the
British mission in the region as a whole and in the compelling need to
exclude Soviet influence from it: to most of them, the question of
Palestine seemed a distraction from the wider issues that they perceived
as vital,[69] in the literal sense of the word. The British were, however,
unable to fulfill either of the two essential conditions of an enduring
post-war Anglo-Arab partnership: the ability to allocate substantial
sums of British economic aid to Arab countries and the willingness to
redeploy, on sovereign British territory, the forces whose presence no
Arab government, with the single Transjordanian exception, could be
expected to tolerate in their own countries for long after the end of the
Second World War. By contrast, the US Government pursued its
policies without committing a single American soldier. Its economic aid
to Israel in the first year of the State's existence amounted to
US$100m.: about the same as the total of all British economic aid
throughout the world at the time.

Within the wider British projection of the future at that time, the
fear that Israel would become a communist bastion in the region has
been confounded by history. The American view, reinforced by the
Israeli military victories of 1948 and increasingly shared by non-
Zionists in the United States thereafter, that Israel was destined to
become 'the most dynamic, efficient and vigorous Government in the
Near East',[70] has been amply borne out by the country's subsequent
development. The only British prediction—but it is one of critical
importance to an historical verdict—that subsequent events have

confirmed is the depth and ferocity of Arab, and especially Palestinian Arab, resentment at what happened in 1948. Moreover—by a supreme irony—one of the consequences of the Anglo-American failure to agree more than temporarily about the Middle East (which continued into the next decade) was that the very power which both governments above all wanted to keep out of it—the Soviet Union—in the end came forward as the champion of the Arab cause. And today the whole region has become the Balkans of the late twentieth century.

CHAPTER 11

The Coda: China

To the four preceding chapters, which have described Anglo-American discords that persisted through the years covered by Part III of the book, this chapter is added as a coda, in the sense that until the concluding months of 1949 there was little Anglo-American interaction related to China; and what there was, was not of much significance. By contrast, in the following year (which forms the subject of Part V) this particular subject of discord would move to the forefront of the relationship between the United States and Britain, culminating in the crisis of confidence between the two countries that arose in the last days of November 1950: a dramatic event, which can be fully understood only against the background of the chain of developments in, and related to, China over the previous five years. This in turn requires some account, however brief, of the American and the British attitudes towards China during the Second World War.

During the war Roosevelt and Churchill were at their worst over Asia. Their racial views, if expressed today, would be distasteful to most people in both countries. In spite of Roosevelt's support for Indian independence, Mahatma Gandhi did not let him forget the racial problem in the United States.[1] Churchill's defiance of Asian change was epitomized in one of his most famous remarks, made during his speech at the Mansion House, London on 10 November 1942: 'I have not been appointed the King's First Minister in order to preside over the liquidation of the British Empire'. On one thing, however, they did agree in Asia: that the China Theatre was a US responsibility—a recognition not only of the military consequences of the initial Japanese victories in 1941–2 (which in fact destroyed Britain's predominant position in East Asia for ever), but also of the impact on US policy of the explosion of American enthusiasm for China in the early 1940s.

This enthusiasm was a phenomenon whose intensity far exceeded anything that the closeness of traditional Sino-American ties[2] in itself logically warranted. The paean of American praise for China, Chinese democracy, and Chinese resistance against the Japanese invaders, was already reaching crescendo pitch even before Pearl Harbor, in spite of the evidence of objective observers on the spot. Nor was the orchestra confined to Henry Luce's *Time* and *Life* magazines and the China Lobby,[3] which exerted a powerful influence on American public opinion. Even newspapers like the *New York Times* and the *Christian Science Monitor* wrote nonsense about China. 'How many', an editorial in the latter newspaper asked its readers in mid-1942, 'have considered what a different balance the world might have today were not the Generalissimo a Christian and his wife American-educated?'[4]

The British, while they too regarded China as an ally in the Far Eastern campaign who could not be allowed to founder, did not suffer from the American national hallucination where China was concerned. Some British leading articles[5] of this period about China make absurd reading, as do some reports from the British Embassy in China.[6] But the pragmatism of the British, against the background of their longer experience of Asia, preserved them from the wilder excesses about China that were reached on the other side of the Atlantic. Nevertheless, as the junior partner in the Far East, the British Government went along willy-nilly* with the Rooseveltian concept of China as the fourth 'world policeman', agreed at the Tehran Conference. So it was that China ended the Second World War in an international position unexpected by anyone in 1939: a permanent member of the UN Security Council, a member of the Council of Foreign Ministers established at Potsdam, the holder of large† reserves of gold and foreign exchange, and with the prospect of the recovery of Manchuria and Formosa (Taiwan). By then, however, the other side of the coin was equally evident: the size of the task of economic reconstruction, the corruption and incompetence of the Nationalist Government, and—as the year drew to its close—the imminence of civil war between Chiang Kai-shek's Nationalist forces and the Communist forces of Mao Tse-tung.

Estimates of the size of Maoist forces at the end of the Second World War varied widely, according to whether or not irregular troops were included in the numbers. But—from the 20,000[8] that had reached Yenan at the end of the Long March in 1935—they had increased to a

---

* Under US pressure, it gave up British extra-territorial rights (except for Hong Kong) in China; it even sought to match the US$500m. loan given to China in 1942 with an offer of a £50m. loan of its own; and, finally, at Yalta Churchill put his signature to the US-Soviet secret agreement on the Far East, which led in August 1945 to the conclusion of the Sino-Soviet Treaty of Alliance, under whose terms the Soviet Government pledged exclusive support in China to the Nationalist Government.

† Over US$900m. on 31 December 1945, in addition to the 'very substantial foreign exchange assets' in private Chinese hands at the time, conservatively estimated as 'at least several hundred million United States dollars' on VJ Day.[7] These combined Chinese holdings of foreign exchange assets may be compared with the £1bn. held by the Bank of England at the end of the Second World War.

strength somewhere between 500,000 and one million by 1945.*
Nevertheless, according to US estimates the numerical superiority of
the Nationalist forces remained of the order of 5 to 1; and the
Nationalists certainly had a virtual monopoly of heavy equipment,
transport, and aircraft. The Maoists' military advantage lay in
geography: they were situated in northern, central, and coastal China.
In order to counterbalance this, US forces transported some 400,000 to
500,000 Nationalist troops by air and sea after the surrender of Japan;
and 50,000 US marines were landed in North China.[10] True, the
Maoist forces were given captured Japanese equipment by the Soviet
Army before it withdrew from Manchuria, in April 1946; but even so,
as late as the close of 1946, by which time the strategic initiative was
about to pass to the Maoist forces, the numerical advantage remained
with the Nationalists.[11]

Two weeks after the death of Roosevelt (who, at the Yalta
Conference had spoken of the 'so-called'[12] Chinese Communists), the
State Department submitted this sober report to Truman:

While favouring no political faction, we continue to support the existing
government of China, headed by Chiang Kai-shek, as the still generally
recognised central authority which thus far offers the best hope for unification
and for avoidance of chaos in China's war effort. However, with regard to our
long-term objective and against the possible disintegration of the existing
government, it is our proposal to maintain a degree of flexibility to permit
collaboration with any other leadership in China which may give greater
promise of achievement of unity and contribution to peace and security in east
Asia.[13]

Had the Truman Administration been able to adhere to this attitude
of detached scepticism, largely shared at the time by the British,[14]
things might have ended differently. Unfortunately, the Americans
had for too long allowed themselves to superimpose an alien, American
image on what was in reality not in any western sense a democracy, but
the unique culture of an ancient people, still undecided on what form
of government should replace their own semi-feudal system. The
Americans were, therefore, reluctant to believe that compromise
between the two sides in China was no longer possible; nor could they
bring themselves to accept the fact that the United States would now
become no more than a marginal factor in whatever Chinese settlement might eventually be reached.

For their part, the British had more than enough to do simply re-

---

* Of Mao's policy during the war years, a contemporary Soviet sinologue has observed drily
that his 'group . . . taking after Chiang Kai-shek, thought more about an alliance with the USA in
order to obtain American aid than about the struggle with the Japanese'.[9]

establishing, as best they could, the Asian positions that they had lost. (The British reoccupation of Hong Kong on 16 September 1945, however, although it required a brisk triangular exchange of tele-grams,[15] did not give rise to the kind of Anglo-American argument that would have been probable three years earlier.) The idea of some kind of British initiative in China was mooted, but no more, in 1945.[16] Even if it had been pursued, the chances that it would have been well received in Washington were remote. The US Government did not even think it necessary to consult the British Government about the negotiations that ended in the conclusion of the Sino-Soviet Treaty of 1945, despite the fact that Britain was a co-signatory of the Yalta Agreement from which these negotiations stemmed. True, Britain still retained major interests in China, not only Hong Kong, but also substantial investments, whose value has been estimated at more than US$1bn.[17] Nevertheless (in strong contrast to South-East Asia, where in 1945 the British not only reoccupied their own colonial territories, but also assumed military responsibility for Indonesia, southern Indo-China, and Siam)* China, like Japan, remained first and foremost a US responsibility.

When the Second World War ended, the United States Government saw itself confronted with three possible options in China. In Acheson's words to the President four years afterwards:

(1) it could have pulled out, lock, stock and barrel;
(2) it could have intervened militarily on a major scale to assist the Nationalists to destroy the Communists;
(3) it could, while assisting the Nationalists to assert their authority over as much of China as possible, endeavour to avoid a civil war by working for a compromise between the two sides.[18]

Truman chose the third course. Neither he nor Byrnes knew anything about Asia; Acheson—then Under Secretary, who was to return to the Department as Secretary of State in January 1949—knew little more.[19] And uncomfortably close to the Chinese scene loomed the formidable proconsular figure of General MacArthur, now Supreme Commander in Japan, and closely linked to the right wing of the Republican Party.

In an effort to give the third option the best conceivable chance of success, Truman appointed Marshall to mediate in China, as his Special Representative, at the close of 1945. Yet even this appointment of the man most likely to be universally acceptable was not enough to

---

* The second and third of these three responsibilities did not last long, but British troops were not withdrawn from Indonesia until the very end of 1946. It was another three years before the Dutch Government concluded an Indonesian settlement. Although the Indonesian problem was sometimes an important item on the Anglo-American agenda during these years, on balance it has seemed preferable not to include it in this study of the major issues of the period.

divert what has well been described as 'the political fault line that ran underneath the Truman administration'[20] in Asia. After spending twelve months trying to deal with a fundamentally intractable situation,[21] Marshall recommended that the US marines should be withdrawn from China; that the attempt to mediate should be abandoned; and that he himself should be recalled. He was indeed recalled; such was his national standing that the Senate then unanimously confirmed his nomination as Secretary of State on the day that Truman submitted it; and if the US Government succeeded in avoiding deeper involvement in China for the next three years, this was in large measure due to Marshall's moderating influence.[22] Nevertheless, the ambivalence of US policy towards China was not resolved. As the annual review of 1948 sent by the Washington Embassy to the Foreign Office observed: 'critics of the State Department could, and did, rightly say, just as they had in 1947, that it still had no China policy worthy of the name'.[23]

It was only eight weeks before the proclamation of the People's Republic in Peking that the US Government at last attempted fully to enlighten American public opinion about what had gone wrong. This attempt took a remarkable form. 'The China White Paper', as it came to be called, published on 5 August 1949, was a 1,054-page apologia for *United States relations with China, with special reference to the period 1944–1949*, transmitted by Acheson to the President in a letter that concluded in the ringing tones of the cold war:

> We continue to believe that, however tragic may be the immediate future of China and however ruthlessly a major portion of this great people may be exploited by a party in the interest of a foreign imperialism, ultimately . . . she will throw off the foreign yoke . . . we should encourage all developments in China which now and in the future work towards this end.[24]

The China White Paper, intended by its authors as a shield for the US Administration, quickly became a boomerang. US post-war policy towards Asia, unlike US policy towards Europe, had never been bipartisan. The so-called loss of China was therefore bound to cause a political storm in the United States. Far from lessening the force of this storm, the White Paper helped to make it more violent. American adulation of the China of the early 1940s had been such that the China of the end of the decade evoked an equally extreme American reaction. To some Americans, treachery* and conspiracy among their own

---

* By the end of 1949 the China Lobby was already in full cry. McCarthy's speech, describing the State Department as 'a nest of communists and communist sympathisers who are helping to shape our foreign policy', was delivered at Wheeling, West Virginia in February 1950. Among the casualties of the 'loyalty programme' that McCarthy inspired were four distinguished members of the US Foreign Service who had been stationed in China.

compatriots seemed the only explanation of such a lamentable end to years of American benevolence towards China, which included the expenditure of US$2bn. since the end of the Second World War. The storm over China at the turn of the year 1949/50 soon merged with the subsequent controversy over the Korean War and Truman's dismissal of MacArthur in 1951. It thus became a major factor contributing to the end of the Democratic Party's long rule.

Perhaps Walter Lippmann came nearest to hitting the nail on the head, when he said: 'Chiang's destiny became our destiny in China and so, when he went down, he took us down with him'. At the time that Lippmann wrote these words, in September 1949,[25] Chiang had not yet left the Chinese mainland. Two months later he fled to Formosa (Taiwan), pledged to recapture China from there. Thus the Sino-American idyll of the 1940s ended with continued US recognition of the irredentist Nationalist Government in Formosa as the Government of China and—in consequence—with an international ostracism of China, inspired by the United States, which lasted for almost a quarter of a century. Worst of all, perhaps, this oscillation from intimate friendship to outraged hostility across the Pacific contributed to the formation of what became known in the west as the Sino-Soviet bloc: a temporary accommodation of Soviet and Chinese interests, as we can now see, but in the circumstances of the time a concept so potent that it retained its hold on western imaginations, especially in the United States, long after the Sino-Soviet alliance of 1950 had become a dead letter.* This alliance, concluded on 14 February 1950 after two months' negotiations in Moscow conducted by Mao himself (and recorded in an extraordinarily gloomy photograph in *Pravda*)[28] broke down in 1957–8; the Sino-Soviet rift at first found mainly indirect expression; in October 1961 for the first time it became overt; and from 1963 onwards the dispute was publicly conducted with increasing bitterness on both sides.

In the same article Lippmann pointed out that the White Paper gave no indication that the US Government had 'ever consulted with any friendly allied power with a view to sharing responsibility for China's future'. The post-war British Government, unlike the US Administration, included men—Attlee, Bevin, and Cripps—with some Asian experience and with ideas about Asia far removed from the Churchillian prejudices that had helped to sour the Anglo-American relationship in Asia during the war. Indeed, there was a reversal of roles in the

---

* As late as October 1961,[26] the US Secretary of State, Dean Rusk, was still referring to the 'Sino-Soviet bloc' in his speeches. Similarly, Senator Henry Jackson, who in the next decade became a prominent sinophile, refused at that time to accept that the differences between the Soviet and the Chinese Communist parties were real.[27]

later 1940s. A 'fundamental unlikeness' in Asia between the Americans and the British remained; but it was now the British who saw themselves as the ideological front-runners for the west in that continent. In August 1949, for example, a British planning paper on 'The United Kingdom in South-East Asia and the Far East' followed the statement that 'the United States does not enjoy the same degree of prestige in the area as the United Kingdom' with this sentence:

It is not a great exaggeration to say that, when Asian Nationalists consider the future political and economic structure of their independent countries, they contemplate only the alternatives of democratic socialism and communism, which they sometimes simplify into a choice between the British and the Russian ways of life.[29]

Such Anglo-American consultation as there was suffered from several handicaps. To begin with, the natural interlocutor for the United States over China in 1945 was not Britain, but the Soviet Union, whose policy towards China then—and arguably for the next four years— was one of caution.* As the United States' relations with the Soviet Union worsened in the late 1940s, it was Europe—not Asia—that became the primary focus of western anxiety and hence of Anglo-American co-operation. And, by the time that the Far East found its place on the Anglo-American agenda, the British were themselves preoccupied with communist penetration, not so much of China as of South-East Asia. The Malayan 'emergency' campaign, which began in June 1948 and lasted well into the next decade, necessitated the deployment of 40,000 troops and nearly twice as many police.

The first initiative, in 1948, was British, as it had been in 1947 in relation to the Middle East. Esler Dening,[31] the Assistant Under-Secretary in the Foreign Office responsible for Far Eastern and South-East Asian affairs, then visited Washington, after a tour of Commonwealth countries, under instructions to sound the Americans about the possibility of secret talks on the Far East. 'The Americans did not respond to this suggestion.' Recalling this rebuff in a letter to his opposite number in the Commonwealth Relations Office a year later, Dening wrote:

Much has happened since then, and in the field of United States foreign policy there are some rather disquieting indications that the Americans in the first place are without any clear policy in regard to the Far East and South-East Asia, and secondly that they are inclined to decrease rather than extend their commitments. We had hoped that we might succeed in directing American

---

* The Soviet action which probably had the greatest impact on Chinese developments after the Second World War was the Red Army's systematic looting—War Booty 'Removals'—of the industrial assets of Manchuria in 1945–6.[30]

attention towards the essentials of the situation by discussing with them the problem outlined in the Cabinet paper on China ... But though they displayed a certain interest and somewhat desultory inter-changes have taken place, no real progress has been made.[32]

By the time that Bevin went to Washington for the signature of the Atlantic Treaty, he had decided to return to the charge personally with Acheson, 'with a view to encouraging the Americans to take a more active interest in the South-East Asia and Far Eastern regions'. The brief prepared in the Foreign Office for the discussion between the two Secretaries of State was drafted to take account, on the one hand, of 'American reluctance to engage in any further commitments' and, on the other, of the 'wish to deter the Americans from embarking upon any policy which would offend the susceptibilities of Asiatic countries and in particular those of the Commonwealth'. It is also clear, both from the brief itself and from the covering minute with which Dening submitted it to Bevin on 23 March 1949, that the main British concern was not China and Korea, about both of which the Foreign Office was profoundly pessimistic,[33] but South-East Asia. Bevin's talk was to be 'in effect an initial step with a view to enlisting American support in principle for the policy we hope to pursue'.[34]

Bevin left Acheson a copy of this brief on 30 March 1949, but neither this meeting nor Dening's subsequent talks with the State Department achieved any significant result in 1949.[35] Another year was to go by before the Korean War induced a sea change in American attitudes towards South-East Asia. The British brief of March 1949 retains a historical importance, however, as one of the earliest signposts pointing in the direction of the South-East Asian policy that the US was, in the following decade, to adopt lock, stock, and barrel. ('South-East Asia' was then very broadly defined, to include not only South-East, but also South, Asia, and even Afghanistan.) Its conclusion was:

... while the strategic necessities of Europe and the Middle East are greater and should have priority, the requirements of South East Asia, though in a different category, are of vital importance. We should, therefore, parallel with our efforts in Europe and the Middle East, do our utmost to encourage a spirit of co-operation and self-reliance in South East Asia with a view to the creation of a common front against Russian expansion in that area.

Even so, Bevin did not follow the advice offered by the British Commissioner-General for South-East Asia, who, in a lengthy des-patch sent from Singapore in March 1949, had strongly argued the need for 'Asian equivalents of the Marshall Plan and the Atlantic Pact' and expounded in its earliest form the doctrine that was later to be called the Domino Theory in South-East Asia.[36] A 'Pacific Pact' was indeed being advocated by the Australian Government; predictably, Jawaharlal Nehru disagreed with this proposal, as did Bevin; and Acheson put his rejection of the idea on public record.[37]*

It was unfortunate that it was not until the last two years of his life that Bevin was able to turn his attention to China. Had he given the Far East half the time that he gave to the Middle East from 1945 onwards, would he have made any impact on the issues? However that may be, by the time that he did address them, all that was practicable was an Anglo-American agreement to differ for the time being. That their difference over China was by then profound was recognized by Bevin on 23 August 1949, when he described the two governments' conflicting views to his colleagues in a memorandum written immediately after the publication of the US China White Paper. Only four months earlier, Acheson had told Bevin during a meeting in Washington that the State Department had 'abandoned the idea of supporting the [Chiang] regime' and that the United States would 'henceforth pursue a more realistic policy regarding China'.[38] In August, in Bevin's words, 'US policy had, without any prior warning, taken a sharp turn in the direction of retreat'. It was 'difficult to understand what the present trend of American policy denotes', but there was 'understood to be a school of thought which considers that Communist China should be allowed to relapse into complete chaos, which will encourage the Chinese people to overthrow the Communist regime'. This was 'diametrically opposite' to the British view that they must do their utmost to maintain China's western contacts 'if we are not to drive Communist China into the arms of Moscow'. Instead, the Americans regarded China as already in the Soviet embrace.[39]

The British Cabinet finally reached the decision to recognize the new Chinese Government on 15 December 1949.† That the British judgement of the new China was sound has been demonstrated by the subsequent course of history. A cool calculation of the nature of the broad western interest in East Asia, based on an understanding of the new China, was possible in London. In Washington it was precluded by the high political temperature being run there as 1949 drew to its close. Nevertheless, as Bevin admitted in the personal message that he sent to Acheson on 16 December 1949, informing him of the British Government's decision to give the new government in Peking *de jure* recognition, there were some special factors which affected the British,

* This was one of many aspects of the US Government's Pacific policy that changed radically under the impact of the Korean War. By the end of 1950 the Australian Foreign Minister, Percy Spender, was well on the way to persuading the US Government to accept the commitment which later became the ANZUS Pact.

† The timing of recognition was left to Bevin's discretion. Formal notification to Chou En-lai was made by the British Embassy on 5 January 1950.

'not only our interests in China but the position in Hong Kong, and also in Malaya and Singapore where there are vast Chinese communities ...'

The message went on:

We also take the view that to withhold recognition indefinitely is to play straight into the hands of the Soviet Union. We feel that the only counter to Russian influence is that Communist China should have contacts with the West, and that the sooner these contacts are established the better ... I had hoped that we might be able to take action together in this matter ... we want to keep in close association with you, but we have to be careful not to lose our grip of the situation in Asia and to take into account the views of our Asian friends.[40]

By far the most important 'Asian friend' was independent India, whose government recognized the Chinese Government on 30 December 1949, and whose views had, since independence, become a major input for the formation of British foreign policy. The Indian example was, therefore, one which it would have been very difficult for the British Government to ignore, in the context of its overall policy towards Asia* at the time.

By contrast, the US Government decided not to recognize the Chinese Government in Peking until 'the dust settled' (a decision defined as meaning until it was 'clearly in the interest of the United States to do so').[41] A whole year was lost, by which time the United States and China found themselves in effect, though not formally, engaged in a limited war with each other—paradoxically, in a territory from which the last US troops had been withdrawn on 29 June 1949: the Korean peninsula. Thus the American attempt to fend off the communist victory in China immediately after the Second World War, their refusal to recognize that victory as an irreversible fact in 1949, followed later by the decision to support Chiang's regime, proved to be the first act of a prolonged tragedy, whose second act was the Korean War, and the third the war in Vietnam.[12]

Chapter 4 has noted the circumstances in which the British signature came to be placed on the secret Yalta Agreement on the Far East. Five years later, the British Government publicly disassociated itself, in effect, from a US policy that had become bankrupt. But this British decision to disagree with the US Government about China had no more influence on US policy in East Asia than had Churchill's decision at Yalta to go along with what Roosevelt and Stalin had agreed about the region bilaterally among themselves. Partly because British recognition of the new China came too late, partly because it was not

* The Colombo Conference, at which Bevin helped to launch the Colombo Plan, opened on 10 January 1950.

followed through to its logical conclusion at the United Nations, and partly because it was soon to be overtaken by the onset of the Korean War, the British did not thereby secure privileged access in Peking. In the autumn of 1950, when the Chinese Government needed an intermediary to convey its warnings to the US Government about Korea, it was to the Indian—not to the British—Embassy that Chou En-lai turned. In November 1950, when Dening tried to visit Peking, his application for a visa was not granted. When the US Government wanted to warn the Chinese Government over two years later, it was again the Indian Embassy that acted as intermediary.[43] And the course of events from the turn of 1949/50 onwards was to confound the British belief, confidently expressed three months earlier, that the US Administration had accepted the British concept of a shared partnership in the direction of world affairs.

PART IV

THE REMAKING OF EUROPE, 1947–49

CHAPTER 12

The Anglo-American Relationship Revived

HALF way through these years of Anglo-American discord (nuclear, financial, Middle Eastern, and Far Eastern) the Americans and the British found themselves thrown together in a great enterprise: the remaking of Europe. As the common endeavour gathered momentum from 1947 onwards, relations as a whole between the United States and Britain were gradually permeated by the spirit of intimate co-operation between the two governments that developed in relation to Europe. Although this harmony could not resolve the discords in the other areas of policy that have formed the subject of the five preceding chapters—nor did the two governments see entirely eye to eye over Europe itself—it softened their impact on the Anglo-American relationship across the board.

The reason why the take-off point for this co-operative effort over Europe was delayed until 1947 was not only that, at the end of the war, the task of economic reconstruction had been underestimated almost by orders of magnitude, but also that until then the European policies both of the US and of the British Government remained to a large extent conditional upon the framework established by the three victorious powers at Yalta, Potsdam, and Moscow in 1945. Through 1946, therefore, the governments of Britain, France, the United States, and the Soviet Union followed—ponderously—the negotiating procedures relating to Europe agreed in the previous year. The second meeting of the Council of Foreign Ministers, held from April–July 1946 in Paris, was followed by the Peace Conference, which opened there on 29 July, and the third meeting of the Council, which was held in New York at the end of the year.

The only agreement reached in all these months of deliberation (the last six of which coincided with the abortive US-Soviet atomic discussions at the UN) was on the terms of the peace treaties with Italy and with Germany's former satellites, which were to be signed, again in Paris, on 10 February 1947. In effect what was finally agreed in New York in December 1946 was an East–West bargain: the Italian treaty virtually on western terms in return for the other treaties virtually on

Soviet terms, coupled with the decision to leave over the central problem of Germany, together with the subsidiary problem of Austria, until the fourth meeting of the Council of Foreign Ministers. This critical meeting, which was held in Moscow from 10 March to 25 April 1947, failed to reach any agreement.[1]* The formal deadlock over Germany came eight months later, at the fifth meeting of the Council at Lancaster House, London, which adjourned on 15 December without even fixing a date for its next meeting.

From August 1945 until December 1947 the United States and the three major European powers had together attempted to deal with the post-war problems of Europe, and above all those of Germany, as a whole. Whatever the final verdict of history on the cold war may be, we know enough already to realize that the sequence of events that led up to the division of Europe was far from clear cut; that it was sometimes fortuitous; and that in the new western approach to European problems there was more than one dimension. The heart of the matter was Germany—'this eventual reservoir of power', as Bevin called it in November 1945[2]—initially in economic, but ultimately in political, terms. Yet in the Anglo-French Treaty,[3] signed at Dunkirk on 4 March 1947, it was the contingency of hostilities with Germany against which the two signatories assumed the obligation to come to each other's aid with 'all the military and other support and assistance' in their power;[4] Bevin believed that Germany 'still constituted the biggest menace to peace in Europe'[5] (a greater threat than the Soviet Union, he said to Bidault even in December 1947);[6] both he and Attlee 'viewed with disfavour a report by the Chiefs of Staff basing British defence on the possibility of war with Russia' in June 1947;[7] and as late as November 1949† British and French fears of Germany remained close to the surface.[8] During the first half of 1947 the British and Soviet governments were even discussing revision of their (1942) treaty of alliance;[9] and in the following year, when Stalin and Molotov met the three western ambassadors in Moscow at the height of the Berlin blockade, the Soviet leaders may not have been speaking ironically when they continued to describe the Soviet Union and the Western powers as allies and 'partners in Germany'.[10]

If the onset of the cold war was a gradual process, so too was the renewal of Anglo-American understanding that accompanied it. For this renewal the wartime Anglo-American relationship served as a foundation, but it was not a question simply of the two countries

---

* On Austria, however, it was agreed to proceed with the preparation of a draft treaty.

† This month is significant because it was on 22 November 1949 that the Petersberg Agreements were signed in Bonn. Their stated primary objective was to incorporate the German Federal Republic in the Western European community.

resuming where they had left off in 1945. Even though the first major Anglo-American response to a European problem in 1947 was made in the field of defence—and, paradoxically, the defence of South-East Europe—the process that developed from 1947 onwards was at first much more economic than military, above all in the joint Anglo-American reaction to the 'economic consequences of the peace', or rather, of the occupation of Germany.

Intertwined with the economic and military strands of the remaking of Western Europe was the political: not only the crude, though potent, imperatives of the cold war, but the vision, shared by many of the statesmen of continental Europe after the Second World War, of a new Europe. This was to be a Europe not just liberated from the threat of German aggression, but qualitatively different from the group of nationalistic rivals that had resorted to two civil wars in the span of a quarter of a century. This fresh approach to the ideal of European unity, alien to the British tradition of pragmatism, evoked a powerful response in the United States.*

Six months after his speech at Fulton, Missouri, Churchill delivered a shorter, perhaps more memorable, speech at Zurich University. Although he made clear his view that Britain would not be a member of this new Europe, but rather, like the United States, one of its 'friends and sponsors' who 'must champion its right to live', the essence of his message to prostrate Europe was this:

I am now going to say something that will astonish you. The first step in the recreation of the European family must be a partnership between France and Germany. In this way only can France recover the moral and cultural leadership of Europe. There can be no revival of Europe without a spiritually great France and a spiritually great Germany ... we must recreate the European family in a regional structure called—it may be—the United States of Europe, and the first practical step will be to form a Council of Europe.[11]

What Churchill, as Leader of the Opposition, said to the students of Zurich University, only a year after the end of the war, was far ahead of governmental thinking in any of the capitals concerned. For *The Times*, which carried his speech in full, British policy 'could not yet afford to despair of avoiding the division of Europe'—a view repeated in a *Times* leading article written as late as May 1947.[12] In Paris, too, the French Government was then pursuing diametrically opposed objectives in its German policy; but the idea of Franco-German

---

* As early as 21 March 1947 the Fulbright-Boggs resolution in Congress had called for the creation of a 'United States of Europe'; and in 1949 the European Cooperation Act was amended to include the words: 'It is further declared to be the policy of the United States to encourage the unification of Europe'.

reconciliation formed an essential part of the vision of a unique Frenchman, Jean Monnet, who, though not himself a politician,* was to exert a profound influence on the post-war political development of Europe. He later summed up his own European belief in the words: 'nous ne coalisons pas les Etats, nous unissons des hommes'. When he came to prepare the first draft of the Schuman Plan in 1950, a Franco-German union was its 'préoccupation centrale'.[13] Churchill's Zurich speech and Monnet's ideas illustrate the fact that, by the time that Marshall delivered his speech at Harvard University nine months later, the seeds which he scattered fell on European ground that was not wholly unprepared. As time went by the different strands—economic, military, and political—in the remaking of Europe became intertwined, so that their interaction on each other cannot always be disentangled. For purposes of clarity, however, they are analysed separately in what follows.

GERMANY

Germany was the central issue in the years that followed the Potsdam Conference. It will be recalled that one of the two component parts of the US-Soviet deal struck at Potsdam, as the conference ground to its close, was an agreement on German reparations based on a complicated formula (see Appendix 2) governing the transfer of industrial capital equipment, over a period of two years, to the Soviet Union from the Western Zones of Germany; that the exact amount of equipment to be removed from the Western Zones was to be determined at the latest within six months; and that this formula foreshadowed the division of Germany, even though the Potsdam Protocol expressly recorded the agreement of the three Heads of Government that the Allied Council would govern the country as a single administrative and economic unit. So indeed it proved.

On 18 March 1946 the Allied Council agreed on the reduction of German industry to approximately seventy per cent of its pre-war level: a decision on which not only France and the Soviet Union but also the United States at that time insisted. Britain, the occupant of the German zone with the most industry and the highest unemployment, reluctantly acquiesced. The French proposals for the Ruhr and the Rhineland were also bound to frustrate the unitary intentions of the Potsdam Protocol. In October 1945 the French Representative on the

---

* Underrated (on occasion even high-hatted) by some of his British contemporaries at the time, he is still sometimes described by British writers simply as a technocrat. He was in fact an empirical visionary.

Allied Council vetoed proposals for the setting up of five German central administrative departments. Moreover, Byrnes' proposal, put forward in May 1946, for a twenty-five year treaty to enforce German demilitarization was rejected by France no less forcefully than by the Soviet Union.

From the British point of view, the Potsdam Agreement—in Bevin's words nearly two years later—'forced on him in his first day of office, left much to be desired'.[14] For its part, the United States disliked the British plan* to nationalize the Ruhr coal-mines (Britain's own coal industry was nationalized on 1 January 1947). Nevertheless, in the second half of 1946, British and US policy began to share a common economic interest: both governments found themselves obliged to subsidize heavily the import of food into their zones of occupation, even to enable the inhabitants to eke out a 1,000-calorie existence.† For the British, this was a cost paid in dollars. In May 1946 the United States issued a warning to both the Soviet Union and France: until the other governments treated the whole of Germany as a single economic unit, no further reparations would be delivered from the US zone. On 2 December the US and British governments signed an agreement to fuse the economies of their zones of occupation in Germany.

It was against this sombre background that the four Foreign Ministers, of whom Marshall was now one (having replaced Byrnes at the State Department on his return from China), travelled in March 1947 to Moscow. There they spent over six weeks locked in fruitless argument over the application of the Potsdam Agreement to Germany. In Stalin's parting words, the failure of this meeting was 'only the first skirmishes and brushes of reconnaissance forces'. Did he perhaps have at the back of his mind Roosevelt's statement, made at Yalta just over two years earlier, that Congress would not support the presence of the American forces in Europe for more than two years? (Since November 1946 the Republican Party had commanded a majority in both Houses of Congress.) Marshall, on the other hand, drew the opposite conclusion from the Moscow meeting of the Council of Foreign Ministers: that Europe could not be left to rot any longer. As he put it in the broadcast that he made on 28 April 1947, after his return to the United States, 'the patient is dying while the doctors deliberate'.[15] His instruction to Kennan to address the European problem at once and afresh‡ followed on the next day.

That the four powers continued—at the end of the grim winter of

* Shelved in November 1947.
† In Vienna, the ration fell even lower in the winter of 1945–6. In April 1947 the Ruhr miners went on strike as a protest against their rations.
‡ See p. 88.

1947—to disagree on how to interpret and how to implement the abstruse document to which, nearly two years earlier, three of their Heads of Government had put their signatures at Potsdam, is hardly surprising. The French, not having been invited either to Yalta or to Potsdam, were able to act as mavericks. As Stalin made clear to Marshall, during his farewell call on 15 April 1947, with particular emphasis, the Russians never lost sight of their goal, almost secured at Yalta, of US$10bn. worth of German reparations to help restore their ravaged economy.[16] For their part, the Americans and the British became increasingly determined to avoid a repetition of the 'economic consequences of the peace'. Nevertheless, the timing of the break between East and West was also influenced by a confused chain of Anglo-American events not directly connected with Germany at all.

THE TRUMAN DOCTRINE

Without access to the Soviet archive we cannot say how much the US President's message delivered to a joint session of Congress on 12 March 1947—two days after the Council of Foreign Ministers had convened in Moscow—affected the way that Molotov played the Soviet hand during the six weeks of quadripartite negotiations that followed. No reference to the Truman Doctrine seems to have been made at any of these meetings in Moscow. There is, however, no doubt about the effect on Congressional opinion at the time of the Administration's adoption of a dramatically new policy towards the Eastern Mediterranean; and in the Kremlin it was certainly regarded as 'a turning point in American foreign policy, a clear departure from Roosevelt's policies'.[17] The significance of the Truman Doctrine for the subsequent evolution of American foreign policy has been much discussed. The basic facts of what happened in the early weeks of 1947—in contrast to the causes of those events—are straightforward. On 21 February the State Department received formal British notification[18] that, from the end of the British financial year (31 March), the British Government would suspend financial assistance to Greece and Turkey. The British estimate of Greece's 'civilian and military needs' during 1947 was 'between £60 and £70 million'; British troops were also to be withdrawn from Greece. No figure was given for Turkish needs, which were clearly not regarded with anything approaching the same degree of urgency. On 12 March Truman asked Congress to approve aid of US$300m. for Greece and 100m. for Turkey, and to authorize the despatch of US civilian and military personnel and equipment to both countries, and the training of selected Greek and Turkish personnel. Truman received a standing ovation. On 22 May the President signed the aid bill for Greece and Turkey into law. Rather more than half the aid given to both countries was military.

On both sides of the Atlantic the claims advanced for what had been achieved during these three months were pitched very high indeed. Attlee's verdict was that 'by giving America notice at the right moment that we couldn't afford to stay and intended to pull out we made the Americans face up to the facts in the Mediterrenean';[19] Hector McNeil, Bevin's deputy at the Foreign Office, recalled (in a minute to the Foreign Secretary written two years later) that it had seemed 'a triumph'.* Truman himself wrote to his daughter on 13 March 1947 that at Potsdam he had realised that 'there is no difference in totalitarian or police states, call them what you will, Nazi, Fascist, Communist ... Your Pop had to tell the world just that in polite language'.[20] What his message to Congress had expressed on the previous day, was his belief that 'it must be the policy of the United States to support free peoples who are resisting attempted subjugation by armed minorities or by outside pressures'[21]—a doctrine whose echoes were to be heard repeatedly in presidential and official statements from Washington for more than twenty years afterwards.

The story told by the documents themselves is more prosaic. In the purely Anglo-American context, two questions call for an answer. Why did Britain choose to present as beyond its means this particular item (some £60–70m.) out of its entire programme of overseas expenditure? And why did the United States decide to accept this particular European commitment—far from the strategic epicentre of the continent—as the first step in what later became a fresh approach to the European crisis? In the event, British financial support for the precarious Greek regime was not wholly withdrawn; and the last British troops did not leave Greece until three years later. Moreover, at the beginning of 1948, as it turned out, the Balkan political alignment was transformed by the Soviet–Yugoslav schism:† an example, perhaps, of Proust's epigram that definite resolutions are taken only on the basis of a state of affairs that is destined not to last.

Ever since the autumn of 1944 the British Government had been deeply involved in Greek politics. In the words of the Englishman who came as close as anyone could to understanding its complexities during and after the war, 'up to 1947 the British Government appointed and dismissed Greek Prime Ministers with the barest attention to constitutional formalities'.[23] On 1 December 1946, a fortnight after a debate

---

* See Chapter 10.

† According to Djilas, Stalin in February 1948 said to the Yugoslav Foreign Minister: 'the uprising in Greece must be stopped and as quickly as possible'[22] (the guerrillas in northern Greece had been able to operate from bases in Yugoslavia).

in the House of Commons in which Bevin's 'drift into the American camp' had come under sharp attack from members of his own party,[24] Attlee wrote a letter, which he typed himself and marked 'Private and Personal', to Bevin, who was then still in New York. In this letter Attlee singled out the Greek commitment as a 'financial and military burden' that he could not 'contemplate ... with equanimity', continuing: 'I feel that we are backing a very lame horse ... I am beginning to doubt whether the Greek game is worth the candle'.[25] Simultaneously forewarned by a telegram from McNeil that the Cabinet was to discuss the whole question of policy towards Greece and Turkey, Bevin at first reacted strongly:

The policy of the Government has hitherto been based on the assumption that Greece and Turkey are essential to our political and strategical position in the world ... Am I to understand that we may now abandon this position? I really do not know where I stand.[26]

Attlee waited for Bevin's return to London before making what amounted to a frontal attack on his foreign policy.* This time the Prime Minister went far beyond the confines of Greece and Turkey. In a memorandum[27] of much broader scope that he circulated to his colleagues on 5 January 1947, he not only returned to the charge on the whole question of what he described as the Chiefs of Staffs' 'strategy of despair' in the Middle East, deploying arguments of the same kind as those that he had used in his similar attempt in March 1946, but he also questioned British ability 'to keep our influence over this congeries of weak, backward and reactionary States' in the face of a 'USSR organised under an iron discipline, equipped with the weapons of a revolutionary doctrine liable to attract the masses, strategically well placed for penetration or attack and with only a limited number of its key points open to our attack'. Attlee's own conclusion was:

Unless we are persuaded that the USSR is irrevocably committed to a policy of world domination and that there is no possibility of her alteration, I think that before being committed to this strategy we should seek to come to an agreement with the USSR after consideration with Stalin of all our points of conflict.

Bevin counter-attacked immediately. 'If we speak to Stalin as you propose, ... we shall get as much of Stalin's goodwill as we got of Hitler's after Munich'; and once again he succeeded so far as his Middle Eastern policy overall was concerned.[28] Over Greece, however,

---

* This attack, however, did not spill over into the crucial field of atomic policy. On the contrary, five days later Attlee took the chair at the meeting of the six ministers (including Bevin and himself) who then took the formal decision to make the British atomic bomb.

faced with the combined opposition of the Prime Minister (Attlee and Bevin had already discussed military withdrawal from Greece together at Chequers on 27 December 1946) and of the Chancellor of the Exchequer, for once the Foreign Secretary gave way.[29] The delivery to the State Department of the two *aides-mémoire* on 21 February 1947 by the British Embassy in Washington was the outcome.

From the reports of the US Ambassador in Athens, Lincoln Mac-Veagh, the State Department was already well aware of how badly things were going in Greece, where an undeclared civil war was being fought (in December 1947 the Greek Communist Party proclaimed a Provisional Democratic Government of Free Greece). This was not the central point of the shock caused in Washington on 21 February 1947. To understand the reasons for the atmosphere of crisis that swiftly developed there, account has to be taken of several other factors. This was the low point of Bevin's foreign secretaryship. Even three months later Acheson was asking the US Embassy in London whether it was 'safe' to assume that he was 'likely to remain in the Foreign Office for the remainder of the year' and whether he was 'making any progress in lining up the Labour back-benchers in support of British foreign policy'.[30] The notification of 21 February also followed hard on the heels of the British decisions to transfer power in India and to refer the Palestine mandate to the UN. These three decisions, taken against the background of the British industrial shutdown in the great freeze of January–February 1947, were alarming enough. But, in addition, the decision to suspend British aid to Greece and Turkey was interpreted in Washington, not in a European context, but as 'tantamount to British abdication from the Middle East with obvious implications as to their successor'.[31] Had Attlee's view prevailed in Cabinet six weeks earlier, this interpretation would have been correct. In the event, it was mistaken. By then the British had—Palestine apart—no intention whatever of leaving the Middle East. Indeed (as has been observed in Chapter 10) the main reason why the involvement of the United States in Greece and Turkey was, in the end, regarded as a triumph in London was that it seemed to the British to have led to an American underwriting of their policy in the region as a whole.

Although Truman never mentioned the Soviet Union by name in his special address to Congress, rhetorical Pelion was piled on Ossa. It was Acheson who had from the outset of the crisis believed that only if he personally laid it on with a trowel, could the Republican leadership in Congress be convinced;[32] and in the subsequent Congressional hearings, he continued to do just this. At a hearing of the Senate Foreign Relations Committee on 1 April 1947—while Marshall was himself at the negotiating table in Moscow—Acheson gave the following reply to a Senator who asked whether the aid programme would press the Soviet Union 'to really sit down with us and settle some of these differences':

I think it is a mistake to believe that you can, at any time, sit down with the Russians and solve questions. I do not think that is the way our problems are going to be worked out with the Russians. I think they will have to be worked out over a long period of time and by always indicating to the Russians that we are quite aware of what our own interests are and that we are quite firm about them and quite prepared to take necessary action. Then I think solutions will become possible.

Asked whether he was 'planning any early participation for the settlement of the issue', Acheson replied 'you cannot sit down with them'.[33]

Both in the United States and in Britain there were some[34] who doubted at the time whether the need to secure Congressional approval for the aid bill for Greece and Turkey justified this amount of hyperbole.* This controversial question is a matter of American, or of US-Soviet, rather than of Anglo-American history. For the purposes of the present study of the Anglo-American relationship, what matters is the fact that the aid bill was approved; and that by then the relatively modest needs of Greece and Turkey were already beginning to be seen by influential members of the Administration in a much wider context, requiring the appropriation by Congress not of millions but of billions of dollars. Two weeks after the bill was signed into law by the President, Marshall delivered, at Harvard, the speech outlining the concept for which his name is most remembered in history, and which gave Bevin the opportunity that he immediately seized.[36]

---

* It is also remarkable that one of the three British Cabinet Ministers involved in the British decision of February, Hugh Dalton, seems not to have realized the significance of Truman's speech to Congress until he heard a broadcast about it by Joseph Harsch two days later.[35]

CHAPTER 13

The Marshall Plan

THE recipients of honorary degrees at Harvard on 5 June 1947 included T. S. Eliot and the US Secretary of State. The short speech that Marshall delivered at this ceremony was not even given the distinction of lead story by *The Times* which published a few lines about it on the following day under a heading 'Mr Marshall's Hint'.[1] The British Embassy thriftily reported it by diplomatic bag, instead of by telegram. Although the speech was not at first given much publicity in the United States either—even the State Department press release simply called it 'Remarks by the Honorable George C. Marshall'—the BBC correspondent in Washington had been briefed beforehand by Acheson. It was, therefore, from his bedside radio that Bevin received a report of Marshall's speech for the first time, in the BBC's American commentary.[2] The man of whom Dalton had written, little more than three months earlier, 'It is quite on the cards that he will collapse completely ... At any rate he has a doctor with him,'[3] was now galvanized. Then, and in the critical months that followed, Bevin was wholly in his element and at his formidable best. Seizing on a single, short passage in Marshall's speech, Bevin instructed the British Ambassador in Paris on 9 June to go and see Georges Bidault at once and tell him that Bevin was 'anxious to cooperate with them at the earliest possible date in studying the new American approach to Europe'. This telegram, which bears Bevin's personal imprint, went on: 'If we are to get advantage out of Mr Marshall's offer, the most complete and constructive programme we can put forward for European countries is of vital importance'.[4] Five days later he telegraphed the proposal, which the French Government accepted, that he should visit Paris the following week.

The passage in Marshall's speech which captured Bevin's imagination and which led, three months later, to the signature in Paris of the first, provisional report on European economic reconstruction, contained little more than a dozen words: 'This is the business of the Europeans. The initiative, I think, must come from Europe'. These critically significant words were preceded by the statement that:

There must be some agreement among the countries of Europe as to the requirements of the situation and the part those countries themselves will take in order to give proper effect to whatever action might be taken by this Government. It would be neither fitting nor efficacious for the Government to undertake to draw up unilaterally a programme designed to place Europe on its feet economically.[5]

Both at Harvard and at a subsequent press conference, Marshall made two other things clear: the the proposed remedy was intended to be root and branch, and that the US offer was being extended to all the nations of Europe:

Any assistance that this Government may render in the future should provide a cure rather than a palliative. Any government that is willing to assist in the task of recovery will find full cooperation, I am sure, on the part of the United States Government. Any government which manoeuvres to block the recovery of other countries cannot expect help from us.

In the course of the negotiations that followed, the Marshall Plan became the European Recovery Program; and on 3 April 1948 it was under this name that Congress voted US$4bn. as the first year's allocation of aid for the countries participating in the programme. But the Marshall Plan is the name that has been remembered on both sides of the Atlantic. It has, therefore, seemed best to use it in this chapter, although it is important to bear in mind that, at its inception, there was no 'plan' as such. Rather, it was an idea that came more or less simultaneously to a small group of men[6] in the US Administration in the spring of 1947. Stamped with Marshall's personal authority at Harvard, it was grasped in Europe—as Bevin was to describe it in his speech to the National Press Club in Washington two years later—like a lifeline to sinking men. It was then converted into the programme that led to the economic recovery of Western Europe, while at the same time—once the Plan had been rejected by the Soviet Union— reinforcing the division of the continent: a consequence, which may now appear inevitable, but which, as the documents show, was not part of the Marshall Plan in its original form in Washington.

THE CONCEPT

The concept behind the Marshall Plan had been under discussion in Washington for two months before the Harvard speech. On 5 March 1947 Acheson* was writing about Greece and Turkey as 'only part of a much larger problem growing out of the change in Great Britain's strength'; and on his initiative the State–War–Navy Coordinating

* Who stayed on as Under Secretary of State for the first six months of Marshall's tenure of office.

Committee set up an *ad hoc* committee to study, in Acheson's words, 'situations elsewhere in the world which may require analogous financial, technical and military aid on our part'.[7] A month later, he devoted a speech at Cleveland, Mississippi, to the themes of 'The Requirements of Reconstruction', Europe's acute dollar deficit, and what 'those facts of international life mean for the United States and for United States foreign policy'.[8] By the middle of May there were rumours of an American initiative. The first draft of the Harvard speech reached Acheson on 20 May, and a week later he submitted a memorandum to Marshall suggesting that 'within the next two or three weeks' the Secretary of State should 'make a speech which would not undertake to lay down any solution but would state the problem and that the great immediate problem is not an ideological one, but a material one'.[9] By that time Marshall had before him two definitions of the problem: Clayton's short memorandum, dated 27 May,[10] on 'The European Crisis', and Kennan's much longer 'first recommendation' of his newly formed Policy Planning Staff, dated 23 May 1947, on 'Policy with respect to American aid to Western Europe'.[11] Clayton had just returned from Europe; Kennan had been carrying out the directive on Europe which Marshall had given immediately after his return from the Council of Foreign Ministers in Moscow.

Beginning with the premise that the extent of European destruction by the war had been 'grossly underestimated', Clayton assessed the combined current annual balance of payments deficit of Britain, France, Italy, and the Anglo-American Zone of Germany as US$5bn. This represented 'an absolute minimum standard of living'; with less than that, he wrote, 'there will be revolution'. British and French resources of gold and dollars would be exhausted by the end of the year. It would, therefore, be 'necessary for the President and the Secretary of State to make a strong spiritual appeal to the American people . . . in order to save Europe from starvation and chaos (*not* from the Russians)'. Europe, Clayton went on, 'must have from us, as a grant, 6 or 7 billion dollars' worth of goods a year for three years'; and this grant 'should be based on a European plan which the principal European nations, headed by the UK, France and Italy, should work out', the plan being 'based on a European economic federation'. The memorandum ended with the words 'we must avoid getting into another UNRRA. *The United States must run this show.*'[12]

Kennan's paper of 23 May shared Clayton's premise: that 'the root of the difficulties of western Europe' was not 'communist activities', but 'the disruptive effect of the war on the economic, political, and social structure of Europe' and 'a profound exhaustion of physical plant and of spiritual vigour'. Indeed he concluded his paper with a section recommending that 'what the press has unfortunately come to identify as the "Truman Doctrine" ' should be clarified; in particular, the 'effort to restore sound economic conditions in other countries' was not to be regarded as 'only a by-product' of a 'defensive reaction to communist pressure'; nor was the Truman Doctrine 'a blank cheque to give economic and military aid to any area in the world where the communists show signs of being successful'. The key passage in Marshall's speech at Harvard, on which Bevin seized with such alacrity, was drawn almost word for word from paragraph 6(a)[13] of Kennan's paper. In addition, Kennan believed that the aid programme 'must embrace, or be linked to, some sort of plan for dealing with the economic plight of Britain'. In an earlier memorandum he had suggested that this plan 'must be agreed in advance with the British'; that it 'should be designed to encourage and contribute to some form of regional political association of western Europe states; and that 'occupational policies in Germany and Austria must be shaped towards enabling the western zones of those countries to make the maximum contribution to economic restoration in western Europe in general'.[14] In his 'first recommendation' paper of 23 May, Kennan went on to select 'the production of coal in the Rhine Valley and its movement to the places of consumption in Europe' as the short-term object (in contradistinction to the long-term programme) of 'effective and dramatic action' which 'should be taken in the immediate future to halt the economic disintegration of western Europe'.

Thus there were several layers to the Marshall concept, even before the Harvard speech. Essentially, it was a programme to close Europe's yawning dollar gap by the early 1950s. But the method by which this programme was to be established and carried out was intended not only to restore the European economy, but also to revive European self-confidence. Care was taken, at this early stage, to avoid a repetition of the ballyhoo of the Truman Doctrine, although the Planning Staff already recognized in its paper of 23 May that 'it may be necessary for the key countries of western Europe to find means of conferring together without the presence of the Russians and Russian satellites'. Finally, the selection of German coal production (with its implications for the level of German industry) as the top priority for the immediate future touched the most sensitive international political nerve of all, since any rational solution of the problem was bound to hit the three rocks on which the meetings of the Council of Foreign Ministers had foundered: Soviet policy on reparations, French policy on the Ruhr and the Rhineland, and British policy on the ownership of the German coal industry. Moreover, this dissension was not only international. The State Department and the War Department had

7. Marshall testifying before the Senate Armed Services Committee, March 1948.

8. Franks looks on as Bevin signs the OEEC agreement, Paris, September 1947: the European response to the Marshall Plan.

long been at loggerheads with each other on German policy—the former lending full weight to the need to carry France along with whatever decisions might eventually be reached over Germany, the latter concerned almost entirely with the revival of the German economy: a controversy that was not settled until August 1947, by an agreement between the two Departments which 'looked remarkably like a treaty between sovereign states' and still did not prevent the Secretary of War, Kenneth Royall, from almost upsetting the apple-cart by an ill-considered remark made during a press conference that he gave in Berlin.[15]

THE BRITISH ROLE IN THE PLAN

On the day that Bevin went to Paris—17 June 1947—for his meeting with Bidault, what became the European Recovery Program still lay far in the future. All that the two men had to go on was the broad brush strokes with which Marshall had sketched the plan at Harvard only eleven days earlier. In spite of this uncertainty* (or perhaps because of it?) and the hostility of Soviet press comment towards the 'Marshall Doctrine'[17] notwithstanding, Molotov accepted an invitation from Bevin and Bidault to meet them in Paris, where the three Foreign Ministers convened on 27 June: arguably the most seminal international meeting since Potsdam. This meeting was followed, on 12 July, by the opening of the Conference on European Reconstruction in Paris. The conference, which was attended neither by the Soviet Union nor (thanks to Soviet pressure) by the Eastern European countries, ended with the signature by the Foreign Ministers of the sixteen participating countries of the preliminary report on the economic reconstruction of Europe, at a ceremony in the Quai d'Orsay presided over by Bevin, on 22 September 1947.

This study of the Anglo-American relationship does not require an account either of this conference, or of the complex of other negotiations directly concerned with or relevant to the Marshall Plan—bilateral, tripartite, and multilateral—that occupied the energies of western ministers and officials during the remaining months of 1947.† Of the importance of Bevin's personal contribution to the launching of the Plan in its early days, there can be no doubt. It was also in large

* At the end of July the Marshall Plan was still being compared in a Washington witticism to 'a flying saucer—nobody knows that it looks like, how big it is, in what direction it is moving, or whether it really exists'.[16]

† In Paris Oliver Franks, persuaded by Bevin to leave Queen's College, Oxford, became Chairman of the Committee of European Economic Cooperation and hence a central figure in the negotiations, both intra-European and translantic, that followed. A recent account of these negotiations is to be found in Bullock, *Bevin*, Chapters 10 and 11.

measure thanks to him that three months later the outcome of the Paris Conference was such that before the year was out, Congress had passed an Interim Aid Bill—the turning-point in the economic recovery of Europe.[18] At a moment when British leadership in Western Europe was essential if the Marshall Plan was to succeed, Bevin and his advisers rose to the occasion. Nevertheless, it was far from plain sailing. The records of the meetings (classified Top Secret at the time), held in London in June to consider Britain's role in the Marshall Plan, throw a brilliant light on the state of the Anglo-American relationship at this juncture, as well as foreshadowing much that was to follow.

As has been noted above, the earlier of Kennan's two memoranda recognized the need to reach prior agreement with the British Government, while at the same time recommending that the Plan should further 'some form' of Western European political unity. Similarly, Clayton had predicated 'a European economic federation'. Thus it came about that, during the brief interval between Bevin's meeting with Bidault in Paris in mid-June and the Anglo-Franco-Soviet meeting held there at the turn of the month, in London five Anglo-American meetings were held on 24–26 June 1947, mainly at 10 Downing Street. The US team was led by Clayton and Douglas, while the British team shifted between Cabinet Ministers and senior officials, with Bevin in the lead. Their long argument related not only to its immediate context: the eve of Bevin's and Bidault's crucial meeting with Molotov. It also has to be assessed within the wider framework of British decisions, commitments, and problems that have been described in earlier chapters—in January, the British decision to make the atomic bomb; in February, the Government's decisions on India and Palestine, and its suspension of aid for Greece and Turkey; in the early weeks of the year, the industrial shutdown; the US loan to Britain, which was fast running out; and—only three weeks ahead—the date of the convertibility of sterling, which the terms of the loan had fixed as 15 July 1947. Nevertheless, Britain remained a world power, with over a million men under arms,* in bases spanning the globe; it was the one major European country whose economy had not been devastated by the war; and London was the centre not only of the British Commonwealth and Empire, but of the world's second international trading system.

At the initial meeting[20] on 24 June, Bevin opened with the proposition that 'Europe can contribute materially to the solution' of 'the chronic troubles of Europe', but that 'Britain with an Empire is on a different basis'. It would, therefore, 'pay the US and the world for the

---

* The strengths of the US and British armies in the latter half of 1947 were almost identical.[19]

US and the UK to establish a financial partnership'. After a tribute to the speed with which Bevin had reacted to Marshall's speech three weeks before, Clayton firmly stated the US position: 'No further piecemeal assistance was feasible for Europe'. He could not 'visualise the Administration going to Congress regarding new proposals for any one country'. A 'European plan must be worked out'; and in this Bevin's 'continued leadership would be welcomed'. Bevin did not give up. If Britain was considered 'just another European country', this would 'fit in with Russian strategy . . . the Russians, in command of the Continent, could deal with Britain in due course'. Clayton stuck to his guns: 'the UK, as a partner in the Marshall Plan rather than as a part of Europe, with special assistance to the UK partner would violate the principle that no piecemeal approach to the European problem would be undertaken'. He 'saw no possibility of interim arrangements for the UK as part of the European approach'.

At the second meeting held in the afternoon at the Treasury, it was argued by the Board of Trade that 'the pooling of assistance would reduce the UK position to that of the "lowest" in Europe' (a proposition repeated by the Cabinet Secretary[21] on the following day). It was even suggested that 'the UK might be better outside of the plan since the British position could be maintained by bilateral deals'. And at the second ministerial meeting, held on 25 June, Bevin again maintained that 'we, as the British Empire, could assist materially', but that the British 'did not want to go into the programme and not do anything—that would sacrifice "the little bit of dignity we have left" '. He got nowhere. Later in the meeting, moreover, Clayton put the British on notice that 'European integration' would have to be included in a firm plan for Europe in order to convince Congress. To this, however, the British response, recorded in writing on the following day, was that 'any proposal that went so far as asking for assurances even in principle that the European countries would constitute themselves into a customs union would present great difficulties and would almost certainly involve delay which in present circumstances would be disastrous'.

In the end it was the British who yielded on the central issue. In doing so, they adopted a device that was to become familiar in Anglo-American bureaucratic dealings over the years: a British draft expressing a policy that was largely American. On 26 June 1947 the British side produced a draft paper, which was then amended in further discussion between officials of the two sides, this time at the US Embassy. The paper, entitled 'Summary of Discussions with Mr Clayton', and expressly described as not to 'be regarded as in any sense a commitment', conformed to the American view of Britain's position within the Plan, while recording British reservations, one of which was to become the source of constant Anglo-American friction over the next three years.[22] The paper was intended as a brief for the Foreign Secretary in Paris; and at the final meeting Bevin is recorded as saying that he thought that he could 'use the memorandum with reasonable safety as an approach in the Paris discussions'. The 'need for great speed' was also underlined. The following day he flew to Paris, to meet Bidault and Molotov.

THE SOVIET REJECTION OF THE PLAN

It has been suggested [23] that, at this point, counsels in the Kremlin were divided; and that it was Stalin's personal decision, overruling Molotov, which decreed the breakdown of the meeting on 2 July. Certainly Molotov's conduct at the conference in Paris was not as sure-footed as usual. The argument subsequently used by Stalin a week later, in order to persuade the Czechoslovak Government to withdraw its acceptance of the Anglo-French invitation to attend the European Conference on Reconstruction, was less sophisticated than the two questions that Molotov himself put to Bidault in Paris on 1 July, which got close to the nub of the matter:

has the French Government changed its views on German reparations and does it favour turning over increased German production for use in European reconstruction before reparations are made? ... Does the French proposal mean that the French Government now favour raising the level of German industrial production?

On the following day, having received his final instructions by telegram from Moscow, Molotov rejected the Marshall Plan on the dual ground that it infringed the national sovereignty of European states and that it ignored the issue of German reparations. Whatever the mainspring of the Soviet rejection on 2 July 1947 may have been— and without the Soviet archive, we cannot be certain—none of the three Foreign Ministers was in any doubt at the time that the breakdown in Paris on that day had divided Europe; and today most historians, asked to name the most plausible date for the beginning of the cold war, would suggest 2 July 1947; Molotov blamed his two western European colleagues, just as they blamed him for what had happened. And Bevin, who told the US Ambassador in Paris that he had 'anticipated and even wished for'[25] the breakdown, was conscious, even while it was taking place in Paris, that history was being made there. He was also well aware that it was French steadfastness—or, seen from the Kremlin, the French Government's U-turn—that had been critical to this outcome. His report to Marshall on the final meeting of the conference, sent via the US Ambassador in London on his return from Paris, ended with these words:

The most satisfactory feature of these talks has been that the French have been quite unwavering in their attitude on the basic issue and I am sure that we can count on the full collaboration of the present Government in the work which we are now setting in hand together.[26]

THE EAST–WEST DEADLOCK OVER GERMANY

The breakdown in Paris on 2 July was followed by the opening of the sixteen-nation European Reconstruction Conference there ten days later. The speed of the political timetable in Western Europe at this point is hard to imagine today, in our age of cumbersome, elaborately prepared, summit meetings. In Eastern Europe things also moved fast, not so much in the economic* field, as in the onset of the final stage of the Stalinization of Eastern Europe: a political process, to which, although it had begun in 1945, full throttle was not applied until the second half of 1947. The liquidation of the leadership of the non-communist parties that had formed part of the coalition governments established in these countries at the end of the war, and the fusion of socialists with communists in single 'united workers'' parties, was followed over the next two years by political show trials, executions, and the grisly paraphernalia of the Great Terror imported from the Soviet Union. In September 1947, the Cominform[27] was established, with its seat in Belgrade. The keynote speech delivered at its inaugural meeting by Andrei Zhdanov proclaimed, in uncompromising language, the division of the world into 'two camps', pointing to the restoration of the industrial areas of Western Germany as 'the cornerstone of the Marshall Plan', and exhorting the communist parties of western Europe to take the lead in the 'struggle against the attempts to enthral their countries economically and politically'.[28] Within a few months the Cominform had been 'deprived ... of any real reason for existence'[29] by the Soviet-Yugoslav schism and the Cominform's failure to resolve it, but the Stalinist wave that engulfed Eastern Europe provided the worst possible backdrop for the fifth meeting of the Council of Foreign Ministers which gathered at Lancaster House, London on 25 November 1947.

This time the Foreign Ministers broke up, on 15 December 1947 without even agreeing on a date for their next meeting.[30] The breaking

* The 'Molotov Plan', such as it was, consisted only of a series of bilateral trade agreements between the Soviet Union and Eastern European countries.

point then was the same as it had been throughout the previous two years—German reparations—although in this one issue there was now visibly subsumed the whole question of the political future of Germany. Looking back, we see this parting of the ways as a logical sequel to events earlier in the year, both in Western and in Eastern Europe. By the time that the Council met, the East–West competition for the political allegiance of Germany was right out in the open. Yet both Bevin and Marshall remained reluctant to rule out, until they absolutely had to, the possibility of a last-minute change in the Soviet position.[31] The State Department, moreover, turned down the British request for preliminary informal discussion of the organization of Western Germany in advance of the meeting of the Council of Foreign Ministers, insisting that there must be a 'clear break between the quadripartite negotiations' and—assuming these broke down—subsequent tripartite 'discussions for a reinforcement of western zonal fusion'.[32]

Once the breakdown had at last taken place, the two Secretaries of State lost no time in meeting[33] to discuss their next moves over Germany. And on the previous day, 17 December 1947, Bevin outlined (separately) to Marshall and Bidault his first broad 'idea' of a system of security:

the problem should not be isolated into a mere quarrel between the western Powers and the Soviet Union ... we must devise some democratic system comprising the Americans, ourselves, France, Italy, etc., and of course the Dominions. This would not be a formal alliance, but an understanding backed by power, money and resolute action. It would be a sort of spiritual federation of the west.

To Marshall, Bevin said that he 'would like to follow the pattern of the recent Middle East talks* using an official team which could either start with us and the Americans alone, or include the French from the start'. Marshall's reply was equally vague; and although he was sympathetic, he was also cautious. He was very willing for discussions at official level to be held 'with a view to arriving at ... an understanding ... They had to reach an understanding. They must take events at the flood stream and produce a coordinated effort'.[34] These discussions lasted, off and on, for the next eighteen months. In its final form Bevin's idea found expression in the North Atlantic Alliance.

* The 'Pentagon Talks': see Chapter 10.

CHAPTER 14

The Security of Western Europe

WESTERN UNION

To launch an idea in general terms in the way that Bevin did in December 1947 reflected his preferred method of working. But, even if this had not been so, in the circumstances of the turn of the year 1947/8 he would have found it difficult—perhaps even counterproductive—to attempt to put forward more definitive defence proposals than those described at the end of the last chapter.

In one form or another, the idea of a Western European grouping had been discussed in Whitehall from 1944 onwards. (The US reaction was then chilly. The State Department briefing book paper, written for the Potsdam Conference, on 'The British plan for a Western European Bloc', concluded with the recommendation that until the Soviet post-war attitude was clear the United States 'should not take a positive stand one way or the other towards this proposal to draw the nations of Western Europe into closer association'.)[1] Three months before Bevin's initiative, moreover, the Canadian Minister of External Affairs, Louis St Laurent, had made a speech suggesting 'some form of Western security association'.[2] What was new in 1948 was the personal stand publicly taken by the British Foreign Secretary in January and—as the year went forward—the gradual involvement of the US Government.

For this four-year gap there were several reasons. First and foremost, it was only at the very end of 1947 that the Yalta–Potsdam mechanism for dealing with the German problem ground to a halt; and, as has been observed in Chapter 12, for the first two years after the war the threat of a German resurgence outweighed the Soviet military threat in the eyes of British ministers. On both sides of the Atlantic, when the immediate post-war military planners addressed the question of Soviet military capability, their general assessments were not just pessimistic; they regarded the Red Army as simply unstoppable anywhere in continental Europe short of the Channel or the Pyrenees. Whatever view may be taken of the relative sizes, after two years of post-war demobilization, of the Soviet armed forces and of the total strength of the American and British armed forces combined, there is no doubt that—leaving the atomic bomb out of the military balance—the equation of ground forces was heavily weighted against the West in Central Europe, where western divisions were deployed not as a shield to deter aggression, but in the posture of an army of occupation.[3]

Of the three western occupying powers in Germany, none was well placed at the beginning of 1948 to redress the military balance on the central front. The United States was entering an electoral year; the Administration's first priority was to secure Congressional approval for appropriations, not for increased military spending or military aid, but for the Marshall Plan; and within the State Department itself there were important divisions of opinion about the US role in European defence.[4] French sensitivity over Germany was still acute. The French Government would therefore insist on German aggression being covered by any new treaty (as it already was by the Treaty of Dunkirk), whereas the three Benelux countries would urge that it should include the possibility of a new Germany re-entering the Western European group of countries. Finally—Bevin's initiative of December 1947 notwithstanding and in spite of the British commitment to France under the terms of the Dunkirk Treaty—the British Chiefs of Staff had not yet reconciled themselves to the conclusion that an alliance with continental Europe must involve a willingness to commit British land forces to Europe once again, in the event of war. Until May 1948 Montgomery's advocacy of defence on the Rhine was still being resisted by his Naval and Air Force colleagues in the British Chiefs of Staff Committee; indeed the formal commitment to send reinforcements to the Continent in the event of war was not accepted by British ministers until March 1950.[5] Two years earlier the primary area of British security concern was the Middle East, as the Pacific area was for the United States.

Into the midst of this confusion Bevin threw the major speech that he delivered in the House of Commons on 22 January 1948. Having first secured Cabinet approval, in his memorandum on 'The First Aim of British Foreign Policy', for the creation of 'some form of union in Western Europe, whether of a formal or informal character, backed by the Americans and the Dominions', he then put the concept of a Western Union to Parliament. The 'free nations of Western Europe must now draw together'; the time was 'ripe for consolidation'; and then he went on:

I hope that treaties will be signed with our near neighbours, the Benelux countries, making our treaty with France an important nucleus in Western Europe. We have then to go beyond the circle of our immediate neighbours. We shall have to consider the question of associating other historic members of European civilisation, including the new Italy, in this great conception ... Britain cannot stand outside Europe and regard her problems as quite separate from those of her European neighbours.[6]

Writing to the Foreign Secretary after the debate, the Prime Minister complimented Bevin on having 'recovered the initiative in European affairs'; in the United States, Vandenberg thought the initiative 'terrific'; and Marshall was 'deeply interested and moved'.[7] Talks between Britain, France, and the Benelux countries began at once.

The reasons why Bevin's initiative was welcomed in this way were not purely military. True, 'newsreels showing valiant Soviet emancipators marching across Red Square were now re-run to the accompaniment of dramatic warnings about the Soviet threat'.[8] But this threat was perceived in London and Washington as no less psychological than military. Bevin's phrase 'a spiritual federation' may have an odd ring today, but at that time it corresponded to a deeply felt fear of what the communist parties of Western Europe might be able to achieve, politically and industrially, in response to the Kremlin's directives. This fear related above all to the two largest communist parties, in France and in Italy, both of which had left the wartime coalition governments and now formed the core of the opposition. We now know that the communist movement in Western Europe had already passed its meridian. The French trade union federation split in December 1947, following the collapse of a wave of strikes; in the critical Italian general election of April 1948, it was Alcide De Gasperi's Christian Democrat Party that triumphed; and the Soviet–Yugoslav schism in mid-1948 destroyed the last hopes of the Greek communists, whose forces were defeated by the Greek national army in the following year. To this catalogue of communist failure, however, there was one major exception: Czechoslovakia. Here, at the end of February 1948, 'the whole character of the State had been changed in less than a hundred hours'.[9] Jan Masaryk, the Czech Foreign Minister, was found dead* outside the window of his offices on 10 March. The President, Edvard Beneš, who capitulated to the Communist leader, Klement Gottwald, in the crisis, died six months later, a broken man.

The Communist take-over of power in Prague in 1948 was the beginning of the conversion of a country which was then, to a large extent, a western democracy into one which—1968 apart—was to become one of the closest and most conformist followers of the Soviet model in Eastern Europe and the one least open to the forces of change.† This take-over did not come as a bolt from the blue (Bevin[10] had foreseen it at least six weeks beforehand). In the Kremlin

---

* The (implausible) official verdict at the time was suicide, but the question was re-opened twenty years later. The enquiry had not been completed by the time Soviet forces invaded the country in August 1968.

† The huge statue of Stalin remained standing in Prague for several years after Khrushchev's revelations about him at the 20th CPSU Congress in 1956.

perspective it was understandable. The Soviet Union wanted to be sure of the advanced, undamaged industries of the Czech lands and of the uranium reserves and the military lines of communication that a close hold on Czechoslovakia would secure. It also needed to make a further riposte to the Marshall Plan; the Cominform was a poor thing. Yet, however 'defensive'[11] the Prague take-over may have seemed to the Kremlin, its timing—from the Soviet point of view—could scarcely have been worse. Western memories of the Munich Agreement and of the subsequent German occupation of Czechoslovakia at the end of the previous decade were still fresh. Together with the Soviet decision to invade Afghanistan nearly thirty-two years later, the coup of 1948 must rank as one of the cardinal errors of communist policy since the Second World War.

THE BRUSSELS AND NORTH ATLANTIC TREATIES

The events in Prague, partly considered in themselves for what they visibly were and partly conceived as a foretaste of what might follow in Scandinavia,* powerfully influenced four sets of western discussions— between Britain and the four Western European states on a regional defence pact; between Britain, Canada, and the United States on a wider system of defence; between Britain, France, and the United States (joined later by Belgium and the Netherlands) on the future of Germany; and between the sixteen countries participating in the Second Conference on European Economic Cooperation, which opened in Paris on 15 March 1948. Of these, the last two need not detain us at present, although their outcome was of lasting significance (the tripartite recommendations of 7 March that helped to pave the way for West German rehabilitation; and the establishment of the Organization of European Economic Cooperation (OEEC), whose members were to receive a total of US$13bn. in Marshall Aid from 1948–52). What is relevant here is the impact on the defence talks on both sides of the Atlantic, which was almost instantaneous.

The force of this impact may be measured by the language used by Bevin in his talks with Douglas immediately after the Prague *coup d'état*: 'the critical period of 6–8 weeks which ... would decide the future of Europe ... the last chance for saving the West'.[12] (A paper circulated to his Cabinet colleagues on 3 March 1948, entitled 'The Threat to Western Civilization', went even further: 'physical control of the Eurasian land mass and eventual control of the whole World Island is

---

* The main western fear was for Norway. In the event, Soviet terms for Finland, announced in the Soviet-Finnish Treaty of April 1948, were surprisingly mild.

what the Politburo is aiming at—no less a thing than that'.)[13] On 11 March the British Ambassador in Washington delivered a message from Bevin strongly urging 'very early steps ... to conclude ... a regional Atlantic Approaches Pact of Mutual Assistance'. Within twenty four hours Marshall, in a positive reply to the British Ambassador, suggested 'the prompt arrival of the British representative early next week'.[14] Yet up to that point the transatlantic defence dialogue, which Bevin had sought to open almost three months earlier in London, had been fruitless. Indeed, the American attitude towards a western defence system had, if anything, receded. In spite of Marshall's warm reaction to Bevin's Western Union initiative, the furthest that his Under Secretary, Robert Lovett, had been prepared to go on 2 February was that, once the United States saw evidence of the Western European countries' 'firm determination ... to act in concert to defend themselves', then it would 'carefully consider the part it might appropriately play in support of such a Western European Union'.[15] And Bevin's proposal, made on a personal basis to the US Ambassador on 25 February 1948, for conversations 'either in Washington or at some point in Europe' evoked from Douglas the cagey comment that he 'could not determine whether Bevin made this suggestion as a slanting effort to entangle us at the moment in European quasi-military alliances or agreements'.[16]

Meanwhile, British defence discussions with France and the Benelux countries moved forward. These talks faced two problems. Was the model to be the bilateral Dunkirk Treaty, centred on the threat of German aggression, or the multilateral Rio de Janeiro Treaty of Inter-American Defense? And, even though it was the State Department who had suggested the Rio model at an early stage, might not a regional pact, as the British at first feared, in practice encourage not American participation in Western defence, but just the opposite? In the end, impelled by what had happened in Prague, the British and the French both accepted a multilateral pact (but including a specific reference in Germany, in Article 7); the Foreign Ministers of the five countries signed the Treaty of Brussels[17] on 17 March 1948,* and two days later they sent Marshall a message proposing discussions in Washington.

Although four months passed before seven-power talks began, by 9 September the Seven-Power Ambassadors' committee had agreed on preliminary conclusions; on 24 December 1948 they agreed on a report embodying the draft text of the North Atlantic Treaty; and on 4 April

* The Western Union Defence Organization was not established for another six months. Montgomery was appointed Chairman of the Commanders-in-Chief of the Western Union forces on 27 September 1948.[18]

1949 the Treaty was signed in Washington. Its key provision, Article 5, read (and still reads) as follows:

The Parties agree that an armed attack against one or more of them in Europe or North America shall be considered an attack against them all; and consequently they agree that, if such an armed attack occurs, each of them, in exercise of the right of individual or collective self-defence recognized by Article 51 of the Charter of the United Nations, will assist the Party or Parties so attacked by taking forthwith, individually and in concert with other Parties, such action as it deems necessary, including the use of armed force, to restore and maintain international peace and security.[19]

Recounted in this way, the agreement to establish the North Atlantic Alliance sounds almost a foregone conclusion, even if the American readiness, formally expressed in Article 5, 'to enter into a commitment to defend Europe in time of peace' (as Bevin described it in a later report to the Cabinet) 'marked a revolutionary step in their policy'.[20] An eyewitness of the Washington end of the transatlantic negotiations recorded at the time his belief that 'the ultimate achievement in some form or another of this main purpose was in little doubt', although his recent view is expressed in less categorical terms.[21]

In the circumstances of 1948/9, could the United States have acted very differently? And to what extent should the outcome be regarded as an Anglo-American achievement? The documents do not support the myth that the North Atlantic Alliance, like the Marshall Plan, was a matter of wily Greeks luring simple Romans. In both cases the Americans entered into new and expensive* commitments with their eyes open and after careful debate, in order to secure an objective which came gradually to be perceived as a US national interest: the revival of European confidence and the fostering of Western European integration. To form a new alliance in the year of a US presidential election—and, moreover, an election that few people expected the Democratic Party to win—was a tall order. Even after the Brussels Treaty was in place, the furthest that Congress was able to go was to pass the Vandenberg Resolution—by 64 votes to 4—in the Senate, on 11 June 1948. The convoluted wording of this resolution, drafted by Vandenberg and Lovett (and approved by Truman), was indeed the prelude to bipartisan support in the following year for US participation in a system of collective Western defence, but it did not commit the United States in any way.[23]

The seven-power committee in Washington also suffered from ebbs

---

* The North Atlantic Treaty was followed, in October 1949, by the Mutual Defense Assistance Act, which became the major source of dollar aid to Europe after the Marshall Plan had come to an end.[22]

and flows. Many arguments were thrown up as the talks went forward. Should Italy be a signatory? (Until the last lap of the negotiations the British Government remained opposed 'on the whole'[24] to Italian participation.) Should Norway and Denmark? Should the Algerian Departments of France? What assurances might be offered to Greece and Turkey? And—in the light of the later problems of alliance management (which remain unsolved today)—we may now wonder whether Kennan's concept of a pact, 'which the European countries would join as one unit and the US and Canada as another', in 'the shape of a dumb-bell with its two weighted ends joined together by the bar of the Atlantic',[25] might not have served the West better than the form that was in the event agreed. Finally, in February 1949, there was the last minute 'cat among the pigeons', as Acheson called it in his memoirs, that arose in Congress over the wording of the heart of the treaty—Article 5. The language of this article is less compelling than that of Article 4 of the Brussels Treaty,[26] which had been modelled on Article 11 of the Dunkirk Treaty. Thanks to Acheson (by now Secretary of State), after consultation with Franks, the retention of the word 'forthwith', coupled with the phrase 'including the use of armed force', saved the day.[27]

Bevin's biographer has described the beginning of April 1949, the climax of his career as Foreign Minister, as 'the greatest ten days of his life'.[28] Although Bevin and Marshall had a common purpose, shared by St Laurent, from the outset, the original initiative was British. As with the Marshall Plan in 1947, there had to be a European initiative before the United States could act. This took the form of the Brussels Treaty. In the long intervals between the conclusion of this treaty and the opening of seven-power negotiations in Washington, not only did the British maintain the momentum, they also formed the hinge between the United States and Europe; a function that they often found themselves fulfilling in the negotiations themselves. Through these months of negotiation, moreover, the American and the British representatives each had at the back of his mind 'the secret code' of the Pentagon proposals.[29] These were the non-binding recommendations agreed at working level during the last ten days of March 1948 in Washington. Those talks, which Bevin had proposed and Marshall at once accepted, in the aftermath of the Prague coup, were secret and tripartite (Anglo-American-Canadian), the US Government having insisted on the exclusion of the French. The recommendations[30] then put forward were to form the core of the treaty drafted by the seven in the second half of the year. Beyond all this, the prospects of success in the North Atlantic negotiations were also substantially increased by the way in which, for almost twelve months, the Americans and the British worked together in countering the most explosive episode in the cold war in Europe.

THE BERLIN BLOCKADE

Earlier in this chapter, reference has been made to the western agreement of principle on Germany, reached in London on 7 March 1948. The London Conference of six powers (the United States, Britain, France, and the Benelux countries) reconvened on 20 April. On 7 June the participants issued a statement, based on the preliminary agreement reached two months earlier, authorizing the Ministers-President of the German *Länder* to convene a Constituent Assembly by 1 September 1948, with the purpose of drawing up a constitution for 'a federal form of government which adequately protects the right of the respective States'; this would be submitted to the German people in a referendum. The Ruhr was to remain a part of Germany, but a six-power International Authority* was to allocate the products of its coal, coke, and steel industries between German domestic consumption and export. The three western occupying powers were to undertake not to withdraw their forces from Germany until the peace of Europe was secured or without prior consultation.[31] On 17 June, by a narrow majority, the French National Assembly gave the government the support that it needed for what amounted to a reversal—with the single exception of the Saar—of French policy towards Germany over the previous three years. On the following day a currency reform in the three western zones of Germany was announced.

The Soviet response to this sequence of events was graduated. A Soviet note on 6 March 1948 recorded the Soviet view that the London Conference violated the Potsdam Agreement and that its recommendations were invalid; on 20 March Marshal Sokolovskii, Head of the Soviet delegation in Berlin, walked out of the Allied Control Council; and from the end of March onwards the first Soviet measures to cut off the western sectors of the City of Berlin began, culminating in the suspension, on the night of 23/4 June, of all rail traffic between Berlin and Helmstedt. Soviet participation in the Allied *Kommandatura* in Berlin ceased on 1 July. By August the Soviet land blockade of the Western sectors of Berlin was complete.

The Soviet Government had chosen for its German counter-stroke a grey area, where the western position was uncertain and vulnerable— uncertain, because the western official record of what had been agreed with Marshal Zhukov in 1945 about access to the western sectors of

---

* The Ruhr Authority never functioned, being replaced in 1950 by the Coal and Steel Authority set up under the Schuman Plan.

Berlin was hazy; and vulnerable, because in June 1948 stocks held in these sectors were enough to supply food for little more than a month and coal for power-stations for little more than six weeks. The blockade lasted until 12 May 1949.

The Anglo-American airlift, which began almost at once, was not at first conceived as much more than a morale-raising, temporary remedy. In the end it brought in enough supplies to the western sectors and so defeated the Soviet blockade. General Lucius Clay, the outstanding western representative in Germany, who would himself have preferred to force the issue running through an armed convoy of trucks, was still thinking only in terms of an airlift of '600 or 700 tons a day' on 27 June.[32] In the early months of the blockade the proportion of US to British flights was 60:40. By the turn of the year the US preponderance had increased; and, on a clear day, Clay's figure had been multiplied by ten. The millionth ton of supplies was flown into Berlin on 16 February 1949 (by a British aircraft).

So far as Berlin was concerned, it has been said that 'in the summer of 1948 western policy was made in London in the constant exchange'[33] between three men: Bevin, Douglas, and the French Ambassador, René Massigli.* In spite of the remarkable degree of trust established between them, the spirit of the people of Berlin (which soon became a political factor in its own right), and the determination of the western air forces to beat the blockade, there were moments when it was nip and tuck. In May 1948 the Soviet Government had drawn maximum propaganda advantage by publicizing the fact that, without informing—let alone consulting—the British, the US Ambassador in Moscow had been talking in a conciliatory manner to Molotov.[34] By 11 August Douglas was concerned enough about the 'undercurrent of feeling here against the US both in and out of government' to address a long personal telegram to the State Department in which he sought to explain 'the sensitive, neurotic behaviour of HMG'. His conclusion, however, was that 'the British appreciate the imperative need for the closest US–UK relationship and on the whole are anxious to accommodate their views to ours'.[35] Worse was to follow in October—three weeks before the election—when the American press leaked Truman's idea of sending Fred Vinson (by now Chief Justice) to Moscow to 'plead with Stalin the aspirations of the American people for peace'.[36] This bizarre proposal was scotched by Marshall.

If Stalin had not overplayed his hand in September 1948, he might have made significant political gains. Even as it was, without lifting the blockade (originally a western precondition for talking at all), he

---

* France was without a government for two months from mid-July 1948 onwards.

succeeded in engaging the three powers in negotiations in Moscow that lasted almost a month[37] and in a subsequent fortnight of negotiations in Berlin, followed by a further written exchange. In the negotiations in Moscow the western representatives made a major concession over currency in West Berlin; the agreement provisionally reached there (criticized by Clay and Douglas, but supported by Bevin and General Robertson)[38] might have caused serious transatlantic dissension; in the event, however, it came apart in the talks that followed in Berlin. Thus the blockade continued into 1949, when it was lifted eleven days before the sixth meeting of the Council of Foreign Ministers, convened at the Palais Rose in Paris. In the end the single concession that the western governments were obliged to make in return for the lifting of the blockade was the holding of this final quadripartite meeting:[39] the last such occasion for nearly five years. Looking back on the shared Berlin airlift six years later, one of those most closely involved at the time on the British side (and a man not given to over-statement) described it, with good reason, as 'a turning-point in the post-war history of Europe'.[40]

'AIRSTRIP ONE'

As well as the airlift, the British and US governments agreed at the outset of the Berlin crisis on a measure of great significance: the transfer of an American strategic bomber force to British bases—that is to say, within striking distance of the Soviet Union. The paternity of the first idea that led to this major decision is disputed.[41] Perhaps it occurred more or less simultaneously to Bevin, Clay, and James Forrestal? In any event, on 28 June 1948 Bevin told Douglas that 'the Cabinet had agreed to the landing [in Britain] of three groups of B-29 aircraft. Two groups would go on to Germany and one would remain here';[42] he also authorized eighty-two US F8 fighters to land in Britain on their way to be redeployed in Germany. The first two groups of B-29s arrived in mid-July; a third arrived in August; Parliament was told that they were in Britain 'to carry out long-distance flying training in Western Europe'. Although the bombers were publicly described in US Government press releases as atomic-capable, in fact they had not yet been modified to carry atomic bombs (it was not until July 1950 that Truman allowed even the non-nuclear components of atomic bombs to be sent overseas).[43]

However much this 'quick and uncomplicated acceptance of the American bombers during the Berlin crisis' may have surprised Forrestal and Lovett, who perhaps wondered whether Bevin had fully considered its implications,[44] a study of the British documents of 1948[45] reveals not a scintilla of doubt. When Douglas asked Bevin on 30 June 1948 whether 'from the political point of view, this transfer was appropriate at the present time', Bevin's immediate reply was that he 'agreed in principle'. Two days later he told Douglas that 'Ministers had given their full approval ... There were no political differences from the United Kingdom side.' Only the question of timing was left open. This was resolved on 13 July, when the US Government was asked to send 'two squadrons at once'. By early August three groups of B29s were based in Britain, and, in the words of Bevin's memorandum to the Cabinet Defence Committee a month later:

the intention was that all three groups of aircraft should remain in the United Kingdom as long as the Berlin crisis continued. No definite proposal that one or more of the groups should be stationed in this country on a more permanent basis has yet been made, and ... this is a question which can only be considered at a later date in view of the political situation then existing. It has always been made clear that it is no part of the intention of the United States Government to maintain their Air Force in this country if their presence is not desired.[46]

Without anticipating a later chapter of this book, however, a postscript must be added to these exchanges. One of Bevin's last actions before leaving the Foreign Office two and a half years later was to send a personal message to Acheson, in an attempt to clarify 'a position which it would be impossible to justify to Parliament or to public opinion if anything went wrong'. It had been, in the words of this message of 13 January 1951, 'implicit in the many talks ... with Ambassador Douglas ... that we should be consulted about any plans for the use of these aircraft. In fact that understanding has been the basis of our agreement to their presence here'. Bevin's message continued:

As I conducted many of the negotiations with Douglas, I feel a personal responsibility in making sure that no misunderstanding exists in the use to which the US air forces in the UK might be put. I cannot feel that I have discharged that responsibility while the British Government has no information as to the strategic plan in support of which these aircraft might be used at very short notice nor how far its plan accords with our own.[47]

This issue had not been resolved by the time of Bevin's death. Thirty years were to pass before it became an important question in British politics.*

---

* And—since this chapter was written—in April 1986 it has become a burning transatlantic issue as well, in relation to the Libyan crisis.

CHAPTER 15

European Integration

AT the ceremony before the North Atlantic Treaty was signed in Washington, the Marine Band played tunes from *Porgy and Bess*. The first week of April 1949, wrote Acheson in his memoirs ten years afterwards, 'was one for which none of us ever needed to feel apologetic'.[1] In the tumultuous history of the first five years following the surrender of Germany this stands out as the high point of the Anglo-American relationship.

Truman, triumphantly elected President of the United States (his party also regained control of Congress) in the preceding autumn, had appointed Acheson, whose 'attitude had long been, and was known to have been, pro-British',[2] to succeed Marshall as Secretary of State. The political position of Bevin, the chief advocate within the British Cabinet of the revived Anglo-American relationship, was, after nearly four years at the Foreign Office, unassailable. The Berlin blockade had been defeated. Many of the stormy problems that had buffeted the relationship between the United States and Britain from the end of the war up to that point, notably in the Middle East, no longer seemed as threatening as they once had. Others—the Soviet atomic explosion, China, the crisis of sterling, the rupture of the Anglo-American atomic dialogue—still lay just out of sight, over the political horizon.

In fact, the world was about to become an even more complicated place than it had already seemed to be since the cold war began. In the course of the next few months a new international environment would have been formed, to which the light-hearted songs played by the band in Washington on 4 April 1949 ('I've got plenty of nothin'' and 'It ain't necessarily so') would seem less appropriate. And yet, even in the halcyon early days of the new alliance, there was still one persistent source of dissension between the US and British governments, which the intimate co-operation between the ministers, officials, and—increasingly—officers of both countries, stimulated by their common effort to remake Europe from 1947 onwards, was not strong enough to overcome. Paradoxically, what the two governments remained at odds about was the future structure of Western Europe itself.

It will be recalled[3] that one of the premises on which the Marshall Plan was based from its inception was variously described in Washington at the time as 'some kind of regional political association of European states', 'European integration', and 'European economic federation'; and that, at the end of the Anglo-American talks held in London in June 1947, the British, reluctantly acquiescing in the American insistence that a privileged position for Britain within the Marshall Plan could not be contemplated, none the less entered a caveat that the US 'continental approach' to European problems presented Britain with 'very special difficulties' and prophesied dire consequences for any American request for 'assurances even in principle that the European countries would constitute themselves into a customs union'. Two and a half years later, Acheson had formed the personal view that the early integration of Europe, including the creation of supranational institutions, must go ahead 'even if the UK finds that its participation must be less than complete'.[4] The present chapter examines what happened in this critical field of European policy during the time that passed between the genesis of the Marshall Plan and Acheson's melancholy conclusion.

In one sense, as the British repeatedly argued, when accused by the Americans of European 'foot-dragging', there was good reason to be proud of the pace at which the countries of Western Europe had been drawn together during this period, initially under British leadership, in the process overcoming seemingly irreconcilable conflicts of national interest. The Brussels Treaty was signed on 17 March 1948. On 3 April 1948 the first fruit of the Marshall Plan, the Economic Cooperation Act, was signed into law by Truman. On the following day he appointed a Republican industrialist, Paul Hoffman, Economic Co-operation Administrator, with Cabinet rank. In Paris the Organization of European Economic Cooperation (OEEC) was then set up as the ECA's opposite number; Paul-Henri Spaak, the Belgian Prime Minister, became the first Chairman of its Council; and Robert Marjolin was released from the French Plan de Modernisation et d'Equipement (where he had been Monnet's deputy) to become the OEEC's first Secretary-General. In 1949, the North Atlantic Treaty was passed by the House of Commons[5] in May and by the Senate in July. At the first session of the North Atlantic Council, held in Washington in September, it was decided to set up a Military Committee, to sit permanently in Washington, whose executive agency would be the North Atlantic Standing Group, consisting of representatives of Britain, France, and the United States.* On 6 October 1949 Truman signed the Mutual Defense Assistance Programme into law.

---

* It is important to realize that, at this early stage of the alliance, what we now know as NATO did not yet exist. The only inter-allied command in being in Western Europe was that of the (Brussels Pact) Western Union Forces.

The first days of April 1949 in Washington had been equally remarkable for 'almost prodigies of agreement on Germany' achieved in the tripartite talks that began immediately after the signature of the North Atlantic Treaty.[6] The key to this achievement was the fresh outlook that Robert Schuman (Foreign Minister from July 1948–January 1953) brought to French foreign policy. For Schuman—he had grown up and been educated at Metz when it was under German rule—the moment had now come to 'place the whole German question on a new and higher plane'.[7] As he put it, in conversation with one of his advisers[8] in Washington in April 1949:

On a tout refusé aux allemands quand on aurait dû leur donner quelque chose et on leur a tout donné quand on aurait dû tout leur refuser: je voudrais, moi, faire autre chose.

Acheson, who had met neither Bevin nor Schuman before April 1949, got on with both of them from the outset. Bevin, who 'had never trusted Byrnes and had never succeeded in penetrating Marshall's reserve', soon became 'Ernie'; Acheson became 'me lad'.[9] The result of the three Foreign Ministers' discussions from 6–8 April was a comprehensive agreement on a whole range of German questions that had long been the subject of inter-allied controversy. This led to the promulgation of the Basic Law of the German Federal Republic on 23 May 1949; at the Palais Rose in Paris, the Council of Foreign Ministers, having once again agreed on nothing so far as Germany was concerned, adjourned *sine die* on 20 June;[10] West German elections followed on 14 August; and Konrad Adenauer was elected the first Chancellor of the Federal Republic by the Bundestag on 15 September 1949.*

Yet the sum of these achievements still did not add up to the full benefit that the United States had intended the Marshall Plan to confer on Europe. In spite of British objections, Marshall continued to insist a year after his Harvard speech that:

Purpose and scope of E.R.P. and C.E.E.C. are far beyond trade relations. Economic cooperation sought under E.R.P., and of which C.E.E.C. is a vehicle, has as ultimate objective closer integration of Western Europe.[12]

Lovett, believing it to be the British objective to revive the wartime 'special relationship' between the two countries, expressly ruled this out, as 'inconsistent with concept of Western European integration and other objectives of E.R.P.'[13]

---

* Two months later the Petersberg Agreement, signed by Adenauer and the three western High Commissioners, finally opened the door for the Federal Republic of Germany to take up its modern place in Western Europe.[11]

At this point in the post-war evolution of the Anglo-American relationship two questions have to be addressed. In the face of the British caveat about the Marshall Plan, forcefully expressed at the London meetings in June 1947, why did the US Government persist in pursuing the goal of European integration? And why did the British Government carry its opposition to US policy in this field to such lengths? The first question is perhaps easier than the second. In part, the answer lies in the conjunction of power and idealism that characterized the United States at the end of the Second World War, which has been discussed in Chapter 8. In part, it may be explained by the traditional contrast between the American preference for setting up formal institutions[14] and the British preference for proceeding with informal gradualism (an attitude epitomized on the British side, by the remark about the Council of Europe attributed to Bevin: 'If you open that Pandora's Box you never know what Trojan horses will jump out'.)[15] In the United States, moreover, the strong desire to bring American power and new ideas together, by means of innovative institutions in Europe, was shared in equal measure by the Administration and by Congress. When the Marshall Plan legislation lay before Congress, therefore, the Administration could argue, with some reason, that Congress expected a new Europe to emerge as the result of American generosity.

To those two explanatory factors on the American side of the Atlantic a third, on the other side, must be added. At least in the early months of 1948, the signals about European unity being received from London were far from clear. The first transatlantic bone of contention in regard to Europe was the question of forming a European customs union. It was Bevin himself who had long argued the case for such a union, beginning at a trades union conference twenty years earlier. As Foreign Secretary after the war, he broached the idea, though without success, in Cabinet.[16] Elements of it are visible in Bevin's cabinet memorandum entitled 'The First Aim of British Foreign Policy' (which also contained echoes of his 1945 concept of the 'Third Monroe': if the resources of European colonial territories could be added to those of Western Europe, 'it should be possible to develop our own power and influence to equal that of the United States of America and of the USSR . . . in a way which will show clearly that we are not subservient to the United States of America or to the Soviet Union').[17] A summary of this paper, which formed the basis of his Western Union speech in the House of Commons a fortnight later, was actually presented to Marshall by the British Ambassador in Washington before the speech was delivered.[18]

Bevin's speech included the crucial sentences: 'We should do all we

can to foster both the spirit and the machinery of cooperation ...
Britain cannot stand outside Europe and regard her problems as quite
separate from those of her European neighbours'. As Spaak was later
to observe in his memoirs, 'never again was he to show up in this light.
On the contrary, he seemed surprised and even worried when he saw
the ideas, which he himself had pioneered, being put into practice'. If
Spaak 'never understood why Bevin changed his view as he did',[19] it is
small wonder that British ideas about Europe also confused policy-
makers in Washington at the time; and that, even after the meetings
held in London in June 1947, they did not resign themselves to
accepting as final the British warnings about the problems that
European integration would present.

What followed may be regarded as a characteristically British
rearguard action, all the more phlegmatically fought because, on the
British side of the Channel, the inspirer of the European Movement
was the Leader of the Opposition. At a rally in the Albert Hall,
London, in 1947 Churchill launched the international lobby that
became, at the Congress of Europe held at The Hague on 7–8 May
1948, the European Unity movement. The Hague conference, whose
eight hundred delegates included nearly fifty former Prime Ministers
or Foreign Ministers (and also a then little known German politician,
Konrad Adenauer) adopted Churchill's Zurich proposal of a Council
of Europe, which was to have its seat in Strasbourg, with an elected
European parliament. The conference was 'mainly (though not exclusively) a personal triumph for Mr Churchill who, through the prestige
of his personality, was able by timely intervention to steer it towards
the limited objectives he had set for it'.[20] On his return to London, he
led Harold Macmillan and others 'in solemn procession from his rooms
in Westminster up Whitehall ... to No. 10', where his proposal, that the
British Government should itself take the initiative in establishing a
European Assembly, fell on exceedingly deaf ears.[21]

In the specific matter of the Council of Europe, the fact that its
leading British advocate was Churchill was certainly an important
contributing factor; Bevin's personal jealousy of Churchill in these
years is well attested.[22] Nevertheless, given the benign subsequent
history of the Council of Europe (whose Assembly was elected and met
for the first time in August 1949), the tenacity with which Labour
ministers and British officials sought to stifle or emasculate it looks
strange to a modern eye. It requires analysis in a study of the Anglo-
American relationship because, in British minds, their attitude towards
Europe was linked with their attitude towards the United States. Both
in London and in Washington[23] British governmental hostility to the
Council of Europe was perceived in a wider context at that time.

The stages in the evolution of British thinking about Europe after the Foreign Secretary's Western Union speech in the House of Commons may be measured by three examples taken from the official records: two documents of March and one of October 1948. Two months after Bevin delivered this speech, he and Cripps jointly submitted to their Cabinet colleagues a memorandum[24] whose ambivalence foreshadowed the frustration, transatlantic as well as intra-European, that was to follow. On the one hand, 'changes of a radical nature in our industrial and agricultural structure may become necessary to secure the economic independence of the Western European countries as a whole and to use our collective resources to the best advantage'. On the other, 'we shall of course have it in our power to moderate the process and to steer it with this in mind. Our influence in the European Organization will be very great and should be decisive on major issues'. So much, therefore, for 'hobby-horses' such as customs unions and transferability schemes. Nevertheless, the European slant was preponderant both in the memorandum and in the conclusion reached in Cabinet after 'a full discussion'. In the words of the memorandum, there was simply 'no option ... We must link ourselves more closely with western Europe'. And the Cabinet concluded that 'the policy outlined' in the memorandum 'was the only means by which the United Kingdom and other participating countries could establish themselves in a position in which they were economically dependent neither on the Soviet Union nor on the United States'.[25]

A few days earlier 'Otto' Clarke, Under-Secretary at the Treasury, had written to his opposite number at the Foreign Office: 'I must say I think our two hours with Mr Healey were very well spent'. The reason for his satisfaction was a memorandum on 'European Cooperation within the framework of the Recovery Programme' written by Denis Healey, then at the Labour Party's International Department at Transport House; it was 'non-commital' on the question of a customs union.[26] In October 1948 this was followed up by a pamphlet from the same source, whose insular title—'Feet on the Ground: a Study of Western Union'—speaks for itself.

Among the last letters that Duff Cooper wrote from the Paris Embassy before his retirement were some addressed personally to Bevin. They vividly express his mounting frustration at the lack of progress—or, as he himself regarded it, Whitehall's persistent obstruction—following the conclusion of the Anglo-French Treaty of Dunkirk. In one of them[27] he attacked as 'a most melancholy document' the record of a meeting held in the Foreign Office to consider matters that had just been discussed in Paris between Bevin and the socialist Prime

Minister, Paul Ramadier. Yet even a year later, in the very different political circumstances of October 1948 (by which time Ramadier had become Defence Minister), Bevin went so far as to tell him that he had been 'frightened' by the Hague resolution, which could not 'be translated very effectively into practical politics'. When Ramadier argued 'the need for some illusions and dreams', Bevin retorted that 'he had dreams himself but as Foreign Secretary he had to keep wide awake'—not a random remark, because he repeated it in London to Schuman in almost identical terms three months later.[28] Whether Bevin changed his mind about Europe in 1948, and if he did, at what point, is debatable. His biographer has stressed the deep and depressing impression made on Bevin by the two-day meeting of the Brussels Pact (ministerial) Consultative Committee, which he attended in July 1948, and from which he returned to London 'convinced . . . that any purely European organisation could not stand on its own feet unless linked with the United States . . . Britain, however much she became . . . involved with Europe, must also—like the United States—retain her independence and look for her security in an Atlantic rather than European framework'.[29]

What is certain is that, in the course of 1948, British ministerial thinking about the security of the very defence organization that Britain had launched at the beginning of the year—Western Union— underwent a change. At the end of December 1948 the British Chiefs of Staff (by now 'impressed by the fact that the defence of a line at least as far East as the Rhine is vital to the security of these islands') recommended to the Defence Committee of the Cabinet that a 'firm and immediate promise of a token force', to be sent to the Continent in the event of war, should be given to Western Union. The Committee decided that 'for the time being at any rate, we should not undertake on the outbreak of war to send to the Continent an infantry brigade group'. It was over a year before this decision was reversed.

As the Chiefs of Staff argued in March 1950, in a memorandum[30] whose passionate language is striking even to a reader thirty-five years afterwards, 'It must seem to our Allies that we have employed every device to avoid giving any answer. We cannot hold back much longer'. During the fourteen months that had passed since the ministerial refusal, they went on:

all sources are agreed that the French morale and will to resist, not only in the nation at large but also in the Armed Forces, have deteriorated and are continuing to deteriorate, largely because of a belief that we are not sincere in our professions of support for Western Union. Many Frenchmen today are saying that we and indeed the Americans are using Western Union only for our own selfish ends and that we intend to make no effective contribution to the common cause except to 'liberate' France after she has been assassinated. We believe that Western Union cannot be placed on a firm basis, politically or militarily, until these French doubts have been allayed.

This time the Defence Committee agreed that Britain's military partners in Western Europe should be given 'the promise to provide a Corps of two Infantry divisions as a reinforcement of the British Army of the Rhine in the event of war with Russia'. British ministerial reluctance to take a decision whose military and political logic now seems obvious, cannot be entirely explained by the fact that, as the Chiefs of Staff themselves pointed out, it meant 'a more realistic acceptance of the fact that ... while the Middle East is crucially important to Allied strategy and must be held if humanly possible, the allocation of reserves to it must not be allowed fatally to compromise' the 'first pillar, the defence of the United Kingdom' itself. It was also relevant that, even as late as the British decision of 1950, the US Chiefs of Staff were 'not at present prepared to commit additional forces to France now', although 'they would consider sending them should the Rhine be holding'. Both these considerations, though important, are not, I suggest, enough in themselves to explain the British reluctance to come to grips with European reality after the Second World War: a reluctance in which they received no encouragement from the United States—quite the reverse.

If Bevin was disenchanted with the European idea, its advocates in Whitehall had little chance. And even in the (British) Foreign Service, where they were mainly to be found, they would have had to contend with other views, powerfully expressed by Makins[31] from a pivotal position after he returned to the Foreign Office from Washington. Not that these views were atypical. It would have required a major effort on Bevin's part—in which he would almost certainly not have been backed by Attlee (who was far more interested in Commonwalth links)—to have overcome the broad consensus of most British ministers, Members of Parliament, and officials about Europe at the time. Paradoxically, the more the Americans sought to persuade the British that European integration was not inconsistent with Britain's position within an Atlantic system of defence, the more entrenched became the British conviction that the opposite was true. At the heart of this conviction lay the opinion 'formed after long deliberation, that neither the Commonwealth alone, nor Western Europe alone, nor even the Commonwealth plus Western Europe' were 'strong enough, either economically or militarily, to hold out against the forces actively opposing them'.[32]

The mainspring of this British consensus was, I suggest, as much emotional as intellectual. The British view of Europe after the Second

World War was an amalgam: atavistic distrust—a legacy of history—and national pride—largely a consequence of the war, particularly the events of 1940 and the British part in the liberation of Europe five years later—coupled with a complacency about Britain's future which sometimes bordered on hubris. Both Bevin and Churchill, each in his own distinctive way, pointed the way forward, away from the historical legacy and towards a new relationship between Britain and continental Europe. (The practical difference between what Bevin at first proposed, in his Western Union speech, and what Churchill urged upon the government, was not really very wide.) But the feelings of pride made it hard for any but a few in Britain to judge what was happening in Western Europe in 1948–50 objectively,* least of all developments in France, the country likely to have been uppermost in the minds of readers of the sentence in the Bevin–Cripps memorandum (referred to above) that warned 'we shall be associating ourselves with partners in Western Europe whose political condition is unstable and whose actions may be embarrassing to us'. And even when summing up the discussion in the Defence Committee in March 1950, before the final decision to promise to reinforce the British Army of the Rhine in the event of war was taken, Attlee still began by saying that 'grave risks were certainly entailed'.[33]

As Monnet put it many years afterwards,[34] for the British 'the price of victory was the illusion that you could maintain what you had, without change'. The international assets that Britain had inherited from the past were essentially threefold: the Commonwealth, Empire, and imperial position in the Middle East, all of which remained largely intact;† the second largest economy in the western world, the pace of its later industrial decline then unforeseen; and, linking these two assets, the sterling area trading and financial system. In addition, the Second World War had left Britain a fourth asset: the undisputed leadership of Western Europe. This asset was gradually dissipated in the years that followed.

It was the value attached in both Whitehall and the City of London to the third of these assets that made it a relatively simple matter for the British mandarinate to block the proposals for a European customs union, evoking in August 1947 this response from Bevin's senior economic adviser:

* The official records show this to have been true even of some of the most intelligent observers. But some members of the Cabinet would not have disagreed with the *New Statesman*'s description of the Schuman Plan, on 10 June 1950, as the outcome of a conspiracy led by the Vatican, Ruhr industrialists, and the Comité des Forges.

† The continued Commonwealth membership of India and Pakistan, after independence, and their acceptance of the King as Head of the Commonwealth, masked for several years the nature of the fundamental change that began in 1947.

There is a well-established prejudice in Whitehall against a European Customs Union. It goes back a long way and is rooted in the old days of Free Trade. It is a relic of a world which has disappeared, probably never to return. The Board of Trade is overstating the case against it. One of their most potent arguments is that we have to *choose* between a European Customs Union and the Commonwealth. However that may be, the Board of Trade have successfully blocked for two years our efforts to look at these proposals objectively. As a result of Marshall's proposals [i.e. the OEEC] European imaginations have been fired. It may be possible to integrate in some measure comparable with the vast industrial integration and potential of the United States, which the Russians are trying to emulate. If some such integration does *not* take place Europe will gradually decline in the face of pressure from the United States on the one hand and the USSR on the other.[35]

British opposition to American plans for Europe did not stop at obstructing the formation of a customs union. And such was the preponderance of British influence within Europe during this period that, until mid-1950, it was the British thesis that defeated the American every time. Thus they successfully saw off the American suggestion of giving the OEEC a secretariat with real powers. Instead, the OEEC's secretary was bound by the decisions of the Council; and the body that counted was the Executive Committee, whose chairman was a British official. The British then prevented the appointment (backed by the Americans) of Spaak, himself a leading member of the European movement, as a politically authoritative Director-General of the OEEC. One of Cripps' least defensible actions as Chancellor of the Exchequer was to block, in February 1950, the idea of a multilateral clearing system for European payments (a decision reversed later in the year by Hugh Gaitskell, who became Minister of Economic Affairs before succeeding Cripps as Chancellor). And even though the British Government compromised in the end over the Council of Europe, the grudging spirit in which this was done could scarcely have been better expressed than by the appointment of the xenophobic Dalton as leader of the British delegation to the Council of Europe working party that was set up at the beginning of 1949.[36]

This brief analysis has left out of account much of the light and shade* in this opening phase of a great debate, which was to last for many years and reached its first major crisis, to Bevin's chagrin and Acheson's embarrassment, in May–June 1950. By a coincidence, however, the contrasting attitudes on each side of the Atlantic seven

---

* For example, the views of the true federalists, who in Britain were a very small minority. Of much greater concern to the British Government was the possibility that a Europe led by the Left might create a 'third force', decoupled from the United States, of the kind then advocated by *Le Monde*, or—alternatively—that a Europe led by the Right might dilute the benefits of the British welfare state.

months before that unhappy encounter may now be found delineated in official American and British documents of almost exactly the same date. A stock-taking conference of US Ambassadors in Europe, including John McCloy, the US High Commissioner for Germany, and Harriman and Hoffman for the ECA, spent two days—21–22 October 1949—in Paris discussing US policy in Western Europe. A week later Bevin presented two interrelated memoranda to the Cabinet, one about the Council of Europe and the other about proposals for the economic unification of Europe.

As far as the Council of Europe was concerned, the British Government had by now decided more or less to live and let live (ironically, two months after the State Department had concluded that the Council would be better off without Britain). Thus, the British Government should 'continue to support the Council of Europe and play an active part in its development', while at the same time maintaining 'a very strict reserve in regard to schemes for the pooling of sovereignty or the establishment of European supra-national machinery'. The importance of the Council of Europe in the German context was recognized; as also the fact that 'we should be attacked in the United States if we could be shown to be preventing European unification by what might be represented as a selfish attitude'. So the balanced antitheses rolled on, before reaching the punchline:

Our relationship with the rest of the Commonwealth and, almost equally important, our new relationship with the United States, ensure that we must remain, as we always have in the past, different in character from other European nations and fundamentally incapable of wholehearted integration with them . . .[37]

The second memorandum,[38] jointly submitted by Bevin and Cripps, reaffirmed the policy approved at the beginning of 1949 by the Economic Policy Committee, that 'for the sake of European economic cooperation, we should not run risks which would jeopardise our own chances of survival if the attempt to restore Western Europe should fail, and we should not involve ourselves in the economic affairs of Europe beyond the point at which we could, if we wished, disengage ourselves'. Economic relations with Europe must yield priority to the 'new relationship with the United States and with the rest of the Commonwealth', which, 'as the result of the Washington talks and of the Commonwealth Finance Ministers' meetings, we have established, or are establishing'. The Cabinet approved both these papers; and a telegram was sent to Washington instructing the Ambassador to tell the US Government, among other things, that the British Government wished to 'do nothing which is incompatible with the objectives of the communiqué issued after the ... economic talks in Washington in September'.

No neat and tidy comparison can be made between the views expressed in these British memoranda and those expressed by the Americans who met in Paris a week earlier (including some of the most illustrious names in post-war US diplomacy), especially since on one important point the conclusions reached at the conference took issue with Acheson's own views, which had been telegraphed in advance to the US Embassy in Paris. Acheson's telegram, however, read in conjunction with the record of the conference, vividly illuminates both the gulf that separated the US and British governments on the future of European economic integration and the dilemma of US policy, to which the conference addressed itself and which occupied much of its time. On the American side, everyone agreed that European integration, economic and political, was essential and that West Germany must be included. Acheson himself had already reached the conclusion—rightly, as later events were to prove—that 'the key to progress towards integration is in French hands ... France and France alone can take the decisive leadership in integrating Western Germany into Western Europe'. On this premise, the policy that he suggested was simultaneous

movement along two lines: first, a strengthening and development of cooperative action by the US, British Commonwealth, and Europe, and second, new institutional arrangements within the larger group ... The needs of the continental countries are in some respects more urgent and more compelling and seem to me to require such action, even if the UK finds that its participation must be less than complete. In some fields and for some purposes, substantial progress towards the establishment of supra-national institutions, as well as arrangements for the freer movement of persons, are needed soon.[39]

On the other hand, for the assembled US representatives, in the words of one of them,[40] 'the central event of the meeting' was 'the complete agreement that European integration without the UK was impossible'. Not that any of their interventions in the discussion showed any trace of softness towards the British stance over Europe. Douglas himself spoke of 'the gradual disintegration of Britain as a world economic power': he described Britain as 'in the worst financial condition since the close of the Napoleonic wars'; and, in effect, he rested his hopes on the outcome of the next General Election. Harriman, whom 'nothing had disappointed ... more keenly than the British attitude toward ... the appointment ... of M. Spaak', remarked that 'our biggest post-war difficulty was that there were many times when we seemed unable to say "No" to Great Britain to the same degree as we have to other European countries'. The US Ambassador in Paris, David Bruce, even though he believed that 'no Frenchman ... can conceive of the construction of a viable Western European world from which the UK would be absent', none the less also said that the United States had been 'too tender with the British since the war'. Nor is there any sign in this record of awareness by any of the assembled ambassadors that a 'new relationship' had recently been established with Britain, still less of 'shared leadership': two concepts which recur in British records of that time. Only from the chair of the Paris conference was it pointed out that 'considerations other than purely European ones were present in US Government thinking concerning the British Empire. There was a deep conviction that the US needed Britain above everything else. This was consistently true in the Pentagon building'.

The telegram[41] sent to Washington after this conference reported its unanimous view that 'it is not realistic to expect that France will take the leadership in bringing about Western European integration without UK participation'; but it also concluded that 'we have ... very urgently to re-examine our attitude towards the British'. A week afterwards Hoffman publicly harangued the OEEC Council, whom he urged to 'move ahead on a far-reaching programme' that would mean 'nothing less than an integration of the Western European economy'.[42]

This state of Anglo-American deadlock remained unbroken a month later, when Acheson, Bevin, and Schuman met in Paris. For Schuman, 'Europe was inconceivable without Britain'; Bevin 'warned' his colleagues that 'the United Kingdom ... could never become an entirely European country'; Acheson's comment was 'once more unto the breach, dear friends'.[43] In the last foreign affairs debate of a Parliament that had sat too long, on 17 November 1949 Bevin advanced the lack-lustre argument: 'We want to bring about a sound relationship between Europe, the Commonwealth and the USA. We do not want a wedge to be driven between any of them, if we can help it'.[44] The only conceivable 'wedge-driver' in sight at that moment was the US Government. Thus the stage was set first for an Anglo-American confrontation and then for a critical British decision, which was to prove—in Acheson's words—'the first wrong choice'.[45]

PART V

1950: PIVOTAL YEAR

CHAPTER 16

The Domestic Political Scene in the United States and Britain

THE British part in the Anglo-American achievement during the two-year period that began in the summer of 1947 was such that there is a temptation for the British historian to allow this to obscure the fact that in the early post-war period 1950 was the pivotal year, whose events were in large measure to determine the shape of history for the next two decades. True, these events had been preceded by other great events, discussed in earlier chapters—the Marshall Plan, the beginning of the cold war, the signature of the Atlantic Treaty, and, in the late summer and early autumn of 1949, the momentous triple conjuncture: the Soviet atomic explosion, the emergence of the two Germanies, and the proclamation of the Chinese People's Republic. 1950, however, saw a quantum leap into the modern era—the beginning in earnest of the nuclear 'armament race of a rather desperate character' that Henry Stimson had foreseen five years earlier; the beginning also of the West's conventional rearmament; the conclusion of the Sino-Soviet treaty of alliance; in Europe, both the launching of the North Atlantic Treaty Organization and the first great initiative in the process of Franco-German reconciliation; in Asia, the unexpected outbreak of a major war, in which American and British troops again found themselves fighting side by side; and, for the western world as a whole, the onset (once the price rises triggered by the Korean War had been absorbed) of an economic golden age, a period of sustained economic expansion without precedent in the history of the industrialized democracies.

In the history both of the United States and of Britain, the interaction between external events and internal politics in 1950 is such as to warrant the short interlude of the present chapter. In both countries the political parties that had governed since the end of the Second World War retained power into the the next decade—the Labour Party only by a narrow majority, won in the general election of February 1950; the Democratic Party through the astonishing victory gained by Truman in the elections of November 1948.* Neither

* The *Chicago Tribune* actually went to press with the headline 'Dewey defeats Truman'. The Labour Party's overall majority in 1950 was reduced to five, although it polled substantially more votes than in 1945.

Truman's second term nor Attlee's second Administration turned out well. By the time that Truman and Attlee met for the second time in Washington, in December 1950, neither leader was in a comfortable position at home. Attlee was out of office less than a year afterwards; Truman's party lost the presidential election of 1952.

BRITAIN

For the first few months of its existence, the outlook for the second Attlee Administration seemed to be set fair. Indeed, the domestic economic news was the best since the end of the war. The effects of the devaluation of the preceding autumn were working through the British economy, which in 1950 at last registered a balance of payments surplus;[1] even *The Economist* described investors as 'facing a more promising summer than they have known in recent years';[2] and, by the end of the year, it was announced that Britain would make no further demands on Marshall Aid funds. But the outlook for the Government was subject to one essential proviso: that the Prime Minister succeeded in making skilful appointments to fill the two gaps in the Government's inner group which the ill health of Bevin and Cripps was bound to leave before long. (Cripps resigned in October 1950, while Bevin, whose condition shocked Acheson in May 1950,[3] managed to carry on at the Foreign Office somehow until five weeks before his death.)[4] Whatever view one may take of the men whom Attlee chose to succeed his two ailing colleagues, the way in which he filled these two critical posts—the Foreign Office and the Treasury— was far from skilful. In the process he antagonized the most brilliant man in his Cabinet—Aneurin Bevan*—to whom he offered neither portfolio.

At the moment when Attlee formed his second Administration in February 1950, he could hardly have foreseen the two major international events that were to confront it in the summer. Of these two crises, the first—the British response to the launching of the Schuman Plan—was handled by the British Government with plodding insensitivity. But it may well not have done the Labour Party much electoral damage, so ill-prepared—in spite of Churchill's eloquence—was British public opinion for the subsequent course taken by the politico-economic development of Western Europe. The second crisis, however, which followed the first in quick succession—the outbreak of the

---

* Attlee appointed Gaitskell to succeed Cripps and Morrison to succeed Bevin. As Minister of Health from July 1945—January 1951, Bevan had been the architect of the National Health Service. He resigned from the Government in April 1951.

Korean War—was another matter. Attlee's* response to it presents an element of paradox for the British historian. Attlee, earlier the would-be conciliator of the Soviet Union, sceptical about American purposes, who had only three years previously made a frontal attack in Cabinet on the strategy of the British Chiefs of Staff,[6] emerged at the beginning of July 1950 as the advocate of 'the closest possible understanding with the United States Government'—indeed as a 'cold warrior'. (In a broadcast to the nation on 31 July 1950, he described the Korean War as a war against aggression: 'the fire that has started in distant Korea may burn down your house'.)[7] As the result of Attlee's initiative, one month after the outbreak of war, Oliver Franks and Marshal of the Royal Air Force Arthur Tedder embarked on secret talks in the Pentagon with Philip Jessup, Ambassador at large, and the US Joint Chiefs of Staff; and in October the British Chiefs of Staff crossed the Atlantic for what were described as 'US–UK Political Military Conversations' with their American opposite numbers.[8] In British domestic political terms, however, what mattered most was the steeply increased expenditure on the rearmament programme which was the British Government's response to the Korean War.

The initial US request was for an extra £6bn. to be spent by Britain over four years. A four-year programme of £3,400m. was approved on 1 August 1950 by the Cabinet, including Bevan, who none the less expressed his reservations clearly.[9] By the turn of the year Attlee had raised this figure, under renewed US pressure,[10] to £4,700m. Although Bevan himself defended the increase in a remarkable speech in Parliament,[11] in the end the Health Service charges,† on which Gaitskell insisted in his first budget, led to his resignation; Harold Wilson followed suit. The immediate result of the increased defence expenditure, combined with the rise in the cost of imports, which also flowed from the Korean War, was a further prolongation of the distortions of the British economy and yet another British balance of payments crisis, in the summer of 1951. It was left to the Conservative Government, which took office in October 1951, to modify a defence programme whose scale was by then seen to be unrealistic. In the longer term, however, Bevan's resignation, his quarrel with Gaitskell, and the split in the Labour Party, which this epitomized, have been rightly regarded as marking a watershed in British political history.[12]

---

* Bevin was by then such a sick man—his deputy[5] at the Foreign Office was a minor political figure in the Labour Party—that British foreign policy in the final eighteen months of the second Labour Government is best considered as largely Attlee's, even though Bevin had some good days as well as bad, and Attlee himself was in hospital early in 1951.

† The sum at stake (£13m.) was in fact small, just as it was nearly twenty years later, when another Labour Government took a decision in the opposite sense—to withdraw from 'East of Suez'.

The scars left by this first defence-versus-welfare controversy are plainly visible in the Labour Party to this day. And although Bevan's personal resignation statement, made in the House of Commons on 23 April 1951,[13] may have been the worst parliamentary speech of his career, it is also remembered for his description of the British Government as 'dragged behind the wheels of American diplomacy'.

The final months of the second Attlee Administration were a dismal period for Britain. The Government staggered on[14] until October 1951, when Churchill—by now approaching his seventy-eighth year—was returned to office; thereafter, the Labour Party was to remain in opposition for the next thirteen years.

THE UNITED STATES

Truman's 'give 'em hell, Harry' electoral campaign returned him to the White House as President of the United States in his own right in January 1949. For the next four years he gave hell to his enemies, and continued in the process to take some of the most difficult presidential decisions in American history. Truman had plenty of enemies to contend with, both at home and abroad. At home his personal triumph in the presidential election did little to alter his problem on Capitol Hill. Although his own party regained control of both Houses of Congress, it was still a divided party. Senior conservative southern Democrats now took over the chairmanship of key Congressional committees, so that the President faced in his second term essentially the same coalition of southern Democrats and conservative Republicans as before. And the Republican Party minority on Capitol Hill, increasingly frustrated, stood ready to attack at the first available weak point. This was not long in coming.

Truman's election had virtually coincided with two very significant events, one internal and one external. In December 1948 Alger Hiss[15] was indicted for perjury. The day before the election the principal city of Manchuria was captured by Chinese communist forces. And as 1949 went forward, the Truman Administration's Asian fault line became increasingly evident. The man who, together with the President, faced the task of preventing a political earthquake was Dean Acheson, whose appointment as Secretary of State to succeed Marshall[16] was destined to have a major impact both on the foreign policy of the United States during Truman's second term and also on the schism that racked the country's domestic political life in these four years.

The strength of Acheson's personal convictions about communism, already demonstrated at the time of the Congressional hearings on the Truman Doctrine two years earlier, was again put on record in

Congress, this time in two sentences drafted for him by Vandenberg, when the Senate Foreign Relations Committee voted to confirm his appointment as Secretary of State on 18 January 1949: 'Communism as a doctrine is economically fatal to a free society and to human rights and fundamental freedoms. Communism as an aggressive factor in world conquest is fatal to independent governments and to free people.'[17] But Acheson's outstanding strengths of intellect and ability were offset by a weakness: snobbery. In his dealings with Bevin, who was impervious to attitudes of this kind, from whatever quarter, this made no difference at all. With what he himself called the 'primitives' of the Republican Party, it was another matter, however.[18] Identified by them as the Secretary of State who had 'lost' China, Acheson went on to compound the fiasco of the China White Paper with his famous remark, made on 25 January 1950—a few hours after sentence was passed on Hiss—'. . . whatever the outcome of any appeal which Mr Hiss or his lawyers may make in this case I do not intend to turn my back on Alger Hiss'.[19] These words ('the uniquely wrong phaseology by the uniquely wrong man at the uniquely wrong moment')[20] were instantly seized on by McCarthy, who embarked on an anti-communist crusade in his famous speech at Wheeling, West Virginia, in the following month. He followed up this speech by sending on 11 February, an equally famous telegram to the President, whose first two sentences were:

In a Lincoln Day speech in Wheeling Thursday night I stated that the State Department harbors a nest of communists and communist sympathisers who are helping to shape our foreign policy. I further stated that I have in my possession the names of 57 communists who are in the State Department at present.[21]

The campaign in the United States that formed the unhappy sequel to McCarthy's speech is mainly a matter of American history, although it has an indirect bearing on the subject of this book. Seriously underrated by Truman, McCarthy now launched himself on his meteoric course across the domestic political arena of the United States. There is no evidence that McCarthy's campaign in any way influenced Truman's private decision, taken secretly in April 1950,[22] not to run for the presidency again. But McCarthyism provided the worst possible accompaniment to the dramatic sequence of events that began on 24 June 1950. And a year later,* even if Truman had wanted to reverse his decision in public, he would have found it exceedingly difficult to do so. What happened in the Far East, combined with the

---

* By which time, among other things, he had relieved MacArthur of his command.

explosion of the communist issue at home, fatally weakened his Administration. When Eisenhower* led the Republican Party to victory in the 1952 elections, twenty years of Democratic rule in the White House were brought to an end.

This, then, was the troubled state of US and British internal politics: the setting within which Truman and Attlee met at the end of 1950. Bevin was too ill[23] to accompany the Prime Minister to Washington. Marshall's wisdom had been brought back once again, as Defense Secretary, but he was now ailing; no longer the man he had once been. Acheson was embattled, fighting on two fronts. As Attlee reported immediately afterwards to Bevin, 'the course of the talks and their results must inevitably be viewed against the domestic political scene' (and this had much to do, he went on to say, with the Administration's 'emphasis on ... the maintenance of their attitude towards China' during the talks).[24] And, as the US Government was equally aware, from two long telegrams in which the US Embassy in London had described the British scene on the eve of the conference with admirable clarity,[25] the same was true of the British position in reverse. Both sides were handling internal political dynamite.[26]

---

* Truman had done what he could to persuade Eisenhower to stand as the Democratic Party candidate.

CHAPTER 17

1 January–24 June

1950 WAS a pivotal year in several senses, two of which are of particular relevance to a study of the Anglo-American relationship. It was in this year that the modern nuclear arms race between what are now called the two superpowers began.* Nuclear co-operation between the United States and Britain having been consigned, in Acheson's words, to 'the deep-freezer'[1] at the beginning of 1950, there was no British input either to the American decision to develop the hydrogen bomb or to the decision, taken in parallel both in Washington and Moscow, not to attempt a fresh US-Soviet nuclear dialogue before embarking on this momentous development.[2] 1950 was also the year by the end of which, although the British did not renounce their claim to the status of a great power—a claim maintained through the decade that followed, against American scepticism[3] and with varying degrees of success—they did accept their world role as being that of the 'principal ally' of the United States, with whom they 'must be prepared in the last resort to continue the struggle together and alone',[4] while exerting 'at the same time . . . sufficient control over the policy of the well-intentioned but inexperienced colossus on whose cooperation our safety depends'.[5]

The word 'superpower' was not as yet an accepted political term in 1950; indeed many years were to pass before it became a part of current political usage. It had, however, been coined in 1944 by an American academic, who had at that time envisaged not two but three superpowers—the United States, the Soviet Union, and Britain. He had also defined a superpower as 'a great power, whose armed force is so mobile that it can be deployed in any strategic theatre, as opposed to a great power whose interests and influence are confined to a single regional theatre':[6] a precise definition which holds good today, in the age of intercontinental ballistic missiles, provided that the state in question is also a nuclear power. In 1950, the British armed forces, spread out over an empire which, apart from the Indian sub-continent, still covered the same territory as it had done five years earlier, retained

---

* In addition, 1950 marked the beginning of the era of what analysts later called 'bipolarity' and—as a reaction to that—of the Non-Aligned Movement.

what would now be called global reach. But in the nuclear field Britain was trailing behind not only behind the United States, but also the Soviet Union. (The first British atomic explosion did not take place until 1952; the first Soviet thermonuclear explosion took place in 1953.) Thus, as the new decade began, Britain might perhaps have been classified as a great power with nuclear aspirations, but not as a superpower. On the other hand, Soviet inferiority at sea and in the air was then such as to make it questionable whether superpower status, in accordance with the strict definition, could at that time have been accorded to the Soviet Union either; and as events later proved, the Soviet Union did not attain rough strategic parity with the United States until the late 1960s.

Even if analysts on both sides of the Atlantic had been given a glimpse of the course of the history of nuclear armaments over the twenty years that followed, it seems probable that in 1950 their vision would still have been blurred by the size of the Red Army and their hearing dulled by the stridency of Soviet political invective in the final phase of the Stalinist era. Nevertheless—in spite of what had happened in Siberia in August 1949, and the Sino-Soviet treaty notwithstanding—the almost equal parts into which the year 1950 falls afford a striking contrast with each other. Up to 24 June 1950—the day* on which North Korean forces crossed the 38th Parallel—the western pace remained leisurely and the tone of western public statements was sanguine. The outbreak of the Korean War caused an instant switch of 180 degrees. This, the great divide in the year, seemed to many to offer a sudden confirmation of the direst fear felt on both sides of the Atlantic: that what had now begun was a titanic struggle for 'the domination of the Eurasian land-mass'.[7]

Acheson's celebrated omission of Korea and Formosa from the definition of the American defensive perimeter in the Pacific (the Aleutians, Japan, Okinawa, and the Philippines), which he expounded in his speech to the National Press Club in Washington on 12 January 1950,[8] was not an isolated instance. Not long afterwards he told the Senate Foreign Relations Committee that the United States could not 'scatter our shots all over the world. We just haven't got enough shots to do that'. Truman himself told a press conference on 4 May 1950: 'I think the situation now is not nearly so bad as it was in the first half of 1946. I think it is improving'. And four days before the Korean War broke out Dean Rusk told the House Foreign Affairs Committee: 'We see no present indication that the people across the [Korean] border have any intention of fighting a major war for that purpose' (which

* By American and European time it was 24 June, but by Far Eastern time the war began in the early hours of 25 June.

had been defined in his previous sentence as that of 'taking over Southern Korea').[9]

The extent of British pessimism about the prospect in Korea (as over China) in 1949 has been noted in Chapter 11. On the eve of the Korean War, however, there is no evidence that the British Government was any better prepared for its outbreak than the US Government.* Over South-East Asia, on the other hand, the wheels of Anglo-American consultation had moved gradually forward from their hesitant start in March 1949. Following an exchange of papers on this area between the State Department and the Foreign Office, in May 1950 what was described as a close identity of views was reached between US, British and French officials. The primary responsibility in South-East Asia remained with Britain and France; and the main emphasis was to be on economic and cultural policy. Nevertheless, on 1 May Truman allocated US$10m. for military assistance to Indo-China: the first step on the beginning of the slope down which US policy was to travel for the next quarter of a century, although—to put the decision in perspective—this amount was equivalent to one thirtieth of the appropriation for Greece and Turkey voted by Congress three years earlier.[11]

For both the United States and Britain, the security of western Europe remained the overriding priority. Yet even here progress was slow. True, as Harriman observed some years later, the fourth meeting of the North Atlantic Council, held in London in May 1950, 'laid the foundation for NATO: in fact, it put the "O" in NATO'.[12] The assembled Foreign Ministers then stretched out their hands towards the German nettle, which—under the impact of the Korean War—they would find themselves grasping by the end of the year. But the US Government had still not decided to commit additional forces to Europe; and the promised British reinforcement, which had been the subject of the British Defence Committee's agonized decision, in fact consisted of two territorial divisions, which were not expected to reach the Continent until four months after the outbreak of any European war.[13]

THE HYDROGEN BOMB AND NSC 68

As the American and the British drew closer to the greatest crisis that they had faced since the end of the Second World War, each

---

* The first news of the outbreak received by the Foreign Office Resident Clerk (duty officer) came from a journalist. This was followed by a telegram from the British Minister at Seoul reporting that he was burning his cyphers.[10]

government took one major decision; the US decision was military and the British was political; both were of lasting significance. On 31 January 1950 a presidential statement announced that the Atomic Energy Commission had been directed to 'continue its work on all forms of atomic weapons, including the so-called hydrogen or super bomb'; and on 10 March Truman decided on an urgent programme to manufacture a hydrogen bomb.[14] The six-month interval between the detection of the Soviet atomic explosion in Siberia and the President's decision to make the 'super' was largely due to the fact that the opinions of his advisers were divided. On 30 October 1949 six members of the US Atomic Energy Commission's general advisory committee, including James Conant[15] and Robert Oppenheimer, had signed a statement that the weapon, whose use would 'involve a decision to slaughter a vast number of civilians', might 'become a weapon of genocide'.[16] The advisory committee unanimously opposed the high-priority development of thermonuclear weapons and the Commission* itself divided three to two against it. Nevertheless, in the end, the White House scales were tipped by the simple argument, advanced publicly in the American press and secretly by the Pentagon, that a decision not to develop the hydrogen bomb meant running the risk that the United States would later find itself in a reversed Hilaire Belloc position of: 'They have got the Maxim gun and we have not'.[18] An early compromise—research, stopping short of development—was soon overtaken by the pressure of events.[19]

This decision, coupled with the parallel Soviet decision, marked the beginning of the era of the balance of terror. It was a decision that stood on its own. Taken by Truman over three months before the outbreak of the Korean War, at a time when he was still insisting on holding the US defence budget around the US$13bn. mark, it did not form part of the general western rearmament that followed the outbreak of the war. There was, however, a bridge between the two: NSC 68. Simultaneously with the White House announcement of 31 January 1950, Truman also directed Acheson and Johnson to 'undertake a re-examination of our objectives in peace and war and of the effect of these objectives on our strategic plans, in the light of the probable fissile bomb capability and possible thermonuclear capability of the Soviet Union'.[20] This report, NSC 68, drafted swiftly for the National Security Council by a small group of officials led by Paul Nitze (who succeeded Kennan as Head of the Policy Planning Staff),

---

* Lilienthal retired from the chairmanship of the Commission, being replaced, in February 1950, by a hardline member of the Commission, Admiral Lewis Strauss. Kennan left the State Department in June 1950, after arguing the case for international control of atomic energy in a memorandum of seventy-nine pages.[17]

is—from the western end of the telescope—a seminal document of the cold war. Although it was presented to Truman on 7 April, the committee established to consider the recommendations of NSC 68 had met little more than half a dozen times when the Korean War began. It then served the purpose for which, as Acheson later described it, the review had originally been intended: to 'bludgeon' governmental opinion in Washington.[21]

One of the basic assumptions of NSC 68 was that the Second World War had left only two major powers in the world, the United States and the Soviet Union; and the assessment did not have much to say about Britain (described as still facing 'economic problems which may require a moderate but politically difficult decline in the British standard of living').[22] Nevertheless, since the broad geo-political outlook reflected in NSC 68 was—with two notable exceptions— shared in London at the time, and since this goes far to explain the reaction in both capitals to what happened in the last week of June, a summary account is needed here.

In the words of this review, the Kremlin was 'inescapably militant'. Its 'Fundamental Design' was defined as:

... to retain and solidify their absolute power, first in the Soviet Union and second in the areas under their control. In the minds of the Soviet leaders, however, achievement of this design requires the dynamic extension of their authority and the ultimate elimination of any effective opposition to their authorit·· The design, therefore calls for the complete submission or forcible destruction of the machinery of government and structure of society in the countries of the non-Soviet world and their replacement by an apparatus and structure subservient to and controlled from the Kremlin. To this end Soviet efforts are now directed toward the domination of the Eurasian land-mass . . .

In the Joint Chiefs' view, which was quoted, the Soviet armed forces were in a 'sufficiently advanced state of preparation immediately to . . . overrun Western Europe, with the possible exception of the Iberian and Scandinavian peninsulas; to drive toward the oil-bearing areas of the Near and Middle East; and to consolidate Communist gains in the Far East'. In the CIA's view, also quoted, the Soviet stockpile of fission bombs might range from 10–20 by mid-1950 and reach 200 by mid-1954. The Soviet 'atomic bomber capability' was assessed as 'already in excess of that needed to deliver available bombs'. The 'critical date for the United States' would be when the Soviet atomic stockpile reached 200. And it was 'quite possible that in the near future the USSR' would have enough atomic bombs and enough bombers to deliver them to 'raise a question whether Britain . . . could be relied upon as an advance base from which a major portion of the US attack could be launched'.

From these assumptions the deduction was drawn that 'the integrity and vitality' of the US system was 'in greater jeopardy than ever before in our history'. The Conclusions recommended 'a much more rapid and concerted build-up of the actual strength of both the United States and the other nations of the free world'.[23] Once this build-up had begun in earnest, 'it might then be desirable for the United States to take an initiative in seeking negotiation in the hope that it might facilitate the process of accommodation by the Kremlin to the new situation'. As for international control of atomic energy, the section in the review devoted to this subject concluded:

It is impossible to hope that an effective plan for international control can be negotiated unless and until the Kremlin design has been frustrated to a point at which a genuine and drastic change in Soviet policies has taken place.

The review ended with the statement that the success of the proposed programme depended on 'recognition by the Government, the American people, and all free peoples, that the cold war is in fact a real war in which the survival of the free world is at stake'. Nevertheless, the 'bludgeon' did not have an immediate effect in Washington. Truman simply remitted NSC 68 to the National Security Council, which set up an *ad hoc* committee; this body had not costed the proposed programme by the time the Korean War broke out. Indeed, it was not until 29 September 1950 that the Council 'adopted the Conclusions of NSC 68 as a statement of policy to be followed over the next four or five years', with the proviso, however, that 'the specific nature and estimated costs of these programmes will be decided as they are more firmly developed'. Truman gave his approval on the following day.[24]

What made NSC 68 significant was its timing—'the tocsin sounded just before the fire'.[25] Its military estimates are open to question. Even at the time, the assumption about the 'fundamental design of the Kremlin', which was itself an essential premise of the whole review, was not shared by everyone concerned in Washington, including Charles Bohlen, who questioned whether the Kremlin really did plan to dominate the world and pointed out that, if this was indeed the fundamental Soviet design, this 'leads inevitably to the conclusion that war is inevitable'.[26]

In the Anglo-American context, had the Foreign Office been given a sight of NSC 68, the assumption that would have caused the greatest difficulty in London (apart from the stated premise that the Second World War had left only two major powers on the international scene) was the doctrine that communism was not just evil, but also monolithic. With the China Lobby in full cry, McCarthy baying for blood, and the US Secretary of State on the way to becoming a domestic political liability to the Democratic Party,[27] this doctrine, although mistaken, was perhaps understandable in the climate of Washington in 1950. In London, even after Chinese intervention in Korea, Bevin stuck to his guns; and in the last speech that he delivered—painfully— in the House of Commons, on 14 December 1950, he reaffirmed that it was British policy 'not to become obsessed with the Communist conception of China but rather to bear in mind that the mass of Chinese scarcely understood what Communism means, and try if we could to bring them along and keep them in association with the other nations of the world'.[28] It was twenty years before a US President and his Secretary of State took something like this view.

THE SCHUMAN PLAN

The major British decision taken in the first half of 1950 was of a different order of magnitude, but its consequences also lasted for twenty years. On 2 June 1950 what was left* of the British Cabinet, meeting under Morrison's chairmanship, decided[29] not to attend the conference convened in Paris by the French Government for 20 June, to discuss the plan to pool coal and steel production under an International High Authority. To be fair to the seven Cabinet Ministers, two junior ministers and four senior officials present at the meeting on 2 June, in 1950 British GNP was equivalent to nearly half the GNP of the six countries who joined the Coal and Steel Community; Britain produced approximately half the amount of coal and one third the amount of steel that the Six then did; and when, five years later, Britain was presented with its second European choice, the Conservative Government then in power behaved no more wisely than its Labour predecessor. In 1950, the British decision to opt out of an institution that was the first stage in the process both of Franco-German reconciliation and of the formation of the European Economic Community, came as the climax to ten days of Anglo-French argument—in all, eleven notes and 4,000 words were exchanged—which might have lasted even longer if the French Government had not addressed a note to all six† governments asking for a reply to its invitation by 8 p.m. on 2 June. The ins and outs of these ten days are largely a matter of Anglo-French history.[30] But this, the first of what proved to be a series of Anglo-French quarrels spread over the next two decades, was also an

---

* Bevin was again in hospital, for his second operation in two months. Attlee and Cripps were both on holiday (ironically, in France).

† Besides Britain, the governments invited were those of the Federal German Republic, Italy, the Netherlands, Belgium, and Luxembourg.

Anglo-American confrontation. It is this aspect of the events of May–June that the remainder of this chapter will address.

It will be recalled (from Chapter 15) that, at the end of 1949, Acheson's response in Paris to the Anglo-French deadlock on the issue of European integration had been to quote *Henry V*. The return to the breach took the form of yet another round of talks between American and British officials, held in London in April 1950. This time the leader of the British team believed that 'the Americans, or some Americans certainly, were coming round to the British view of *Atlantic Community* development, including of course Europe in this wider grouping';[31] indeed, on the eve of the critical May meetings, the chief British concern was not so much how to resist renewed pressure for European integration—let alone how to respond to a brilliantly imaginative proposal drafted in five weeks, its secret confined to Monnet and a group of eight other Frenchmen—but how to persuade the French Government to take a 'more realistic view of the future place of Western Germany in Western Europe'.[32] Acheson's own view of European integration, however, had not changed during the winter. Indeed, at a meeting (with Douglas, among others, present) at the State Department on 7 March, he is recorded as asking whether there were 'any British who could work at overall matters imaginatively' and he suggested that 'Bevin had distinct limitations along this line'. Douglas agreed about Bevin, adding that Strang was 'also rather limited'.[33] Not only did Acheson demonstrate his 'immediate and intense displeasure' with a paper on the special nature of Anglo-American relations drafted during the talks in London (by ordering 'all copies of the paper that could be found' to be 'collected and burned');[34] he also broke his journey to London with a forty-eight hour stop-over in Paris. As he later recorded in his memoirs, the 'circumstantial evidence' of his 'presence there, in view of what occurred, convinced Bevin of a Franco-American conspiracy against him'.[35]

On 7 May—forty-eight hours before Bevin was informed by the French Ambassador in London—Schuman disclosed his coal and steel plan to Acheson and the US Ambassador in Paris under pledge of secrecy. Acheson was, therefore, obliged to have his first meeting with Bevin in London without saying a word about it.* There is no reason to doubt Acheson's account of the 'innocence' of his motives in going to Paris.[36] On the other hand, it is hard to believe that all the senior officials concerned, without exception, were taken entirely by surprise. In spite of the secrecy in which Monnet and his team had worked, the fact remains that his links with Bruce and Harriman were close; the

---

* Adenauer was also informed in secrecy, by a special envoy; the French Ambassador in Bonn was kept in the dark.

view has recently been expressed that Harriman probably had some advance warning of the Plan;[37] and the US Ambassadors in Paris and London and the High Commissioner in Bonn worked very closely together at this time. Although British officials were given no such inkling, they can hardly have expected a man of Monnet's vision and determination to take 'no' for an answer after the many days that he had spent talking to them about economic plans for the future over the previous twelve months.[38] However that may be, Bevin felt justified in having a furious row[39] with Acheson: a bad beginning to what should have been one of the most serious transatlantic deliberations since the Second World War.

Both Acheson and Truman publicly supported the French with enthusiasm in May.[40] In the developments of the following weeks, the state of Bevin's and Cripps' health, coupled with Attlee's and Morrison's imperfect grasp of the great issues that were at stake, obliged British officials to play a major part. The question arises, therefore, to what extent they realized, as their American counterparts undoubtedly did, the full significance of the Schuman Plan, which embodied not just a new economic concept, but an inspired answer to what the British themselves had defined as the problem of the 'future place of Western Germany in Europe'. For this is what it undoubtedly was. Witness, for example, Adenauer's immediate response—'Das ist unser Durchbruch'—when he was first informed about the Schuman Plan; to the German Chancellor it was at once clear that the plan meant 'the beginning of an independent German existence, an independent German role, and the recovery of some authority from the total collapse of defeat'.[41] Even Bevin, ill and angry, saw (in his initial off the cuff reaction) that 'something' had 'changed' between Britain and France. And Cripps at first took the sound view, and said as much to Monnet, to the distress of senior officials both in the Treasury and in the Foreign Office: that 'we should collaborate from the outset'.[42] The initial reaction of some British officials was positive;[43] but they did not include those who mattered most in Whitehall; and the Noes soon had it. Thereafter almost everything that could go wrong did go wrong, not least the publication on 13 June of a parochial twelve-page statement on *European Unity* by the Labour Party's National Executive Committee, before Parliament had reconvened; this statement (described by Churchill as the 'Dalton Brown Paper') coincided with the publication of a Governmental White Paper, which it partly contradicted; and even the White Paper contained a major inaccuracy, which was pounced on by the press.[44]

Hoffman described the Labour Party statement as 'deplorable';[45] and in the words of the *Christian Science Monitor*:

What has stirred sharp criticism ... is its additional theme that unification with the continent must be opposed in the interests of socialist doctrine ... the Labour Party is all for union, if everybody will be like the British.[46]

On the same day as the two British statements were published, Acheson was speaking of the 'promise of a great new era in Europe'.[47] On 21 June the *New York Times* carried the full English text of a speech made by Schuman on the previous day, in which he said once again that Europe was inconceivable without Britain.[48] The die was, however, cast in London within a week. The Cabinet (this time chaired by the Prime Minister) having taken its formal decision not to take part in the Schuman Plan,[49] the Government then defeated by thirteen votes the motion put down by Churchill in the name of all the Opposition parties in the House of Commons:

to accept the invitation to take part in the discussions on the Schuman Plan, subject to the same condition as that made by the Netherlands Government, namely, that if the discussions show the plan not to be practicable, freedom of action is reserved.[50]

The debate was memorable not only for a devastating[51] speech by Churchill—'les absents', he observed, 'ont toujours tort'[52]—but also for the fact that it had to be interrupted to enable the Prime Minister to make a statement on the outbreak of the Korean War.[53] British policy in Europe had been completely outflanked.* For both the British and the Americans, however, their differences over Europe were now over-shadowed by the sudden shock of what was happening in Asia.

---

* Since this chapter was written, *DBPO*, Series II, vol. i has been published (in 1986), covering the period May 1950–December 1952 in relation to European Integration.

CHAPTER 18

25 June–31 December

THE OUTBREAK OF THE KOREAN WAR

THE Korean War came as a bolt from the blue. The first reports that North Korean troops had launched a heavy attack across the 38th parallel came to both the State Department and the Foreign Office from the press. Everyone who mattered in both capitals was out of town. Such was the degree of suprise both in Washington and in London that the immediate reaction was confined to convening a meeting of the Security Council, to be held on the afternoon of Sunday 25 June. Truman was dissuaded by Acheson from flying back to Washington from Independence, Missouri, the previous night, but when he did take off, on the Sunday afternoon, he was in such a hurry that he left without a navigator.[1]

By a coincidence, two Australian officers, members of the United Nations Temporary Commission on Korea, had spent the previous fortnight as observers on and around the parallel. Although their report, written the day before the North Korean attack, included what were—in retrospect—some ominous signs, it ended with the words: 'No reports . . . had been received of any unusual activity on the part of the North Korean forces that would indicate any imminent change in the general situation on the parallel'.[2] Nevertheless, the report of these officers proved to be a document of considerable importance in the subsequent proceedings of the UN Security Council, because, on the strength of their evidence, UNCOK was able to report from Seoul to the UN Secretary General at once the fact that 'South Korean forces were deployed on a wholly defensive basis in all sectors of the parallel' and 'were taken completely by surprise'.[3]

From the outset the American conviction was that the North Korean attack was the work of 'centrally directed Communist Imperialism'; and in his special message to Congress on 19 July, Truman said: 'The fateful events of the Nineteen-thirties, when aggression unopposed bred more aggression and eventually war, were fresh in our minds'.[4] The British, while in no doubt that the attack was a clear case of aggression, were reluctant[5] to go along with the assumption that

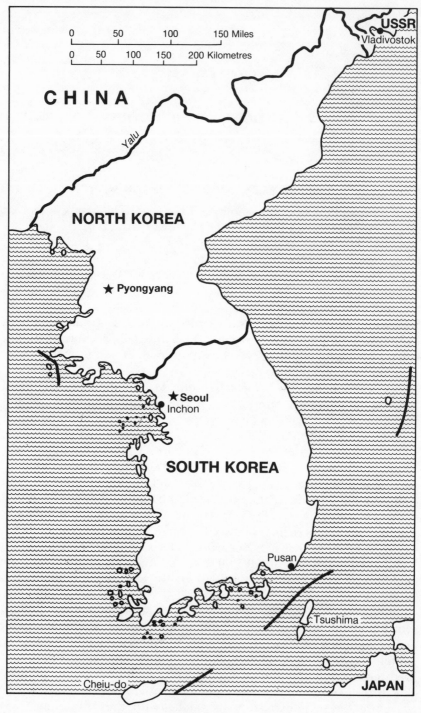

Map 6. Korea

communism was monolithic: the root of the disagreement between the two governments, which was soon to become critical. Today, because the evidence[6] on the origins of the Korean War remains fragmentary, it is still an open question which governments were involved, and to what extent, in the North Korean decision to invade South Korea. On balance, however, the belief (by no means confined to Americans in 1950)[7] that responsibility for the North Korean decision rested in the Kremlin, is hard to sustain—not least because the Soviet Government had tied one arm behind its back in New York. In January 1950 the Soviet Representative in the Security Council, Yakov Malik, had walked out in protest at the Council's refusal to unseat the Chinese Nationalist Representative. When the Security Council took its critical vote in June—not only branding the North Korean attack as 'a breach of the peace' and demanding its 'immediate cessation', but also calling upon all members 'to render every assistance to the United Nations in the execution of this resolution and to refrain from giving assistance to the North Korean authorities'[8]—Malik had still not resumed his seat. His absence from the Security Council, barely explicable if the Soviet Government was the prime mover of the invasion, was to have significant international consequences, the first of which was that the Security Council 'overnight emerged as something surprisingly like the custodian of peace and security which the Charter had envisaged'.[9] On 7 July the Council resolved that UN members providing forces for Korea should make them available to a unified command and asked the United States to designate a commander; Truman appointed MacArthur. Thus it came about that in its first phase, from June–October 1950, the war was fought between North and South Korean troops, the latter aided by forces under UN command. From November 1950 onwards, however, the war became something different and far more dangerous: a conflict that was still limited, but fought between two armies, one of which was, in effect, under US and the other under Chinese leadership. A selective chronology of the first six months of the war will be found at Appendix 6.

THE US RESPONSE

Once the magnitude of the disaster facing the South Korean forces became apparent, as it did in a matter of days,[10] US decisions followed thick and fast. US air and naval forces were committed to the battle, south of the 38th parallel; the Seventh Fleet, which had been instructed to steam north from the Philippines at the outset, was now interposed to seal off Formosa, because (in the words of Truman's statement of 27 June) 'the occupation of Formosa by Communist forces would be a direct threat to the security of the Pacific and to United States forces performing their lawful and necessary functions in that area';[11] and at five o'clock on the morning of 30 June, in response to an urgent telephoned appeal from MacArthur,[12] Truman approved the commitment of 'one regimental combat team' to the battle. Later the same morning he extended this commitment by authorizing MacArthur to deploy in Korea US forces under his command in Japan, 'subject only to requirements of the safety of Japan'.[13]

Thus, in the course of less than one week, the tide of military events in the Korean peninsula (from which US troops had been withdrawn a year earlier) brought about what amounted to a U-turn in US policy in East Asia. The previous ambivalence about the importance to US security of Korea and Formosa was swept away; and this led ultimately both to a new US relationship with Japan and to the conclusion of an alliance with the Chinese Nationalist authorities in Taipei. But the immediate political result was—in the perspective of Peking—to plunge the United States back into the Chinese civil war. Truman did not simply interpose the Seventh Fleet; he stated publicly that the determination of Formosa's status 'must await the restoration of security in the Pacific, a peace settlement with Japan, or consideration by the United Nations'.[14] Predictably, the Chinese response to this action by the US Government (which had nothing to do with the UN) was that the Seventh Fleet's presence in the Formosa (Taiwan) Straits was an act of aggression against Chinese territory:[15] territory, moreover, from which Chiang had offered his troops for deployment in Korea.

THE BRITISH RESPONSE

In the longer term, the most important consequences of the Korean War for Britain lay in the economic field. It was not only that, as has been described in an earlier chapter, the scale of the British rearmament programme (raising defence expenditure from 7.1 per cent in 1950 to 10.7 per cent of GNP in 1952) split Attlee's Cabinet. Looking back on the first post-war decades of the British economy, one of its earliest historians argued that the thrust of Britain's industrial investment was first thrown out of gear by the demands of rearmament in the early 1950s; then, in 1953–4, the needs of the consumer had at long last to be met; and in the following year a fresh balance of payments crisis brought economic expansion to a halt—a period of five years, during which 'Germany, investing at home up to the hilt of its resources, refurbished the whole of its industrial machine'.[16]

At the time, although US pressure had a good deal to do with the size of the British rearmament programme, the primary motive of Attlee, Bevin and Gaitskell (appointed Chancellor in October 1950) was their fear for the safety of Europe, to which the dramatic events of June added a sharper edge, coupled with anxiety lest, impelled by the same events and under the pressure of their internal political debate, the Americans should adopt a 'Pacific first' strategy. Thus the British Government lost no time in deciding to support the United States in the Security Council in June; and British naval forces were at once released from Hong Kong in support of US forces operating in Korea. But at a meeting held on 6 July 1950 the Cabinet Defence Committee 'decided that no land or air forces should be made available for operations in Korea'.[17] It was only after Franks had written Attlee a personal letter in manuscript on 15 July,[18] which he followed up with a powerful telegram to the Foreign Office, that the British Chiefs of Staff 'recognised the strong political arguments' in favour of sending British land forces to Korea as well, even though this was, in their view, 'still militarily unsound'.[19] On 25 July the earlier decision was reversed[20] and announced in the House of Commons on the following day; the first British troops reached Pusan on 29 August; and nine months later there where the three brigades in Korea, which in July 1951 became the 1st Commonwealth Division.*

ANGLO-AMERICAN DISSENSION

In the context of the Anglo-American relationship, the most serious political consequence of the war centered on the question of China. What had been a leisurely gentlemen's agreement to differ six months earlier, now became, in the wake of the US President's public statement regarding the future of Formosa, a burning political issue. The British Government found it all the more difficult to handle both because Bevin was again in the London Clinic and because it was not only an Anglo-American issue, but also an issue of Commonwealth politics, particularly for the Indian Government, which cast itself in the role of mediator. From hospital Bevin sent Acheson a message, based on British discussions both with the Indian and with the Soviet Government, urging the US Government to play down those parts of the President's message of 27 June that did not bear directly on Korea. Acheson sent a long and dusty answer on 10 July, the US Ambassador being instructed to 'leave him in no doubt of seriousness with which I view implications of his message and their possible effect

---

* Two of these brigades were British; the third was composed of Canadian troops and Australian forces of all three services.

on our whole future relationship'.[21] Acheson's message to Bevin reiterated the view that the invasion of South Korea 'was ordered by the Kremlin and is being actively directed by key Soviet personnel in Korea'. He later recorded in his memoirs:

Bevin's written reply warned that we must not drive China into Soviet hands, and stated that although aggression must be repelled and he would not yield to blackmail, he wanted the Soviet Union and the Communist Chinese in the Security Council. The correspondence clearly had no future, so we dropped it.[22],*

As the records show, this last sentence was not accurate. When Bevin returned to the Foreign Office at the end of July, Douglas brought him a message from Acheson suggesting another look at their policies on Formosa. Bevin gave a temporizing reply.[24] The British records also reveal that Attlee wanted to have talks with Truman three months before his dash to Washington in December. Acheson, sounded out by Franks, did not think the moment opportune; 'it would be much best for the visit of the PM to take place after the November elections'.[25] By the second half of September MacArthur's brilliant counter-attack had transformed the tactical situation in the Korean peninsula. It also evoked the first Chinese warnings. On 3 October 1950 the US Chargé d'Affaires in London sent a telegram to Washington (classified for action at any hour of the day or night), reporting the contents of a telegram from the British Embassy in Peking: on that day Chou En-lai had summoned the Indian Ambassador and told him that if UN Command forces crossed the 38th parallel, Chinese troops would cross the frontier to take part in the defence of North Korea. This warning was disregarded in Washington and dismissed as 'pure bluff' in Tokyo. Bevin, on the other hand, took it seriously enough to telegraph from the *Queen Mary* (he was on his way home) his view that 'armed intervention by China would be a great catastrophe'.[26] The crossing of the parallel by the UN Command forces (which included the British 27 Brigade) on 9 October 1950 was followed by the fateful advance towards the Yalu River.

Two days earlier the UN General Assembly had passed a resolution calling for 'all appropriate steps to be taken to ensure conditions of stability throughout Korea'. Without being mentioned in the resolution, the crossing of the 38th parallel was implied. It was the British who sponsored this resolution. This apart, the order, counter-order, and disorder,[27] which entitled MacArthur to testify (after he had been relieved of his command by Truman in the following year) that it had

* Almost simultaneously, the British Cabinet accepted a US offer to move two bomber groups and a fighter group to airfields in the United Kingdom.[23]

been with Washington's complete agreement that he had ordered his forces to cross the parallel, were almost entirely an American responsibility.

GERMAN REARMAMENT

Such was the optimism about the Korean War felt in New York, after the success of MacArthur's landing at Inchon, that the meetings held there in September 1950 between Acheson, Bevin, and Schuman hardly touched on Asian matters at all.* What the three Foreign Ministers were concerned with was Europe. The new US commitments in Asia did not prevent the Administration from accepting further commitments in Europe as well. NSC 68 now came into its own. By 1 December 1950, Congress had approved two supplementary appropriations, which together added a total of US$35.3bn. to the original defence budget of US$13.3bn. for the fiscal year 1951.[29] (By 1953 the overall budget for defence and international affairs had trebled). On the eve of the September 1950 meeting, Truman announced his intention of making substantial increases in the strength of US forces to be stationed in Western Europe, provided that a greater European contribution to the collective defence was forthcoming.[30] The US ground forces in Europe were to be brought up to a strength of six divisions; the British to three. But what the European contribution really meant was West German rearmament. On the insistence of the Pentagon, the United States' offer to its European allies in the autumn of 1950 took the form of a single package: a united North Atlantic command, increased US forces, and German armed participation—a decision of a take-it-or-leave-it kind which Acheson at first resisted and later, in his memoirs, described as a mistake.[31]

Not only was Schuman flatly opposed to the creation of German military units—a matter which he was prepared to discuss only after France and her North Atlantic allies had been rearmed themselves— but initially Bevin himself was not prepared to go beyond the formation of a West German gendarmerie (Adenauer's own proposal). In mid-September, however, faced with the Franco-American deadlock, he changed his mind and extracted, by telegram, grudging approval from the Cabinet that he should give the British Government's 'general agreement to an acceptance in principle of German participation in Western defence'.[32] In so doing, he deployed to Attlee an argument which was to recur when the Prime Minister himself

* In October 1950, however, the US Administration secured Congressional approval for 'approximately one half billion dollars' for military assistance in the Far East. The 'major part of this sum' was earmarked to 'provide military equipment for the armed forces both of France and of the Associated States of Indo-China'.[28]

visited Washington and which was to influence many British decisions over the years: 'Our country is a leading Power and I cannot take part in discussions without giving an opinion. We must either reject the US thesis or accept it and co-operate with them. Otherwise Great Britain will look weak and indecisive.'

In the event no decision on German rearmament was reached in September 1950; and at the beginning of December, while Attlee was in Washington, Bevin changed his mind again [33]—without effect. In the interval much had happened, including the announcement by the French Prime Minister, René Pleven, on 24 October 1950 of another French plan prepared in secrecy:* the Pleven Plan for European defence. The long history of this initiative, which culminated in the rejection of the European Defence Community legislation by the French National Assembly in 1954, lies outside the period on which this book is focused. In the short term, however, it helped to make possible French approval of the so-called Spofford Plan, whereby 'the unit for incorporation of German troops into the unified [North Atlantic] command should be the regimental combat team, a self-contained unit of approximately six thousand men, on condition that this be combined with approval of the longer-term goal of a European Army'.[34] It was this compromise formula that was finally adopted by the North Atlantic Council, at its meeting in Brussels on 19 December 1950. Although four years were to pass before the way to rearming the Federal Republic was opened by that country's admission to the Western European Union, the agreement reached in December 1950 enabled General Eisenhower to be appointed Supreme Commander of the integrated North Atlantic forces: the real beginning of the North Atlantic Treaty Organization,† within which the Brussels Treaty military organization was now merged.[35]

CHINESE INTERVENTION IN KOREA: CRISIS

Meanwhile, the advance towards the Yalu River that had begun on 9 October led MacArthur's army into 'the largest ambush in history',[36] laid by a force of over 300,000 Chinese troops, at the end of November 1950. Faced by these numbers, what had been a 'home by Christmas' advance became a retreat, beyond Seoul. Early on the morning of 28 November Truman received a message from MacArthur (via the Joint Chiefs of Staff) that the United States now faced 'an entirely new war'.[37] On 15 December Truman announced on television that on the

---

* Acheson was informed in advance; Bevin was not.

† This led to a prolonged debate in the Senate in the early weeks of 1951, which ended in approval both of Eisenhower's appointment and of the sending to Europe of 'not more than four divisions'.

9. Bevin and Acheson, Washington, September 1949.

10. Truman greets Attlee, Washington, December 1950.

following day he would proclaim a national emergency; 'all the things we believe in are in great danger. This danger has been created by the rulers of the Soviet Union'. A fortnight earlier, however, he had made an astonishing gaffe: his remarks at his press conference on 30 November were interpreted around the whole world as meaning not only that the United States had the use of the atomic bomb under active consideration, but also that the decision whether to use it would be MacArthur's. Despite a White House denial,[38] the damage was done—not least in the British Parliament, which, at the time of Truman's press conference, was engaged in a foreign affairs debate that showed wide support on both sides of the House for a positive reponse to the Soviet Government's proposal of a meeting of the Council of Foreign Ministers to be held without delay, 'for consideration of the question of the fulfilment of the Potsdam Agreement regarding the demilitarisation of Germany',[39] put forward by Andrei Gromyko in a note handed to the US Ambassador late on the night of 3 November 1950.* There was also deep concern in Parliament lest the United States should yield to the temptation to become involved, in Asia, in what General Omar Bradley described in the following year as 'the wrong war, at the wrong place, at the wrong time and with the wrong enemy'.[40]

This time Attlee could not be fobbed off. After an emergency Cabinet meeting and a meeting with the French Prime Minister and Foreign Minister (who came to London for consultations), he flew to Washington on 3 December 1950 to attend a conference that has recently been described as 'the last great attempt to recreate the kind of relationship based on equality which had existed in the early days of the [Second World] war'.[41] It would be more accurate to say that Anglo-American partnership had long been the British concept of what the post-war relationship between the two countries should be, and that this was Attlee's first personal attempt to put the concept into practice himself (his first visit to Washington five years earlier had had a different aim; and the talks held then were tripartite). Many of the claims advanced, moreover, at the time and since, for what this conference achieved, have been exaggerated.[42] Nevertheless, this was the last occasion on which a British Prime Minister crossed the Atlantic in order to bring out a reversal of US policy on a major issue—China. The records now available show conclusively first, that Atlee's idea of proposing himself as a visitor to Washington in 1950 was not simply a reaction to the general consternation aroused by Truman's

---

* Four months later a meeting was convened at the Palais Rose in Paris, not of the CFM, but of the deputies of the four Foreign Ministers. It adjourned, after seventy-four fruitless sessions, on 21 June 1951.

mishandling of the atomic question put to him by the United Press representative during his press conference on 30 November—on the contrary, the Prime Minister and the Foreign Secretary had seen the need for personal consultation with the President three months earlier—and secondly, that when Atlee was finally received in Washington, he argued the British case on China with tenacity and skill.

Unlike the talks held in Washington in November 1945,[43] the six meetings held there from 4–8 December 1950 were fully documented on both sides. The official American and British prose of these records still conveys a sense of high drama after a lapse of thirty-five years. This, the only Anglo-American summit meeting during the Labour Government's six years in power, was held at the nadir of the Korean War. The possibility that the Korean peninsula might have to be evacuated, and how this should be done, was openly discussed at the talks. On both sides there was much plain speaking. Each side was, at least in part, playing to an unseen internal political gallery.[44]

Except in relation to the principal economic issue—the control of the allocation and prices of raw materials—the conference achieved no concrete result. It was the British view (powerfully expressed in forthright terms in an emergency telegram sent by Bevin to Attlee on 4 November)[45] that 'an attempt should now be made to reach an all-round agreement with the Chinese before it is too late ... the only policy now open to the United States Government in the interests of us all'. The 'essential features' of this settlement were described in the same telegram as:

(a) Acknowledgement of China's position as a great power in the Far East. Agreement by the United States Government that in the event of a satisfactory settlement of Korea they will consider afresh the question of United States recognition of the Central People's Government.

(b) Formosa. Agreement by the United States Government that they stand by the Cairo and Potsdam Declarations.

(c) Agreement on a cease-fire in Korea leading up to restoration of the *status quo* before June 25th i.e. re-establishment of the 38th Parallel as the line dividing United Nations forces and Chinese forces.

Bevin recognized that it might be 'politically very difficult' for the United States Government to accept this programme. It was indeed. The Washington communiqué recorded the two government's continued agreement to differ over this, the central issue, round which most of the discussion revolved. On the atomic issue, which had taken up relatively little time in Washington, Attlee told the House of Commons on his return to London that, in the spirit and against the background of the wartime Anglo-American partnership, he had received assurances which he considered to be 'perfectly satisfactory'. At the end of his report on the meetings, telegraphed to Bevin from Ottawa on his way home, Attlee had said that he believed Truman's 'personal assurances about the use of the atomic bomb' to be 'perfectly sincere'.[46] That they were sincere is undoubted, but—from a British point of view—they were far from satisfactory.

Even without knowing exactly what had happened in Washington, on which Attlee refused to be drawn in Parliament, Churchill's reaction to the discovery that the Quebec Agreement had been abandoned was to write personally to the President.[47] In fact, the minutes of the two sides disagreed on the wording of the presidential assurances. In the words used by Acheson over twenty years afterwards, the US side 'had to unachieve' what Truman had said privately to Attlee—'that he would not consider the use of the bomb without consulting the UK and Canada'—and in the end the penultimate paragraph of the Final Communiqué simply said that the President hoped that 'world conditions would never call for the use of the atomic bomb' and that he had 'told the Prime Minister that it was also his desire to keep the Prime Minister at all times informed of developments which might bring about a change in the situation'.[48] Almost a year was to go by before a formula was at last devised to cover the use in an emergency of US air bases and facilities in Britain.[49]

Looking back on this conference today, it is clear that, although it resolved nothing of substance, it did—for a time—relax transatlantic tension. The former US artillery officer and the former British infantry officer got on well together personally. The claim that they altered history by meeting is untenable. Yet, had they not met when they did, no one can say for certain what might have happened. The use of atomic bombs was *considered* in Washington in the dark, final weeks of 1950 and their use was, by implication, threatened early in 1953, although the fact that they were not used is, I suggest, proof of the paradox which has repeated itself more than once since then, namely, that when it comes to the point, a nuclear power cannot make use of its nuclear arms in a conventional war.

Even as it was, within a month Attlee was confiding his second thoughts to Bevin, whom he had asked to go and see him at 10 Downing Street:

... he had been talking to a number of military and other personalities (he specifically mentioned Mountbatten)* who were thoroughly alarmed about

* These bracketed words have been inked out in the original minute by Bevin's Private Secretary.

the present American attitudes to world affairs which they were afraid would end by dragging us into war.[50]

At the time, however (in his telegram to Bevin sent from Ottawa), Attlee resorted to phraseology that will be familiar from earlier chapters of this book: 'throughout these talks the UK was lifted out of the "European queue" and were treated as partners unequal no doubt in power but still equal in counsel'.[51] By contrast, Acheson's later summing up, in his memoirs, of this, his first summit conference was characteristically acerbic: 'an ungranted prayer that I might be spared another, and ... the grateful recognition of the gap between Mr. Attlee's brief and the consent decree—to use lawyers' terms—that he had signed.' The American and British positions having drawn 'substantially closer', Acheson concluded that 'the furor ... fizzled out and was soon forgotten'.[52]

PART VI

THE SEQUEL, 1951–62

CHAPTER 19

The Mould Broken

OF the many explanations for the evolution of the post-war Anglo-American relationship that have been suggested, one of the least convincing is the Greco-Roman analogy. Attributed to Harold Macmillan (while he was in Algiers during the war), although it is hard to conceive that a classical scholar would have been capable of such a misunderstanding of ancient history,[1] this concept initially took the following form:

These Americans represent the new Roman Empire and we Britons, like the Greeks of old, must teach them how to make it go.[2]

On the other hand, the underlying thought—that even a super-power sometimes needs an interlocutor, especially one who speaks the same language—became increasingly invoked in Whitehall in later years as the intellectual basis for the relationship. With the passage of time, this frequently repeated concept has, I suggest, been superimposed on the earlier period, even though—as the documents now available on both sides of the Atlantic show—the relationship between the United States and Britain during the years 1945–50 was of a different kind. In reality, for this variegated period of the Anglo-American relationship, largely conducted before the mould had set, within the constraints of what each country perceived to be its national interest, in a manner sometimes brilliantly creative, sometimes destructively obtuse, and often paradoxical, history offers no close parallel. By a further parodox, during these years a Labour Government was in power in Britain; by the time that the Labour Party left office in October 1951 the Anglo–American mould was set; and yet five years later it was a Conservative Prime Minister who broke it.

1956 was a tremendous year. Its catalogue of great political events included, in chronological sequence, Khrushchev's denunciation of Stalin at the Twentieth Congress of the Soviet Communist Party; the decision taken by the governments of the Six at the Venice Conference to form a European Customs Union (ineffectively countered by the ill-fated British 'Plan G');* the Polish 'October'; the Hungarian

---

* The Venice decision led to the signature of the (EEC) Treaty of Rome in 1957; the British 'Plan G' attempted—unsuccessfully—to set up a European Industrial Free Trade Area instead.[3]

Revolt; and the final act of the Suez crisis—the cause of the first major Anglo-American schism[4] since the Second World War. Because the inner history of this schism will, it is hoped, soon be open to public scrutiny, no attempt will be made in this chapter to anticipate the release of the 1956 official documents (nor, for the reason given in the Preface,[5] will the remainder of this book be documented in the manner followed in Parts I–V). Instead, the questions that this and the next chapter will seek to address are these. How did the British sustain their claim to be regarded as a world power as effectively as they did during the first five years of Conservative rule? Why did the US Government acquiesce in the exercise of the British claim for as long as it did, before it finally drew the line over Suez? And why, in spite of the bitterness of the schism on both sides of the Atlantic, was the Anglo-American relationship so swiftly re-established—even reinforced—in the six years that followed?

THE FIRST PHASE: 1951–5

The simplistic temptation—to explain what happened in the five years before the Suez crisis largely in terms of personalities—must be resisted. True, personalities mattered greatly in 1956. Eden's unhappy relationship with John Foster Dulles,* whom he had even advised Eisenhower not to appoint as Secretary of State, together with his long known vanity and his increasingly severe bouts of illness, combined—at any rate from 1954 onwards—to constitute what would in today's jargon be called a semi-adversarial factor. But Eden had known Eisenhower for many years, both as wartime Supreme Commander and as the first peacetime Supreme Commander, Europe from the beginning of 1951; and it was Eisenhower who was the ultimate arbiter of US foreign policy.[6] Indeed, to begin with, Churchill's return to power in October 1951, followed by Eisenhower's election to the presidency a year later, seemed to augur well[7] (though perhaps more to British than to American public opinion). Within three months of forming his third Administration Churchill spent nine days in Washington, conferring with the Truman Administration there officially for the first time and addressing a joint session of Congress for the third. And even before Eisenhower was inaugurated as President, Churchill met him privately in New York, in December 1952. First names seemed

---

* Not that the faults in this relationship were all on one side. In an early and perceptive analysis of Foster (as he was always called in conversation) Dulles, both as a man and as Secretary of State, Joseph Harsch observed: 'One of the many curious things about him is the fact that he tends to arouse either approval bordering on veneration, or disapproval ranging close to moral contempt': *Harper's Magazine*, September 1956, p. 27, 'John Foster Dulles: a very complicated man'.

to cross the Atlantic by telegram, telephone, and letter, almost as they had during the Second World War.

Eisenhower's warnings to Churchill, not to overplay his hand, have been much quoted. Thus, when they met in New York, he told Churchill that to give cause for 'concern in world opinion that we meant to form a two-power "coalition" . . . would create jealousies and suspicions that would be harmful in our work towards a world of justice'.[8] In August 1953 the US Ambassador in London told Macmillan that the President was 'embarrassed by Churchill's attempt to revive, by personal correspondence, the old Churchill–Roosevelt relationship'.[9] And in December, at their Bermuda meeting, in Eisenhower's words, 'the last thing the two strongest Western powers should do was to appear before the world as a combination of forces to compel adherence to their announced views'.[10] Nevertheless, the fact that Eisenhower felt obliged to give these warnings is evidence of the potent impact of which Churchill was still capable, certainly until his stroke in June 1953, and on occasions even thereafter; the statement that, following Eisenhower's electoral victory of November 1952, 'American hegemony was asserted in a manner which made it open for all to see'[11] is an example of historical telescoping; this demonstration was in fact delayed until November 1956. In order to understand the course taken by the relationship between the two countries in the early years of Conservative government in the 1950s, what must be examined is not so much the personal relationships between the men who led them at that time, as the ambiguity that was already inherent in the delicate balance of the Anglo-American relationship as a whole. During the post-war quinquennium this ambiguity had been latent, prevented from reaching the surface during the latter half of the period by the continuous pressure of external events. In the new era it gradually became less manageable, until it found its final expression in the Suez crisis.

It will be recalled from Part V that by the end of 1950 the British had accepted their role as the 'principal ally' of the United States. Of this there was no doubt on either side of the Atlantic. But the British concept that, in spite of the disparity of power between the two countries, Britain remained the United States' 'equal in counsel' depended on the validity of the British claim still to be a world power. As Franks had written, in his manuscript letter to Attlee of 15 July 1950:

Three or even two years ago . . . we were one of the queue of European countries. Now with new strength and vitality in our associations with the Commonwealth, a reviving, more flexible and much stronger domestic economy, and great improvement in our overseas payments, we are effectively out of the queue, one of two world powers outside Russia. This has been recognised by the recent exchange of messages ...*

Attlee's and Bevin's successors in Downing Street believed, if anything even more strongly, that this view was right. Their counterparts in Washington, whether in the Democratic or in the Republican Administration, never accepted it. To recognize now the nature of this ambiguity in the early 1950s, however, does not mean, as is often assumed today, that Britain had already declined to its modern position of international weakness (an assumption which recently has been well described as an enormous source of confusion).[14] The balance between the far more powerful United States and Britain was indeed delicate, but there were still some high cards held in the British hand, three of which are described below. A fourth—British influence in the Middle East—though still highly valued by both governments at the beginning of the 1950s, was already on the way to becoming something like a singleton ace.

In the first place, in October 1952 Britain joined the atomic club. Admittedly this was only a month before the first US hydrogen bomb explosion. It was five years before the first British hydrogen bomb explosion followed, but with the advent of the (British made) fleet of V-bombers later in the decade, Britain was set to become the possessor of a nuclear deterrent to be reckoned with. No less important—the world's first nuclear plant designed to feed electricity into the national grid was British.

Secondly, the United States continued to need British bases for its strategic bomber force; and, as the decade went forward, the usefulness of Britain as a base for other kinds of strategic weapons would enter the transatlantic calculus. Although the Anglo-American atomic dialogue still made little progress, in spite of Churchill's personal efforts during his visit to Washington in January 1952, and although the so-called 'Attlee–Churchill understanding' regarding the use of American atomic weapons based in Britain was phrased in very general terms, by the end of that year the British Chiefs of Staff 'had received on a highly personal basis a great deal of information' about the 'strategic and tactical aspects' of the US air plan.[15]

Thirdly, Britain remained—the United States apart—the backbone of NATO through those years. In spite of the agreement on German rearmament reached by the North Atlantic Council at its meeting at the very end of 1950, and although the integrated command structure was established, with Eisenhower as NATO's first SACEUR (Supreme Allied Commander, Europe), the Federal Republic of Germany

---

* In which Attlee had proposed, and Truman had accepted, bilateral talks, to be conducted 'in the greatest secrecy' in Washington.[13]

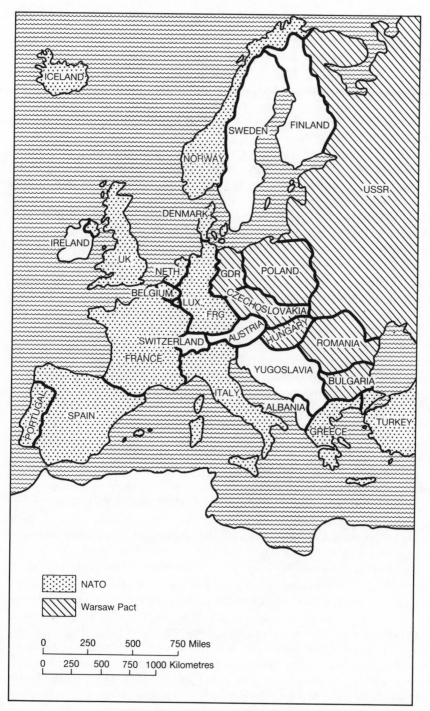

ICELAND

SWEDEN

FINLAND

NORWAY

USSR

DENMARK

IRELAND

UK

NETH

GDR

POLAND

BELGIUM

LUX

CZECHOSLOVAKIA

FRG

AUSTRIA

HUNGARY

SWITZERLAND

ROMANIA

FRANCE

YUGOSLAVIA

BULGARIA

ITALY

PORTUGAL

SPAIN

ALBANIA

TURKEY

GREECE

NATO

Warsaw Pact

| 0 | 250 | 500 | 750 Miles |

| 0 | 250 | 500 | 750 | 1000 Kilometres |

Map 7. The Alliances of Modern Europe

remained unarmed. On 26 May 1952 the Bonn Convention was signed, whereby the Federal Republic of Germany attained what Acheson described at the ceremony as 'the independence in foreign affairs and authority in domestic affairs which befit a free state'. The European Defence Community agreement, however (signed in Paris on the following day), was subsequently ratified by all the signatories except its originator, France, which for the next two years engaged in its greatest debate since the Dreyfus Affair.[16] For this long delay the weakness of the Fourth French Republic, under a succession of unstable governments and with its armed forces more and more embroiled, first in Indo-China, and then in North Africa, was primarily responsible. And when, on 30 August 1954, the French National Assembly finally rejected the EDC, it was Eden—not Dulles—who stepped into the breach, with a skilful round of shuttle diplomacy.[17] The 1948 Brussels Treaty was taken off the shelf, dusted, and became the Western European Union, with its headquarters in London, backed by the British Government's assurance that four British divisions and the British Tactical Air Force would be maintained on the European mainland as long as the majority of the Brussels Treaty powers desired it. Signed in Paris at the end of October, the agreements ratifying Western European Union received the approval of the French Chamber of Deputies on 29 December 1954. On 5 May 1955 the Paris Agreements took effect, thus enabling the Federal Republic of Germany to become a member of NATO at last. The formation of the Warsaw Pact, concluded on 14 May, completed the pattern of military alliances in Europe which has endured over the thirty years that followed.

Thanks to these events in the latter half of 1954, Britain briefly recovered the European initiative that it had thrown away four years earlier—only to repeat the same error, this time with graver consequences, by its cavalier treatment of the two conferences of the Six which preceded the conclusion of the Treaty of Rome: the Messina Conference of 1955 and the Venice Conference of 1956. As before, these moves towards Western European integration had the fullest support in Washington; and 'a very ill-judged and ... unfortunate approach' made by the British to the US Government, in an attempt to enlist US support for the British effort to oppose the formation of a European customs union, received 'an extremely dusty answer'.[18] In the Anglo-American context, however, the two governments' continuing disagreement on the economic integration of Western Europe was overshadowed by a clash of policy on a far broader stage. At the heart of this clash, as in the British disdain towards the efforts of the Six, lay the British conviction that they remained a world power.

On 6 March 1953 Stalin died. At the time, the cautious US response took the form of Eisenhower's 'What is the Soviet Union ready to do?' speech, delivered to the American Society of Newspaper Editors on 16 April.[19] On the other hand, Churchill, without consulting the US Government, delivered his celebrated speech in the House of Commons on 11 May 1953, in the course of which he suggested that:

A conference on the highest level should take place between the leading Powers without long delay. This conference should not be overhung by a ponderous or rigid agenda, or led into mazes of technical details. The conference should be confined to the smallest number of Powers and persons possible. It should meet with a measure of informality and a still greater measure of privacy and seclusion. It might be that no hard-faced agreements would be reached, but there might be a general feeling among those gathered together that they might do something better then tear the human race, including themselves, into bits.[20]

This Anglo-American difference of approach was to have been discussed at a conference of the US, British, and French Heads of Government in mid-June. France was without a government for five weeks. By the time that the Cabinet crisis in Paris had been resolved, Churchill was incapacitated by his stroke (Eden was already ill). Precious months slipped by. When the Three Power Conference finally convened, at Bermuda at the very end of 1953, the atmosphere was far from happy; and its only concrete result was to accept a Soviet proposal for a meeting of the four Foreign Ministers. Churchill still did not give up his summit project. In mid-1954, on the way back (by boat) from another meeting in Washington, he drafted a message to Molotov—without consulting either the US Government or his own Cabinet, and after 'a blood row' with Eden—in which he proposed a bilateral meeting with Georgii Malenkov.[21] By the time that the Soviet Government at last received the message it may well have been aware of Churchill's failing powers. His initiative petered out; and when a Four Power summit meeting was finally held—in July 1955—he had, at the age of eighty, resigned. It was not at all the kind of meeting that he had had in mind two years earlier.

When the four Foreign Ministers met in Berlin from January–February 1954, they achieved nothing in relation to Europe, but they did agree to hold a conference in Geneva, at which China would be represented, relating to the Far East. This Geneva Conference proved to be of great importance not only for one of the two items on its agenda (Indo-China), but also for Anglo-American relations, in particular for the relationship between Eden and Dulles. Arguably, the decisive event of the conference, which was held from April–July 1954, and brought the Franco-Vietnamese war in Indo-China to an end, was the capture of Dien Bien Phu by the Vietminh on 7 May. But for Eden this conference was a personal triumph. When the final declaration of the Geneva Conference was signed on 21 July 1954, the US Government was not prepared to join in it, simply taking note of what had been agreed and undertaking not to use force to disturb the settlement of a colonial war, for which, by the time it ended, the US Government was paying most of the cost with dollars given in aid to the French Government. As Co-Chairman (with the Soviet Foreign Minister) of the conference, Eden dealt—with evident satisfaction—on equal terms with Molotov and with the French and Chinese Prime Ministers, whereas his irritation with the US Secretary of State, and with US diplomatic moves made while the conference was in progress, comes clearly through even the sanitized account of the conference offered in his memoirs.[22] For his part, Dulles had been obliged to watch 'the British delegation' standing 'silently by', while 'the Communists poured invective on the United States' "imperialistic actions" '; he was convinced that Eden had 'repudiated'—without offering an alternative—the understanding reached earlier between the US and British governments in London, that Dulles should 'start to set up conferences in Washington to discuss an alliance* with the nations involved'.[23] In this controversy lay the seeds of future Anglo-American trouble.

SUEZ

Chapter 10 suggested that, just as the Far East turned out to be the Achilles heel of US foreign policy after the Second World War, so in the end did the Middle East for British foreign policy. From the concluding months of the second Attlee Administration the incoming Conservative Government inherited far fewer anxieties about the Far East than might have been expected from the crisis at the turn of the year 1950/1. But almost on the eve of the British general election two major events occurred in the Middle East: the Egyptian Government abrogated both[24] its treaties with Britain; and the British cruiser *Mauritius* evacuated the staff of the world's largest oil refinery, at Abadan—the culmination of the first phase of the Anglo-Iranian oil crisis. Both these events had important Anglo-American overtones.

In September 1951, in an attempt to square the Anglo-Egyptian circle (five years after Bevin's abortive agreement with Sidky, described in an earlier chapter), the British and US governments had agreed on the concept of a 'Middle East Command', under a Supreme Allied Commander Middle East, to be established in Cairo and to be

---

* In the end, this took the form of the SEATO Pact, signed at Manila in September 1954.

responsible for the Suez base, which would be given to Egypt. After French and Turkish approval had been secured, the plan was put to the Egyptian Prime Minister, who 'rejected it without reading it'. In Acheson's words, 'If ever there was a political stillbirth, this was it'.[25] In spite of Acheson's personal scepticism[26] about the eventual outcome of the Middle East Command plan, the extent of Anglo-American co-operation in relation to Egypt in 1951 was considerable; it continued over the next three years; and Anglo-Egyptian agreement on the withdrawal of British troops from the Suez base was at last reached in October 1954.

The first phase of the Anglo-Iranian crisis was a different story. Although this crisis was finally resolved by a combined Anglo-American effort four years later, the early months of the crisis carried some pointers towards the Anglo-American disaster of 1956. In the first place, one of the causes of the Anglo-Iranian crisis was a purely American action: the extraneous decision taken by Aramco to share the company's profits fifty–fifty with Saudi Arabia—'a landmark in the economic affairs of the Middle East'.[27] Secondly, the American and British assessments of the nature of Iranian nationalism were quite different.[28] Thirdly (and perhaps most important of all), although the British Cabinet went a long way down the path leading towards military intervention in Iran, what finally decided them against intervening was the attitude of the United States. In the words of the minutes of the (Labour) Cabinet meeting held on 27 September 1951, 'we could not afford to break with the United States on an issue of this kind'.[29] A month later the new Government went along with the decision of its predecessor. Yet in 1956—in relation to another nationalist leader, this time in Egypt—it did precisely the opposite.

As in the Anglo-Iranian crisis, what triggered the Suez crisis was a US decision: this time, to withdraw its offer to the Egyptian Government to finance the Aswan High Dam.[30] One week later, on 26 July 1956, Gamal Abdel Nasser nationalized the Suez Canal. The Israeli invasion of Sinai began on 29 October. The Anglo-French ultimatum* was delivered on 30 October. British bombing of Egyptian airfields began on the following day. On 5 November the Anglo-French invasion began at Port Said. On 6 November—the day of the US presidential election—Eden accepted a ceasefire. By Christmas the last British troops had been withdrawn, unconditionally, from Port Said.

The question why it was that Eden and his small inner circle of

---

* Nominally, to both the Egyptian and the Israeli governments, in order to 'separate the combatants'. In reality, it was addressed to Egypt.

ministerial colleagues and official advisers miscalculated* to such a
dramatic extent the transatlantic 'correlation of forces'—to use Marxist terminology, which is perhaps more appropriate to what happened
in November 1956 than the bland language of the traditional Anglo-
American communiqué—is best left for the documents of that year,
once they have been released, to answer. So, too, is the parallel
question, how exactly Eisenhower and Dulles assessed the indications,[31]
which must have reached Washington from many sources in the last
days of October, of the degree of consultation between the British,
French, and Israeli governments, and of what was impending. Whatever the official documents may reveal about the diplomatic imbroglio
that occupied the three-month gap between Nasser's announcement of
his decision to nationalize the Canal and the Anglo-French ultimatum,
they are very unlikely to alter the conclusion that the final outcome of
the Suez crisis demonstrated, beyond a shadow of doubt, how the
'forces' between the United States and Britain were really 'correlated'.
Moreover, in New York, the spectacle of the United States and the
Soviet Union acting against Britain in the Security Council, in parallel
if not in concert, was a poignant reminder, for the British Government,
of the coincidence of US and Soviet interests over Palestine in 1947–8.
Remnants of what has been described in an earlier chapter as the
British imperial position would survive in the Middle East for a little
over ten years after the Suez crisis. But from now on the two major
powers that counted in this critical region were the United States and
the Soviet Union.

The Eisenhower Administration brought the Eden Government to
heel in a few days in November 1956 not merely by refusing to have
further dealings with the Prime Minister personally (instead, the US
Ambassador dealt with Eden's colleagues, notably Macmillan), but by
the use of unvarnished economic and financial pressure, exerted both
bilaterally and through the International Monetary Fund. By comparison with the impact of this pressure, the Soviet Prime Minister's
warning to Eden of 5 November 1956[32] was only an additional
embarrassment, shrugged off in London as no more than atom-
rattling. It was the sustained US pressure that was decisive.[33] The
Emperor was seen, for the first time in public, to have no clothes. The
US attitude in November 1956 is perfectly summarized in the words
attributed to Eisenhower when he first telephoned 10 Downing Street:
'Anthony, have you gone out of your mind?'.[34] The British attitude
does not lend itself so easily to half a dozen words. Perhaps uncons-

---

* A critical supplementary question, on the British side of the equation, is which Egyptian
politician was believed, in London, to be capable of taking Nasser's place, and how he was to be
kept in office.

ciously, Eden and his colleagues seem to have trusted to the belief expressed nearly two hundred years earlier by Cardinal Richelieu, that 'of two unequal powers, joined by a treaty, the greater power runs more risk of being abandoned than the other'.[35] Eden was not in the Richelieu class of statesmanship.

The Suez crisis left the Anglo-American relationship temporarily shattered. But by far its worst consequence was not the damage that it did to the relationship, but the fact that the preoccupation with the Middle East that underlay it (by no means confined to the British side of the Atlantic) was fundamentally a diversion from the East–West issue, which was what really mattered. Eisenhower may have been right—without the Soviet archive we cannot be certain—in writing afterwards that, even without the diversion of Suez in November 1956, the western reaction to what happened simultaneously in Budapest would have been no more 'intense'.[36] But Churchill's first instinctive reaction after Stalin's death in the spring of 1953 had been the right one. The full resources of American and British diplomacy should have been employed in the direction that, in his old age, Churchill, the original architect of the Anglo-American relationship, had tried to point in his speech of 11 May 1953, rather than in this preposterous irrelevance.

Today most people would agree with the thrust of the view, expressed nearly a quarter of a century afterwards by an American historian, that the era of Stalin's successors ... 'particularly the years 1955 to 1960, presented what was unquestionably the most favorable situation that had existed since the 1920s for an improvement of relations with Russia'.[37] Whether this belief is indeed wholly beyond question cannot be finally established without more Soviet evidence than is at present available. What is certain is that Stalin's death and the early years of the Soviet leadership that succeeded him offered the western powers a unique opportunity. They did not grasp it. From this failure both West and East have suffered.

CHAPTER 20

The Mould Reset

ON 9 January 1957 Eden—'omnium consensu capax imperii, nisi imperasset'—resigned.[1] When his successor as Prime Minister, Harold Macmillan, had his first audience of the Queen, he warned her 'half in joke, half in earnest' that he 'could not answer for the new Government lasting more than six weeks'.[2] He remained at 10 Downing Street for more than six years: a period that has been regarded as the golden afternoon of the Anglo-American relationship. As the present chapter seeks to show, it is more accurate to describe it as a long parenthesis.* And when this parenthesis finally came to an end, at the turn of 1962/3, the extent of the changes in the balance of world power, which had been proceeding simultaneously during these six years, became dramatically evident.

However this period may be assessed, it was remarkable for the speed with which the two governments mended their fences and for the effort devoted thereafter—especially, but not exclusively, on the British side of the Atlantic—to keeping them in good repair. Once again, personalities counted. But, as in the preceding period, they were not the dominant factor. To take personalities first, however, the fact that Eisenhower and Macmillan had worked together in the Mediterranean theatre of operations during the Second World War, and that for the four years after Macmillan's appointment as Prime Minister each of them was at the head of affairs in his country, made the initial work of fence-mending much easier than it would otherwise have been (the tone of Eisenhower's personal message[4] of congratulation to the new Prime Minister speaks for itself). Within days[5] of Macmillan's appointment he received an invitation to meet the President in March. This invitation led to their conference at Bermuda, the first of the many Anglo-American summit meetings that characterized this period. Even more remarkable was the way in which, in effect, this Anglo-American parenthesis was for a time extended beyond the end of the second Eisenhower Administration. Had the same paper-thin majority of votes brought Richard Nixon, instead of John Kennedy, to the White

---

* 'You see the inconveniency of a long parenthesis; we have forgot the sense that went before'; a 1659 usage, cited in the Shorter Oxford English Dictionary.[3]

House in January 1961, relations between the United States and Britain during 1961–3 would hardly have been the same. As it was, in spite of a difference in age of nearly thirty years between the two heads of government, the final stage of the parenthesis was marked by a personal understanding between Kennedy and Macmillan, which, as it grew over a period of two and a half years, developed a political significance, not to be confused with the hagiographical overtones added after the President's assassination in November 1963. Instead, the relationship between the two men was down to earth. Macmillan does not seem to have harboured any illusions at the time; towards the end of the first year of Kennedy's term of office, for example, he described the President as:

A very sensitive man, very easily pleased and very easily offended. He likes presents—he likes letters, he likes attention. To match this, he is clearly a very effective, even ruthless, operator in the political field ...[6]

In fostering this relationship Macmillan was helped by the shrewd appointment of David Ormsby Gore as Ambassador in Washington.[7]

At the same time, the late 1950s were also years during which the number of 'transnational actors' on the Anglo-American stage in every field—political, economic, financial, and cultural—rapidly multiplied, contributing to an interpenetration[8] between American and British societies on a scale inconceivable before the advent of cheap transatlantic travel, and since then intensified still further by other factors, notably television. What has been described in an earlier chapter as the 'elitist' factor in the first post-war years of the Anglo-American relationship thus gradually became diluted. Nevertheless, for the men serving during this decade in the senior echelons of the bureaucracies, civilian and military, on both sides of the Atlantic, the practice of Anglo-American co-operation became almost a habit; most of them saw the Suez crisis simply as the exception that proved the rule. Thanks in part to the existence of this 'historical generation',[9] a series of agreements between the US and British governments was negotiated in the years 1957–62 with a combination of professional smoothness and personal trust that may perplex later generations. In time this practice of co-operation developed its own momentum. But as the momentum became built in, so too did reluctance to recognize publicly the degree to which the asymmetry of the Anglo-American relationship was simultaneously increasing. This reluctance to come to grips with reality was most marked in London: witness the outcry that greeted Acheson's famous remark in his speech at West Point on 5 December 1962:

Great Britain has lost an empire and has not yet found a role. The attempt to play a separate role—that is, a role apart from Europe, a role based on a 'special relationship' with the United States, a role based on being the head of a 'Commonwealth' which has no political structure, unity or strength ... this role is about to be played out.

All this said, what brought the United States and Britain together again after the Suez schism and what kept them close to each other for the rest of the period assessed in the present chapter, was essentially a perception of shared national security interests, above all in the nuclear field: a perception which took on a particularly keen intensity precisely in 1957. Thus, ironically, the nuclear understanding that had eluded the two governments in the late 1940s, was in the end reached not long after their greatest post-war dispute, the Suez crisis. It was this, above all, that reset the mould.

A number of factors contributed to this, at first sight, paradoxical outcome in 1957. As has been noted in the preceding chapter, Britain was by now a nuclear power in the military sense; and a year before the Suez crisis, the world's first land-based nuclear power plant, producing electricity on an industrial scale, had opened at Calder Hall. Moreover, for the United States Air Force, Britain remained 'Airstrip One'. Even before Eisenhower met Macmillan at Bermuda in March 1957, the British Government had received a US proposal for the deployment of Thor IRBM's in Britain, to be controlled under what later came to be known as a 'dual key' arrangement. At Bermuda, Eisenhower and Dulles at first still 'found it difficult to talk constructively' with Macmillan and Selwyn Lloyd[10] about Suez, 'because of the blinding bitterness they felt toward Nasser', with whose removal the British ministers were still 'obsessed'.[11] On the other hand, the IRBM agreement was concluded and indeed announced in the official communiqué issued at the end of the conference. In the words of Eisenhower's memoirs, 'this agreement would put missiles of intermediate range within striking distance of the heartland of the Soviet Union'.[12]

Both in the United States and in Britain strategic military doctrine was profoundly influenced by the far-reaching deductions drawn in each country from two major events in the 1950s: the Korean War and the Soviet launching of the first sputnik on 4 October 1957. From the aftermath of the Korean War flowed the financial conclusion that national security must be maintained by affordable means; and to both governments the cheapest strategy seemed to be defence based on nuclear deterrence. For Dulles, this became 'maximum deterrence at a bearable cost'.[13] The new British doctrine was embodied in the Defence White Paper of April 1957,[14] accurately described by Macmillan in his memoirs, so far as Britain was concerned, as 'the biggest change in military policy ever made in normal times'.

In the age of 'Star Wars', it is not easy to recapture the international effect of the Soviet announcement of the success of the sputnik at the time. The second meeting between Eisenhower and Macmillan was held in Washington in the wake of this announcement. A single sentence of the communiqué of 25 October 1957, described as a Declaration of Common Purpose, stated that the President would request Congress to amend the McMahon Atomic Energy Act of 1946 'to permit of close and fruitful collaboration of scientists and engineers of Great Britain, the United States, and other friendly countries'. In Macmillan's later words, this was 'the great prize'; in Eisenhower's, 'a great alliance requires, above all, faith and trust on both sides'.[16] On 2 July 1958 the Act was duly amended, to allow the exchange of information about the design and production of nuclear warheads with countries that had already made 'substantial progress in the development of atomic weapons', as well as the transfer of fissile materials to them.[17] The day afterwards, the Anglo-American Atomic Agreement was signed; it was approved by Congress and came into force one month later.[18]

By the end of 1960 nuclear weapons collaboration between the two countries had been taken a stage further. The United States had secured the British Government's agreement to a Scottish base for its new ICBM Polaris submarines and to the establishment of a missile-warning system at Fylingdales; Britain had been promised the US Skybolt air-to-ground missile system (for use with its strategic V-bomber force, whose operational life would thus be prolonged); and in addition there was an understanding that Britain might claim the US Polaris system, if this later became necessary. Meanwhile the British Government had taken a decision which proved to be of lasting significance: the abandonment of its own intermediate range Blue Streak ground-to-ground ballistic missile. Macmillan's memoirs suggest a certain uneasiness, perhaps felt in retrospect, about the 'difficult and even heart-breaking' decision to abandon Blue Streak.[19] By contrast, the French, entering the same field later, were able to avoid many of the pitfalls,[20] which in the late 1950s seemed to the British to justify their decision to give up the development of an indigenous ballistic missile system, and to rely instead on their privileged access to American know-how.

However that may be, by the end of the second Eisenhower Administration the nuclear keystone of the Anglo-American arch was at last visibly in place. As full nuclear partners, the British acquired a unique position within the western alliance.[21] In the Washington perspective, this not only secured an additional measure of anti-Soviet deterrence at little cost. It also provided an insurance against any future British adventure of the Suez kind in the international arena (anti-American feeling in Britain did not by any means disappear overnight at the end of 1956).[22] In the British perspective, it was common ground between the Prime Minister and the Leader of the Opposition that the possession of a British nuclear deterrent, even if to some extent indebted to US technology, gave Britain 'a better position' (Macmillan) or 'a measure of independence' (Gaitskell) in relation to the United States.[23]

In a series of public pronouncements, from the Declaration of Common Purpose onwards, the term of art for this state of Anglo-American affairs was 'inter-dependence'. On 11 December 1962, however, a large crack appeared in the nuclear keystone: the US Defense Secretary* informed his British opposite number that the Skybolt project was dying, if not dead, thus confronting the British Government with a major reassessment of its defence and foreign policy—and indeed a crisis in the Prime Minister's own credibility.[25] The immediate consequence of the US Government's decision about Skybolt was the Nassau Agreement of 21 December 1962, concluded after nearly four days of fierce discussion. (The subsequent Chevaline and Trident agreements are its direct descendants.)[26] Given the importance of the Nassau agreement not just for the Anglo-American defence relationship, but for the overall relationship between the United States and Britain, the relevant extracts from the 'Statement on Nuclear Defense Systems' of 21 December 1962 will be found at Appendix 7.

At Nassau Macmillan invoked his understanding with Eisenhower about Polaris. This was the last occasion when a US President spoke publicly of an American moral obligation to Britain. The agreement finally reached at Nassau, however, embodied a compromise, not unfairly described by one of the participants at the conference as a 'monument of contrived ambiguity'.[27] In return for the provision by the United States of 'Polaris missiles ... (less warheads) for British submarines', British forces 'developed under this plan' and 'at least equal' US forces 'would be made available for inclusion in a NATO mutilateral nuclear force'. The British forces would be used 'for the purposes of international defense of the Western Alliance in all circumstances', except where British 'supreme national interests' were at stake. The US Government offered Polaris on the same terms to the French Government. The Nassau agreement was made public before

---

* Robert McNamara, who (in a speech delivered at Ann Arbor six months earlier) had spoken out strongly against the development of limited independent nuclear deterrents.[24] He made matters worse over Skybolt by talking about it at Gatwick Airport, before he reached London.

the offer was communicated to de Gaulle (who had returned to power in June 1958). He declined it.

The abortive US attempt* to secure allied agreement on the establishment of a multilateral force within NATO lies outside the time-frame of this section of the book. One postscript must be added, however: the Partial Nuclear Test Ban Treaty, signed eight months later by representatives of the United States, British, and Soviet governments in Moscow. Although the date of its signature puts this agreement also outside the time-frame, it represented the outcome of years of transatlantic deliberation, for which a considerable measure of the credit is due to Macmillan. Assisted by able advisers, he did his nuclear homework. The draft taken to Moscow in August 1963 was genuinely Anglo-American: the last example, perhaps, of a major political initiative in East–West relations in which the British Government took part on an equal footing with that of the United States.

Outside the nuclear field, central though this was to Anglo-American relationship, much had changed during the years 1957–62. The British contribution to the North Atlantic Alliance remained a major one; but on land the Federal German Army had become the largest Western European component of NATO by 1960;[28] and the Fifth French Republic under de Gaulle, once the Algerian problem had been solved, was set to become a formidable force not only in European but in world politics during the new decade. In the Anglo-American context these changes remained for a time beneath the surface, partly because of the commitment of many dedicated people on both sides of the Atlantic—and not only mid-Atlantic votaries—to the belief that the peace and prosperity of the world still depended, in large measure, on Anglo-American understanding. An additional reason was the virtuoso role played by the Head of Government who, until illness obliged him to resign in 1963, had been longer in office than any of his opposite numbers.[29]

Like his two Conservative predecessors, Macmillan was seldom absent from the centre of the international stage. He too sometimes gave the President cause for anxiety, notably during his solitary pilgrimage to the Soviet Union in February–March 1959.[30] But the constant stream of long, elegantly drafted messages sent across the Atlantic from 10 Downing Street ensured that he seldom strayed too far from the White House guidelines. And because of the length of Macmillan's ministerial experience and the elaborate care that he devoted to his relationship first with Eisenhower and then with

---

* The idea of a multilateral force, assigned to NATO, had first been put forward at a NATO ministerial meeting in December 1960, by the US Secretary of State.

Kennedy, what he had to say contributed an input to the 'policy of the well-intentioned but inexperienced colossus' (the description used in Bevin's final memorandum–already quoted—which still represented the received Whitehall view of Washington ten years later). Nevertheless, although Macmillan's international performance scintillated, the results were meagre. First his tireless promotion of a Four Power Summit Conference and then his attempt to secure an Anglo-French understanding that would permit British entry* into the EEC ended in fiasco. The summit meeting, held at long last in Paris in May 1960, was stultified by the shooting down of an American U-2 aircraft over Soviet territory just before the conference was due to open. (The initial US disclaimer on 7 May was made to look absurd as soon as the Soviet Government announced that the pilot, Francis Gary Powers, was in Soviet hands.) The British attempt to enter the EEC was stopped dead by de Gaulle's first 'Non', at his famous press conference on 14 January 1963. The belated British conversion to belief in European integration was not helped by what Macmillan had agreed six weeks earlier with Kennedy in Nassau. It will be recalled that it was Bevin who had feared the 'Trojan horses' who might leap out of the Council of Europe. Fifteen years later, it was de Gaulle for whom Britain had become the Trojan horse of the United States. In the words that he himself used at his press conference:

It is to be foreseen that the cohesion of all its members [sc. the Six, plus Great Britain and the members of the European Free Trade Area], who would be very numerous, very different, would not hold out for long and that the outcome would be the emergence of a colossal Atlantic Community dependent on America and under American direction, which would soon absorb the European community. This is a hypothesis which can be justified perfectly well in the eyes of some people, but this is not at all what France has wanted to do and is doing—and that is a strictly European construction.[31]

The greatest change towards the end of this period that affected the Anglo-American relationship, however, was not European, but American. In spite of the bellicose rhetoric[32] that is remembered as the hallmark of Dulles' Secretaryship of State, it was under the next Democratic Administration that this change took visible shape; and the area of the world that provided the setting for the first public demonstration that the United States was now a superpower, answerable in the last resort to none of its allies, Britain included, proved to be one in which Britain had largely lost interest since the Second World War: Latin America (Argentina was the principal exception). This region had indeed not occupied much US governmental attention

* Britain applied to join the EEC in August 1961, and to join the Coal and Steel Community (established under the original Schuman plan) in March 1962.

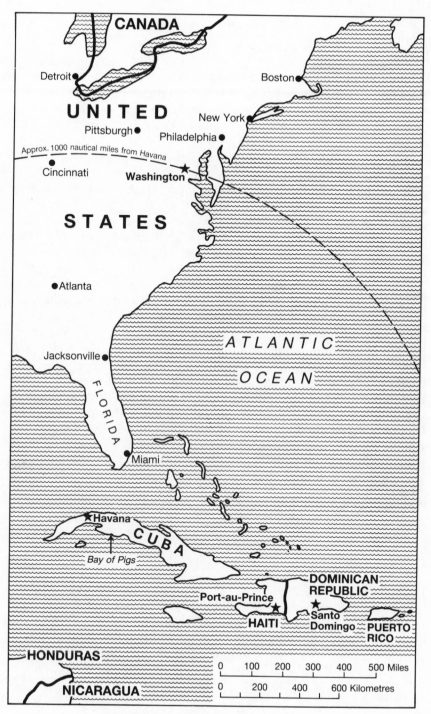

Map 8. Cuba in Relation to the United States

either, since the conclusion of the Treaty of Rio de Janeiro in 1947. The warning lights for the United States in the Latin American sub-continent had, however, already begun to flash towards the end of the Republican Administration (while on an official visit, as Vice-President, Nixon had received an extremely hostile reception in the streets of Caracas). And in 1961 the incoming Democratic Administration may be said to have inherited from its Republican predecessors the tragi-comic planning that ended in the ill-starred invasion of a beach on the southern coast of Cuba,* in April of that year.

What happened in the following year is the first example of superpower military involvement in a country of the developing world, which had forsworn allegiance to one of the two superpowers and assumed the colours of the other.[33] The Cuban missile crisis of October 1962 is also the nearest approach that the world has seen to a thermonuclear exchange. That Kennedy took pains during the critical fortnight to keep the heads of allied governments informed is not in question; and the published[34] excerpts from his exchanges with Mac-millan at the time show the lengths to which he went in discussion with the United States' principal ally. But there is no evidence that at any point during this seminal crisis[35] British influence on the decisions of the US ExCom[36] was more than marginal. The long parenthesis in the Anglo-American relationship was drawing towards its close. With the resignation of Macmillan and the death of Kennedy in the following year, the parenthesis came to an end. So too did a period of over twenty-three years, during most of which the British Government was headed by men who were half-American. They were, however, 'half-American, but all British'.[37] In any case, by the end of 1962, the age of the superpowers had begun.

---

* Fidel Castro entered Havana in January 1959. The landing of Cuban exiles at the Bay of Pigs in 1961 was sponsored by the CIA. Kennedy accepted full responsibility for this fiasco.

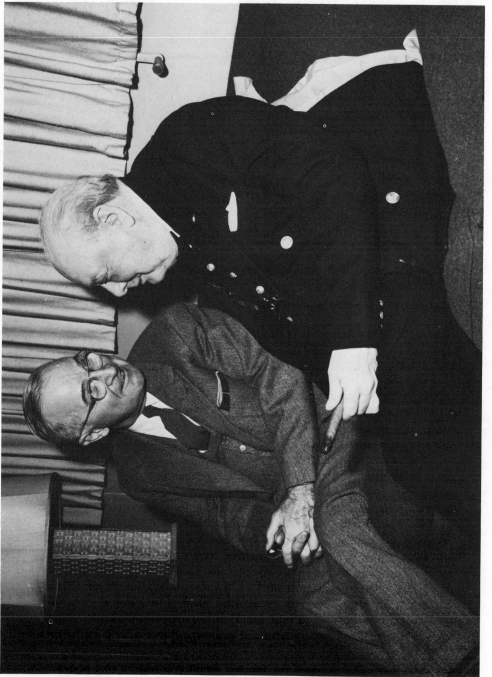

11. Truman and Churchill, on board the *Williamsburg*, January 1952.

12. The Bermuda Conference, March 1957: Lloyd, Eisenhower, Macmillan, Dulles.

13. Kennedy and Macmillan, Nassau, December 1962.

CHAPTER 21

The Balance Sheet

WHAT happened in 1962 brought the relationship between the United States and Britain to a point within striking distance of its contemporary state—just close enough, therefore, to enable this concluding chapter to draw up a balance sheet, based on three questions quoted in the Introduction: irrespective of 'bicentennial oratory', had the Anglo-American relationship 'damaged or strengthened the two countries'? Had 'the one been able to prevent the other from making serious mistakes or to contribute to its learning process'? And, finally, had the two countries 'lured each other into adventures, inspired a false sense of confidence in each other, distorted the other's perspective . . . ?'[1]

Clearly, if the opening date selected for the balance sheet is 1941, the benefits derived by the peoples of both countries from the unprecedented unification of their military effort (to quote Marshall's description once again) must weigh heavily in the scales. Both sides made some serious mistakes during the four years. But wars are usually won by the contestant who makes fewer mistakes than his opponent; a war without mistakes is unknown to history. Nor did either of the two countries lure the other into military adventures in the Second World War. Nevertheless, as has been argued in Part II, the consultation—or rather lack of consultation—between the two governments on the great atomic issues, during the final weeks of the war and immediately after the Japanese surrender, must today be accounted a major international misfortune.

It is only when applied to the four post-war decades that the three questions become pertinent. And the nearer an assessment of the sort that Alastair Buchan had in mind gets to the modern world and to the state of the Anglo-American relationship as it is today, the less easy the questions become to answer, as the language in which he expressed them (in his article posthumously published in 1976) implies. The profound changes that were already under way then have been intensified still further by what has happened since.

CHANGES IN THE ANGLO-AMERICAN RELATIONSHIP

Through the 1960s and the first half of the next decade, the United States first embarked on its disastrous war in Vietnam; then reversed its twenty-three-year-old ostracism of the People's Republic of China; and in 1972 it also began the dialogue with the Soviet superpower that came to be known as the policy of *détente*. 1973 was the so-called 'Year of Europe', of unhappy memory;[2] in the mid-1970s *détente* between the superpowers began to falter; and at the very end of the decade it collapsed. Some of the American rhetoric[3] used in the early 1980s was reminiscent of the language of Dulles thirty years earlier; latterly its tone has softened; and, at the time of writing, the US-Soviet dialogue has just been re-opened by the Joint Statement[4] issued on 21 November 1985, at the conclusion of the Geneva summit meeting between Ronald Reagan and Mikhail Gorbachev—the first such meeting to be held for over six years. The factor common to all these wild swings[5] of US policy was that, whichever party was in office in Washington, the US Government acted—at any rate so far as these swings were concerned—not in the manner of the senior ally that had been typical of the years on which this book is focused, nor even as it did for the most part during the twelve years that followed, but from the position that became publicly demonstrated for the first time in October 1962: the lonely eminence of a superpower.

As the full significance of US superpower became increasingly apparent, so too did that of Britain's decline, in relation not only to the United States, but also to its partners in the European Economic Community, which it succeeded in joining, at the third attempt, in 1973. Today it requires almost an effort of will to recall that in the early weeks of 1947 (dark though they were in Britain, literally as well as metaphorically), the British Empire was still at the largest territorial extent ever reached in its entire history; the sterling area financial system was intact; the size of the British and US armies was almost identical; and in June of the same year some of Whitehall's most senior officials were even ready to question, at any rate for the sake of (secret, bilateral) Anglo-American argument, whether British interests might be better served if Britain were to stay outside the Marshall Plan. In the pivotal post-war year, 1950, British GNP was still the second largest in the world, exceeded (albeit by a large margin) only by that of the United States, and having only recently ceased to equal that of France and the Federal Republic of Germany combined. Eighteen years later, Federal German GNP had overtaken that of Britain, whose GNP now exceeded that of France only by a whisker. Today, per capita income in Britain is comparable with that of Italy.

This remorseless decline, whose pace has been slowed down, not yet reversed, by the policies of Margaret Thatcher's two Administrations, is all the more striking in that Britain ended the Second World War with an advanced technological base (for example, atomic, electronic, and aviation-related) and that since the late 1970s British energy resources have become the richest in Western Europe. Indeed, Britain is now the world's fifth largest producer of oil.[6] To say that, during the world's twenty golden years of expansion from 1950–70, Britain's rate of economic growth was between a half and three-fifths of that of its industrialized competitors is only to express in different terms what has been called the British disease. Its much debated causes are not directly relevant to this chapter, although they form a significant element among the underlying post-war trends of which a contemporary Anglo-American balance sheet must take account.

So far as the relationship between the United States and Britain is concerned, the most obvious economic consequence of British industrial decline has been that, whereas at the end of the Second World War it was still legitimate to draw a broad comparison between the American and British economies, today (save in certain special sectors)* such a comparison has little or no meaning. Not that American resources have proved to be infinite, as they had seemed to be during the war and for most of the period covered in this book. While Britain lurched from one economic crisis to another in the later 1960s and in the 1970s—devaluing the pound sterling still further and finally abandoning its strategic role 'East of Suez' on the way—the United States, which had enjoyed roughly half the world's GNP after the Second World War, generated rather less than one quarter thirty years later.

The Vietnam War (which Lyndon Johnson sought to combine with the establishment of the Great Society) demonstrated the limits not only of the US Government's conventional military power, but also of its financial resources. And the 1980s have witnessed the extraordinary phenomenon of a vast US Federal deficit being financed by an unprecedented inflow of foreign capital. They have also demonstrated a trend which had begun in the previous decade: that of major regional and ethnic shifts within the United States, which have had the effect of reducing the influence of the East Coast (the traditional home of the majority of Britain's interlocutors in the United States and of some of the greatest practitioners of the Anglo-American relationship).†

In a lecture delivered at Chatham House in May 1982, Henry Kissinger observed:

That two countries with such divergent traditions should form a durable

---

* Of which the City of London's financial services are perhaps the best example. Since the City's discovery of the Eurocurrency market in the 1960s, it has hardly looked back.

† However, Ronald Reagan, whose respect for Margaret Thatcher is a matter of public record, is from California.

partnership is remarkable in itself. The period of the close Anglo-American 'special relationship', the object of such nostalgia today, were also times of occasional mutual exasperation.[7]

The principal discords that gave rise to this exasperation on both sides of the Atlantic have been discussed in Part II of this book. What is perhaps still more remarkable is the fact that today—long after the resolution of those post-war discords—at a time when the asymmetry between US and British power and influence has increased to a point which few people on either side of the Atlantic would have prophesied forty years ago, none the less enough of the Anglo-American relationship has survived for it to be the object of continuous analysis and debate. Part of the reason for this—to quote another passage from Kissinger's lecture—is the ease and informality with which the two governments grew accustomed to doing business with each other: 'a source of wonder—and no little resentment—to third countries'. Only during the brief experiment of Edward Heath's administration was an attempt made in London to give European interests, and hence British consultation with European governments, at least equality with, if not priority over, traditional Anglo-American imperatives (an experiment reversed by subsequent British governments, both Labour and Conservative). On the US side, an example of continued confidence in British judgement is the fact that the US-Soviet Agreement on the Prevention of Nuclear War (signed during Leonid Brezhnev's visit to Washington in June 1973) was largely drafted in the Foreign Office.[8]

Instances such as these, however striking, are not enough to reverse the basic hard facts. However good or bad US-Soviet relations may be at any given moment, the superpower relationship is of its very nature unique. Qualitatively different from the relationship of each of the superpowers with any other country, it over-arches their policies towards the international community as a whole. And—in parenthesis—if there is one country in the world today whose unique relationship with the United States extends across the board (politico-military-economic), it is surely Israel. This said, even within NATO Britain is no longer the country most important to the United States. The British conventional contribution to the NATO order of battle has retained great significance and—thanks in part to American technology—Britain is still a strategic nuclear power. In the US perspective, however, the major European member of the North Atlantic Alliance has for long been not Britain, but the Federal Republic of Germany. It is this country, lying at the epicentre of European politics, that is crucial to the whole complex of East–West issues, for many reasons: its geographical position, its industrial strength, the size of its army, and its special relationship with the German Democratic Republic.

In the economic sphere, in spite of the volume of trade between the United States and Britain (vastly increased by comparison with the inter-war period) and the massive size of each country's direct and portfolio investment in the other, it is Japan which now looms largest, in the economic sense, for the United States.* Only in the realm of ideas can Britain, despite its decline in other fields, be said to have held its own. This is indeed an area of the utmost importance. At governmental level, two sets of professional eyes assessing the same flow of shared intelligence, derived from all sources, should still be better than one, however great the difference in size and resources between the two intelligence communities. More broadly, the same ease and informality, not to mention a common language, that has pervaded the relationship between Washington and Whitehall is also a source of vitality in the powerful conceptual flow in the academic field that has developed in both directions across the Atlantic since the war. Even here, however, there are warning signs already evident: to cite only one British example,[9] the restriction of governmental funds for basic scientific (including medical) research.

Finally, full account must be taken of the human factor. This is no longer as potent in either country as it was during the years described in this book. Nevertheless, even at a time when the appearance of British news in the American media is often prompted by events in Britain, or British behaviour abroad,[10] of a kind which in Britain itself most people would rather forget, in the United States the human factor still operates to British advantage. The most recent demonstration of the depth of the reservoir of anglophile feeling in the United States has been the American popular reaction to the Falklands War in 1982. At any rate at its onset, the balance of US national interest in this conflict was by no means self-evident (US interests in the Latin American sub-continent weighed against US relations with Britain). But as the crisis developed, the aid given by the United States to the British task force was unstinting; and it was critically important for the eventual outcome of the Falklands campaign.[11]

THE RELATIONSHIP IN RETROSPECT

In short, the Anglo-American relationship still exists, albeit in an attenuated form when measured against the standard of the vigorous exchanges, both co-operative and combative, that characterized relations between the United States and Britain during the period that

* The influx of Japanese capital into the United States made the largest single foreign contribution to financing the Federal deficit in 1985.

has formed the focus of this book. A balance sheet, therefore, calls for an exercise of contemporary judgement as well as of historical assessment. From the early 1960s onwards, the verdict becomes increasingly an open one, always provided that the historian scrupulously follows Buchan's example in specifically discarding the commonplaces of Anglo-American oratory. For the 1950s, extended into the following decade up to the point at which it was left at the end of Part V, the analogy of the curate's egg,* with 1956 standing out as a particularly rotten year, offers the most objective verdict. So far as the post-war quinquennium is concerned, it is a fact that repeated attempts made by officials to define the relationship between the two countries and to set out its objectives, met with a firm refusal by their Foreign Ministers even to circulate to ministerial colleagues, let alone to publish, the conclusions reached by their advisers: a fact that is all the more striking given that the process of decision in the field of foreign policy was then concentrated in the hands of a small group of men, both in Washington and in London, most of whom either already knew, or got to know, each other very well indeed. In the United States the group of decision-makers was more broadly based than in Britain; but even within the American group, the closeness of the links between, for example, Bruce, Douglas, Harriman, and McCloy was a significant factor.

The most well known of these attempts at definition—because Acheson described its abortive outcome in his memoirs—was made in the spring of 1950. The 'wretched paper' to which Acheson took such strong objection, is presumably the State Department's Secret 'Policy Statement—United Kingdom', dated June 7 1950.[12] Nineteen pages long, it shows greater understanding of the British dilemma at that time than any other US document of the period; it even describes it as 'basic' US policy to consider the United Kingdom as a 'partner', as well as the United States' 'most important ally', in American 'efforts to develop political and economic stability in the free world'. No doubt this was one of the reasons why Acheson ordered that the paper should be destroyed; in any case by 7 June it was in several crucial respects already out of date.[13]

Two earlier attempts, both made by British officials, are less generally known, because they came to light later, in the British Public Record Office. Of these two memoranda (both prepared in the Foreign Office), the first, although it contains some sentences which sound

* This expression—meaning 'good in parts'—is familiar to most British readers, but is hardly known in America, because it was introduced into the language by an ancient edition of *Punch*. The illustration showed a nervous curate at his bishop's breakfast table. Asked whether he likes the (doubtful) egg, with which he has just been served, he replied: 'Parts of it are excellent'.

grotesque today, has the merit of having identified as early as 1944 what became one of the greatest difficulties in the management of Anglo-American relationship during the early post-war years. In the words of this memorandum, the 'special quality' of the 'connexion' between the two nations 'brings with it no automatic guarantee of immediate unity or solidarity . . . on the contrary, the complexity and all-pervading quality of the relationship makes it, on the national plane, a source of endless irritation'. If, however, the British went about 'their business in the right way', the memorandum prophesied:

we can help to steer this great unwieldy barge, the United States of America, into the right harbour. If we don't, it is likely to continue to wallow in the ocean, an isolated menace to navigation.[14]

When this paper, entitled 'The Essentials of an American Policy', was submitted (with the support of his ministerial deputy) to Eden in 1944, he simply minuted: 'I frankly doubt whether the circulation of this paper would be politic now'. Five years later, confronted with a memorandum on 'Anglo-American Relations: Present and Future', Bevin wrote more succinctly: 'I will not circulate this'.[15]

Grandiloquent definitions of the Anglo-American relationship belong, for the most part, to later years, when its global importance had diminished; indeed the height of the rhetoric may be said to have increased with the passage of time in inverse proportion of the real content of the relationship. By the end of the post-war quinquennium the Great Republic had publicly assumed its global mission. Britain had not renounced its claim to the status of a great power—its empire and its imperial position were still largely intact— but it had accepted the role of principal ally of the United States. In the course of this remarkable evolution of their relationship, the two countries' greatest shared achievement was the remaking of Western Europe, which brought them together again from 1947 onwards. Yet this was marred by the British refusal to assume the new European role that at first formed an essential part of the American grand design for the political as well as for the economic recovery of Europe. By contrast, the British regarded this role as the least important of the 'three circles' of their 'life and power',* and as one which, at any rate from 1950 onwards, must yield pride of place to the other two. On this issue, which—British membership of the EEC notwithstanding—is by no means an academic question in British political life even thirty-five years later, there was a paradoxical reversal of traditional philosophi-

---

* The concept of the three concentric circles—the Anglo-American relationship, the British Commonwealth, and British relations with Western Europe—seems to have originated in a remark made by Churchill to Franks.[16]

cal attitudes between the United States and Britain. It was the Americans who joined the French in supporting Monnet's pragmatism (an empirical approach to problems, which Monnet himself had learned to a large extent from his own Anglo-Saxon experience in the two world wars),[17] whereas the British took refuge in a kind of Cartesian fortress. The concept of what the British perceived as the new 'Atlantic' community was paramount for them. Western European continental countries might follow the American advice to 'integrate' their economies if they so chose. But the lessons of the British historical experience still seemed in London to dictate that Britain should, together with the United States, share the leadership of the broader Atlantic community, within whose framework all three of the British 'circles' of power might be secured and reconciled.

Except for Bevin's sortie in 1948 (assessed in Chapter 15), the British stayed within the walls of this conceptual fortress for the first fifteen years after the Second World War, regardless of the holes in its logical defences. Whereas the Atlantic model was precisely expressed, in terms of the national security of the members of the North Atlantic Alliance, once the scope and purpose of this alliance had been laid down in the North Atlantic Treaty of 1949, what it might mean in broader political terms was never clearly defined. Nor could it be. If what was intended was the 'western world', this included countries who were not then members of the North Atlantic Alliance; nor are they today. Moreover, since what Europe needed most in its post-war crisis was a firm sense of political identity, in what meaningful way could Mediterranean countries like Italy, Greece, and Turkey be regarded as 'Atlantic'? Outside the field of defence, to which the term 'Atlantic' had first been applied in 1948–9, the Atlantic community was essentially an artificial concept, which for a time enabled the British to close their eyes to the European choice confronting them and to seek instead to derive the maximum advantage from their relationship with the United States.*

The Americans were unable to dissuade the British from committing this first serious mistake[18] after the war; nor did their repeated efforts to 'contribute to the British learning process' over Europe meet with any success. Indeed, they may even have had the unintended effect of making the British more obstinate. But it remains to the credit of the makers of US policy in the early post-war years that they recognized British policy towards Western Europe, and the British attempt to make the concept of the Atlantic community bear far more political weight than it could reasonably carry, for what they both unfortunat-

---

* The Commonwealth 'circle' was also a powerful constraint on British policy in this respect during the early post-war years. But its force lessened as the process of decolonization accelerated.

ely were: fudged thinking, for which Britain—and Europe—has since had to pay a heavy price.

As for the Anglo-American relationship outside Europe, it will be recalled that this book has excluded from its scope the greatest American and British successes in foreign policy in these areas during the post-war quinquennium—in Japan and the Indian sub-continent respectively—because in the first case the British contribution was minimal, just as the American contribution was minimal in the second. (The fact that the United States had long supported Indian independence is largely irrelevant; the British transferred power in the sub-continent in 1947 for other reasons.) For the purposes of this study, therefore, the assessment is restricted to the two areas which an earlier chapter described as the Achilles heel of each country's foreign policy: for the United States, China, and for Britain, the Middle East. If the second of the three questions is applied to these two areas, the answers cannot be favourable.

Although many other factors have contributed to the Balkanization of the Middle Eastern region since 1945, the initial failure in this area of policy after the Second World War was—as has been argued in an earlier chapter—Anglo-American as well as British. As I have also suggested, it was over China that the most serious mistake during this period was made—by the United States, whose post-war policy towards that country was at first ambivalent and later grievously wrong, whereas British policy, though half-hearted, was right. The responsibility for this great error was American, like the responsibility for the United States' subsequent South-East Asian adventure. At any rate, up to the turn of the year 1950/1, the British made an attempt to represent the voice of reason. Yet, even without the official documents at his disposal, the British historian is left with the impression that, as the decade unfolded—and still more in the 1960s—the ease of British access in Washington could and should have been used more often and more effectively in an attempt to restore realistic dimensions to the American 'distorted perspective' in the Far East and South-East Asia. Instead, the United States took a quarter of a century to complete its 'learning process' in this field largely on its own.

Meanwhile, the British stood by, relying on their belief in the overall beneficence of the Anglo-American relationship to enable them to assume that the replacement of British by United States world power— almost a 'moral duty' for the United States—was not against their own interests.[19] Irrespective of its effect on British interests, what it did damage, as time went by, was intellectual rigour in Whitehall. To this consequence there have always been—and are—honourable exceptions. But even before the Anglo-American mould had set, there was already a British tendency* to react to a problem by minuting a file in the first instance with the two words, 'Consult Washington'. Understandable perhaps in the hectic years immediately after the Second World War, this later evolved into a way of thinking, or rather not thinking a problem through. Once the mould had set completely, a natural consequence in Britain was the growth of a large body† of officials, civilian and military, for whom the ease and informality of doing transatlantic business became not just the professional advantage that Kissinger cited in his 1982 lecture, but also what would in French be called a professional *déformation*: an outlook on the world that made it increasingly difficult for them to think in any other way,[20] and in particular to do business effectively with other countries whose leaders and officials were often neither easy nor informal. France is the prime example; but it was not until twenty-three years after the end of the Second World War that Britain sent a fluent Russian-speaker as its ambassador to Moscow.[21]

For the junior partner in the Anglo-American relationship, the development over the years of what would in American terms be called a bureaucratic lobby, well organized and well intentioned, but avoiding the consideration of any alternative option that might conflict with or call in question the conventional Anglo-American wisdom, gradually contributed—again making use of Buchan's definition—to a 'distortion of perspective'. Of this distortion the British architect of post-war Anglo-American understanding, Ernest Bevin, had no inkling. Nor would he have had much sympathy with it. (On the contrary, when Bevin's biographer asked Attlee what Bevin's greatest contribution as Foreign Secretary had been, his reply was: 'Standing up to the Americans'.)[22] This was a distortion to which the senior partner was far less susceptible, but which, in the longer term, did not serve US interests well either. The opposite side of the coin of ease and informality is a cosy mutual admiration, which in the long run can benefit neither partner in a mature political relationship.

Before the contemporary Anglo-American relationship is considered, the two major functional areas of the post-war relationship (financial and nuclear) must be assessed. In the initial phases of both the strong element of American chauvinism will have been evident from the account in Part III of this book. In neither field, however, can a simple, clear-cut answer be given. On the one hand, the US decision

* In his early years in the Foreign Office, the author was doubtless as prone to this habit as his colleagues were.

† Counter-balanced to a certain extent, since 1973, by the (slower) growth of the number of British officials whose international experience has been primarily European. But a Martian observer of the House of Commons today would still have difficulty in deducing that this more recent growth had been in progress for the past thirteen years.

to terminate Lend-Lease one week after the Japanese war ended, followed by US insistence on the inclusion of the twelve-month sterling convertibility clause in the subsequent Financial Agreement with Britain, inflicted severe damage on the immediate post-war relationship between the two countries. On the other hand, the brilliant simplicity of the Marshall Plan, together with the swiftness of the British response, turned the tide in their post-war relationship, as it did for the recovery of Western Europe as a whole. It did not, however, resolve the Anglo-American conflict of views on how the new international economic order should be managed, nor—although the British at first thought otherwise—did the Anglo-American financial discussions of September 1949.[23] Each side was defending what it perceived to be its vital national economic interests. In spite of everything, however, they did in the end realize what Keynes, in his last great speech in the House of Lords,[24] called the 'world's best hope, an Anglo-American understanding' in the financial field. After a shaky start, the system first designed at Bretton Woods finally prevailed. Future historians may decide that this was the greatest Anglo-American achievement in the years following the Second World War, just as the greatest failure was not to seize the opportunity for a fresh attempt at East–West understanding in the years immediately following Stalin's death.

To the nuclear question, Buchan provided the main answer himself: during the four years' 'complex and sorry muddle' of the early post-war atomic dialogue between the two governments, no sustained Anglo-American attempt was ever made to address the most crucial post-war atomic problem, that of international control.[25] In the bilateral context, however, once the McMahon Act had been amended, the keystone to the Anglo-American arch was at last put in place, by the terms of the Anglo-American Atomic Agreement of July 1958. Taken together with the intelligence interchange (already in force) and the 1962 Nassau Agreement on Polaris (subsequently extended by the Chevaline[26] and Trident agreements) between the US and British governments, this 1958 agreement was, as has been noted in the preceding chapter, 'the great prize' for Macmillan at the time. Up to the end of his premiership the British nuclear deterrent, carried in the RAF's fleet of some 180 medium range bombers, remained formidable. As its deterrent value waned during the rest of the decade, this force was replaced by the Royal Navy's fleet of four submarines equipped with the Polaris missiles that had been the subject of the Nassau Agreement. Can all this fairly be regarded—as it still is by many people in Britain today—as part of the British 're-adjustment'? Or did an American historian of the post-war Anglo-American nuclear relationship come nearer to the mark when he argued, early in the 1970s, that the end result of the whole British attempt to keep up with the American Joneses in this field of military technology was a 'psychology of dependence' on the United States? All the evidence that is public today suggests that he was right.[27]

THE RELATIONSHIP TODAY

Today, as in 1962, the question of a long-term replacement for the British strategic nuclear deterrent has once again become a central point of choice in Britain. It is also an issue of intense party political controversy. Whether or not the verdict suggested at the end of the preceding paragraph is correct, at the time of writing it seems virtually certain that the Trident II agreement with the United States will not survive unless the next British Parliament is returned with a clear Conservative Party majority. Nor is the current debate on this question by any means dictated only by party political allegiance. Within the past year both Britain's best informed newspapers (neither of them by any stretch of the imagination supporters of the Labour Party) have taken up the cudgels against Trident: *The Economist*[28] on grounds of cost, and the *Financial Times* on broader grounds of policy, arguing that 'by its treatment of the opportunity cost of the Trident II project', the British Government will 'show whether or not it is seriously committed . . . to the strengthening of the European pillar of the Alliance and the enhancement of defence cooperation in Europe'.[29]

The relationship between the United States and Britain has survived many slings and arrows over the past forty years, among which only the Suez crisis was powerful enough to break the Anglo-American mould, and that only for a matter of months. Suez apart, there have been articles by writers of repute, with access to unimpeachable sources on both sides of the Atlantic, under titles like 'The Relationship in Disarray', for as far back as any observer of the Anglo-American scene can remember. It may therefore be that from the present erosion of transatlantic consensus on the fundamental concepts of defence, there will eventually emerge a healthier transatlantic relationship, and within whatever new overall framework may be agreed, a healthier Anglo-American relationship as well. That this is the outcome devoutly to be desired, goes without saying. But—judging by developments both in the United States and in Britain during the current decade so far—it is as well not to be too sanguine.[30]

In one sense, the Anglo-American relationship is today more broadly—and, it is therefore to be hoped, more soundly—based than it was during the years 1945–50. The financial, commercial, cultural, and scientific bonds that today link the United States and Britain, to their mutual advantage, do not depend on the residue of the 'special' relationship. They will surely endure, however Britain's governmental relationship with the United States and with its European partners may eventually be redefined. On the other hand, the Anglo-American governmental relationship, by comparison with what it consisted of thirty-five years ago or even twenty-five years ago—has become reduced to a hard core, consisting of three main elements: nuclear technology, intelligence, and those aspects of defence policy with which these two elements are linked.[31]

Given the extent of Britain's decline in other respects and the consequent growth of the imbalance in the Anglo-American relationship overall, sooner or later both countries will, I suggest, be well advised to consider how far the conditions of the late twentieth century still warrant the present degree of British dependence on the United States in this restricted, though critically important, field. Is such dependence—for the Macmillanite concept of 'interdependence' can no longer seriously be sustained—now healthy for either country? Are there really no other options, which might serve the interests of both countries better? Whose interest is served by Whitehall memoranda* beginning with words like 'As the US President has recently observed . . .' or including phrases such as 'The Americans might be embarrassed if we were to . . .'? Has not the time arrived for all the Europeans, particularly the British, to grow up—in general, but above all in the field of defence?[32]

The alternative policy currently advocated for Britain by the Labour Party[33] will certainly win Britain no friends in the United States. But is Labour's the true alternative to the conventional wisdom of past years in Whitehall? Galling though it would be for the British to admit this after so long, may not the American interest be at least as well, if not better, served by the continued existence of a fully independent nuclear deterrent, in the hands of a French government whose President's European credentials are so good that he was able two years ago to address the German Bundestag in person, in order to argue the case for the deployment of Pershing II and cruise missiles in the Federal Republic?[34] Would not Britain be a more useful ally for the United States if it were at last to become whole-heartedly committed to Europe? And would not a western alliance resting on two pillars of more or less equal strength and length, become, in the longer term, a more effective instrument than NATO in its current state?

My answer to these questions has been outlined elsewhere[35] and need

---

* Both are purely imaginary examples—'no resemblance to any living person is intended'.

not be repeated here. As things stand at the time of writing, the parallel between the options that face Britain now and those of 1962 is a warning that can be ignored only at British peril; and if it is ignored, it will not serve US interests in the longer term. In 1962, Macmillan sought and obtained Kennedy's agreement at Nassau to help him in squaring a circle which in reality could not be squared at all. In 1985, there is a danger—in relation not only to the Trident project, but also to a host* of other issues—that Britain may be encouraged by the continued existence of its residual 'special' links with the US Government, to make almost exactly the same mistakes as it did thirty-odd years ago. In 1962, Macmillan had the excuse that he felt himself up against a wall; he was an old man in a hurry. Today we have a little time, not to break the mould in the manner of 1956, but to reset it, in a new form, better suited to the interests both of the United States and Britain in the final years of the twentieth century.

---

* One of which, since this chapter was written, severely buffeted the British Government, causing the resignation of two senior Cabinet Ministers in January 1986: the Westland Affair. Three months later the US air strike against Libya and the British Government's decision to allow American air bases in Britain to be used for this purpose has divided British public opinion. In the view of one usually dispassionate observer of the transatlantic scene, 'nothing has so vehemently separated America from Europe since 1945.'[36]

1945: a selective chronology

11 January	US forces land on Luzon
17 January	Soviet Army captures Warsaw
31 January	Soviet Army crosses Oder
30 January– 2 February	Malta Conference
4–11 February	**Yalta Conference**
11 February	Soviet Army captures Budapest
6 March	Allied forces capture Cologne
1 April	US forces invade Okinawa
5 April	Soviet Government denounces Soviet-Japanese neutrality pact
11 April	Soviet forces capture Vienna
12 April	Death of Roosevelt; Truman becomes US President
19 April	Soviet Government signs Treaty of Friendship with Polish Provisional (Warsaw) Government
25 April	Soviet and US forces meet on Elbe
29 April	German forces in Italy capitulate
30 April	Death of Hitler
2 May	Soviet forces capture Berlin
3 May	British forces capture Rangoon
8/9 May	End of war in Europe
12 May	Soviet forces enter Prague
9 June	Yugoslav forces withdraw from Trieste
26 June	United Nations Charter signed at San Francisco
5 July	US and Britain announce recognition of Polish (Warsaw) Government
6 July	British general election
16 July	Atomic weapon successfully tested at Alamogordo, New Mexico
16 July– 2 August	**Potsdam Conference**
26 July	Churchill defeated in British general election; Attlee becomes Prime Minister
6 August	Atomic bomb dropped on Hiroshima

Appendix 1

8 August	Soviet Union declares war on Japan; invades Manchuria
9 August	Atomic bomb dropped on Nagasaki
14 August	Japan capitulates; end of Second World War
14 August	Sino-Soviet Treaty of Friendship signed
2 September	Formal Japanese surrender in Tokyo Bay
11 September– 2 October	**First meeting of Council of Foreign Ministers, London**
12 September	Formal Japanese surrender in Singapore
13 September	British forces occupy Saigon
16 September	Formal surrender of Hong Kong to British forces
September	50,000 US marines landed in North China; US airlift of Chinese Nationalist troops begins; US forces occupy South Korea (38th parallel divides US and Soviet occupation zones)
10–16 November	Truman-Attlee meeting, Washington
27 November	Truman appoints Marshall as his Special Representative in China (Chinese civil war imminent)
12–31 December	**Meeting of Soviet, US, and British Foreign Ministers in Moscow**

Section III of the Potsdam Protocol: Reparations from Germany

1. Reparation claims of the U.S.S.R. shall be met by removals from the zone of Germany occupied by the U.S.S.R., and from appropriate German external assets.

2. The U.S.S.R. undertakes to settle the reparation claims of Poland from its own share of reparations.

3. The reparations claims of the United States, the United Kingdom and other countries entitled to reparations shall be met from the Western Zones and from appropriate German external assets.

4. In addition to the reparations to be taken by the U.S.S.R. from its own zone of occupation, the U.S.S.R. shall receive additionally from the Western Zones:

   (a) 15 per cent of such usable and complete industrial capital equipment, in the first place from the metallurgical, chemical and machine manufacturing industries, as is unnecessary for the German peace economy and should be removed from the Western Zones of Germany, in exchange for an equivalent value of food, coal, potash, zinc, timber, clay products, petroleum products, and such other commodities as may be agreed upon.

   (b) 10 per cent of such industrial capital equipment as is unnecessary for the German peace economy and should be removed from the Western Zones, to be transferred to the Soviet Government on reparations account without payment or exchange of any kind in return.

Removals of equipment as provided in (a) and (b) above shall be made simultaneously.

5. The amount of equipment to be removed from the Western Zones on account of reparations must be determined within six months from now at the latest.

6. Removals of industrial capital equipment shall begin as soon as possible and shall be completed within two years from the determination specified in paragraph 5. The delivery of products covered by 4(a) above shall begin as soon as possible and shall be made by the U.S.S.R. in agreed instalments within five years of the date hereof. The determination of the amount and

character of the industrial capital equipment unnecessary for the German peace economy and therefore available for reparations shall be made by the Control Council under policies fixed by the Allied Commission on reparations, with the participation of France, subject to the final approval of the Zone Commander in the Zone from which the equipment is to be removed.

7. Prior to the fixing of the total amount of such equipment subject to removal, advance deliveries shall be made in respect of such equipment as will be determined to be eligible for delivery in accordance with the procedure set forth in the last sentence of paragraph 6.

8. The Soviet Government renounces all claims in respect of reparations to shares of German enterprises which are located in the Western Zones of occupation in Germany as well as to German foreign assets in all countries except those specified in paragraph 9 below.

9. The Governments of the United Kingdom and United States renounce all claims in respect of reparations to shares of German enterprises which are located in the Eastern Zone of occupation in Germany, as well as to German foreign assets in Bulgaria, Finland, Hungary, Roumania and Eastern Austria.

10. The Soviet Government makes no claims to gold captured by the Allied troops in Germany.

The eight articles of the *Modus Vivendi*, 7 January 1948

1. All agreements between the three governments or any two of them in the field of atomic energy shall be regarded as null and of no effect, with the following exceptions:

    (a) The Patent Memorandum of 1 October 1943 as modified by subsequent agreement on 19 September 1944 and 8 March 1945.

    (b) The Agreement and Declaration of Trust dated 13 June 1944.

    (c) The exchange of letters between the Acting Secretary of State and the British Ambassador of 19 and 24 September 1945, concerning Brazil.

    (d) The agreed public Declaration by the President of the United States, the Prime Minister of the United Kingdom, and the Prime Minister of Canada of 15 November 1945.

2. The Combined Policy Committee, already established, and subject to the control of the three governments, shall continue as an organ for dealing with atomic energy problems of common concern. The Committee shall consist of three representatives of the United States, two of the United Kingdom, and one of Canada, unless otherwise agreed.

3. The Committee shall *inter alia*:

    (a) Allocate raw materials in accordance with such principles as may be determined from time to time by the Committee, taking into account all supplies available to any of the three governments.

    (b) Consider general questions arising with respect to co-operation among the three governments.

    (c) Supervise the operations and policies of the Combined Development Agency referred to in paragraph 4 below.

4. The Combined Development Trust, created on the thirteenth of June 1944 by the Agreement and Declaration of Trust signed by President Roosevelt and Mr Winston Churchill, shall continue in effect except that it shall henceforward be known as the Combined Development Agency. Of the six persons provided for in Clause 1(2) of the Declaration of Trust, three shall represent the United States, two the United Kingdom, and one Canada.

5. The United States, the United Kingdom and Canada will, within the limits of their respective constitutions and statutes, use every effort to

acquire control of supplies of uranium and thorium situated within their respective territories. The United Kingdom will, in so far as need exists, communicate with the governments of the British Commonwealth for the purpose of ensuring that such governments exercise control of supplies of uranium and thorium situated in their respective territories. The United Kingdom will consult with the Commonwealth Governments concerned with a view to encouraging the greatest possible production of uranium and thorium in the British Commonwealth, and with a view to ensuring that as large a quantity as possible of such supplies is made available to the United States, United Kingdom and Canada.

6. It is recognised that there are areas of information and experience in which co-operation would be mutually beneficial to the three countries. They will therefore co-operate in respect of such areas as may from time to time be agreed upon by the CPC and in so far as this is permitted by the laws of the respective countries.

7. In the interests of mutual security, classified information in the field of atomic energy will not be disclosed to other governments or authorities or persons in other countries without due prior consultation.

8. Policy with respect to international control of atomic energy remains that set forth in the Three-Nations Agreed Declaration of 15 November 1945. Whenever a plan for the international control of atomic energy with appropriate safeguards which would ensure use of atomic energy for peaceful purposes only shall be agreed upon, and shall become fully effective, the relationship of these countries in atomic energy matters will have to be reconsidered in the light thereof.

Selective Middle Eastern Chronology, 1945–49

31 May	British troops intervene against French in Syria
24 July 1945	Truman's first approach to Prime Minister over Palestine
31 August	Truman's request for admission of 100,000 Jews into Palestine
13 November 1945	Anglo-American Committee of Enquiry into Palestine announced
20 December 1945	Egyptian Government demands revision of 1936 Anglo-Egyptian Treaty
22 March 1946	Anglo-Transjordanian Treaty signed
March–April 1946	Azerbaijan Crisis
15 April 1946	Withdrawal of French and British troops from Syria
30 April 1946	Anglo-American Committee's report on Palestine
7 May 1946	British intention to withdraw troops from Egypt announced
4 July 1946	Cairo evacuated by British troops
22 July 1946	Irgun blows up King David Hotel, Jerusalem
25 July 1946	Morrison–Grady Plan leaked to the press
4 October 1946	Truman's Yom Kippur statement
17–25 October 1946	Negotiations between Bevin and Egyptian Prime Minister (Ismail Sidky) concluded
9 December 1946	Sidky resigns
January–February 1947	Palestine Conference in London
18 February 1947	British referral of Palestine issue to UN announced
13 May 1947	UN Special Commission on Palestine formed
11 July 1947	Egyptian Government submits Anglo-Egyptian Treaty issue to UN
31 August 1947	UNSCOP majority favours partition of Palestine
29 November 1947	UN General Assembly votes for partition of Palestine
15 January 1947	Anglo-Iraqi (abortive) Treaty signed
15 March 1947	New Anglo-Transjordanian Treaty signed (Transjordan renamed Jordan on 21 June 1949)
9 April 1948	Massacre of Deir Yassin (Palestine)
14 May 1948	British Mandate in Palestine ends; Israel proclaims independence; recognized *de facto* by US; first Arab-Israeli war begins

17 May 1948	Israel recognized by USSR
17 September 1948	UN Palestine Mediator, Bernadotte, assassinated
25 November 1948	Truman backs Israeli retention of the Negev
December 1948–January 1949	Egyptian Army defeated by Israel
29 January 1949	Britain recognizes Israel *de facto*
31 January 1949	US recognizes Israel *de jure*
24 February–20 July 1949	Armistices between Israel and (successively) Egypt, Lebanon, Jordan, and Syria
11 May 1949	Israel becomes member of UN

Selective Chronology, China, 1945-49

1945

11 February	Secret Yalta Agreement on the Far East
9 August	Soviet Union invades Manchuria
14 August	Sino-Soviet Treaty of Friendship and Alliance; Surrender of Japan
September onwards	US marines land in north China; US transports Chinese Nationalist forces northwards
27 November	General Marshall appointed Special Representative in China of US President (Chinese civil war imminent)

1946

10 January	Chinese cease-fire agreed
April	Soviet forces withdraw from Manchuria; Chinese civil war in Manchuria follows
6-30 June	Truce in Manchuria
July	Nationalist offensive begins
2 November	Sino-Soviet Treaty of Friendship, Commerce, and Navigation
November	Chou En-lai leaves Nanking, returns to Yenan

1947

6 January	Marshall's recall announced
19 March	Yenan occupied by Nationalist troops
15 August	Syngman Rhee inaugurated as first President of Republic of Korea
19 September	General Wedemeyer's report submitted to President Truman
September	US refers Korean question to UN
By December	Nationalist forces in Manchuria cut off

1948 Communist counter-offensive in China

1949

31 January	Peking occupied by Communist forces
25 May	Shanghai occupied by Communist forces
29 June	US occupation forces leave Korea

5 August	China White Paper published in Washington
1 October	People's Republic of China proclaimed in Peking; Chiang Kai-shek flees to Formosa (Taiwan) in December
3 October	Soviet Union recognizes new Chinese Government
15 December	British Government's decision to recognize Chinese Communist government (announced on 5 January 1950)
16 December	Mao Tse-tung arrives in Moscow, to negotiate new Sino-Soviet Treaty

NOTE

For the reason given in the Preface, the Wade–Giles system of transliteration of Chinese has been used in this book. The pinyin equivalents of names of people and places are given below.

Wade–Giles	Pinyin
Chiang Kai-shek	Jiang Jieshi
Chou En-lai	Zhou Enlai
Mao Tse-tung	Mao Zedong
Canton	Guangzhou
Chungking	Chongqing
Kwangtung	Guangdong
Nanking	Nanjing
Peking	Beijing
Tientsin	Tianjin
Yenan	Yan'an

Selective Chronology of the Korean War

1950

24 June*	North Korean forces attack South Korea
26 June	South Korean Cabinet leaves Seoul for Pusan
28 June	News of fall of Seoul reaches Washington
29 June	MacArthur inspects the front in Korea; on return, asks for US ground troops to be committed
1 July	US task force leaves Japan for Korea
31 July	MacArthur visits Chiang in Formosa
1 August	All UN forces ordered to withdraw to the Pusan Perimeter
29 August	First British brigade (27) reaches Pusan
15 September	US Tenth Corps lands at Inchon
15–30 September	North Korean army defeated; retreats north of 38th parallel
7 October	First UN patrols cross parallel
9 October	Invasion of North Korea begins
15 October	Truman and MacArthur confer on Wake Island
19 October	Pyongyang captured
7 November	US aircraft bomb Yalu bridges
24–8 November	MacArthur launches 'Home for Christmas' offensive; Chinese forces counter-attack; retreat of UN forces begins
4 December	Chinese forces recapture Pyongyang
26 December	Chinese forces reach 38th parallel

The Sequel: 1951–3

Having lost Seoul, the US Eighth Army (now commanded by General Matthew Ridgway) began a counter-attack on 25 January 1951; by mid-March the 38th parallel had been reached again. MacArthur was dismissed by Truman on 9 April. On 1 July 1951 armistice negotiations begin, but they lasted for over two years and the war became a stalemate. Having described it as 'the burial ground for twenty thousand American dead',† Eisenhower—as President-elect—flew to Korea in November 1952. After the new US Administration had threatened to widen the war, the Korean Armistice was signed at Panmunjom on 27 July 1953.

* 24 June by US and European time, but 25 June by Korean time.
† The final figures were nearly twice this number.

Extract from the Nassau Agreement: Statement on Nuclear Defense Systems—21 December 1962

... (5) The Prime Minister then turned to the possibility of provision of the Polaris missile to the United Kingdom by the United States. After careful review, the President and the Prime Minister agreed that a decision on Polaris must be considered in the widest context both of the future defense of the Atlantic Alliance and of the safety of the whole Free World. They reached the conclusion that this issue created an opportunity for the development of new and closer arrangements for the organization and control of strategic Western defense and that such arrangements in turn could make a major contribution to political cohesion among the nations of the Alliance.

(6) The Prime Minister suggested and the President agreed, that for the immediate future a start could be made by subscribing to NATO some part of the forces already in existence. This could include allocations from United States Strategic Forces, from United Kingdom Bomber Command, and from tactical nuclear forces now held in Europe. Such forces would be assigned as part of a NATO nuclear force and targeted with NATO plans.

(7) Returning to Polaris the President and the Prime Minister agreed that the purpose of their two governments with respect to the provision of the Polaris missile must be the development of a multilateral NATO nuclear force in the closest consultation with other NATO allies. They will use their best endeavours to this end.

(8) Accordingly, the President and the Prime Minister agreed that the U.S. will make available on a continuing basis Polaris missiles (less warheads) for British submarines.* The U.S. will also study the feasibility of making available certain support facilities for such submarines. The U.K. Government will construct the submarines in which these weapons will be placed and they will also provide the nuclear warheads for the Polaris missiles. British forces developed under this plan will be assigned and targeted in the same way as the forces described in paragraph 6.

These forces, and at least equal U.S. forces, would be made available for inclusion in a NATO multilateral nuclear force. The Prime Minister made it clear that except where Her Majesty's Government may decide that supreme national interests are at stake, these British forces will be used for the purposes of international defense of the Western Alliance in all circumstances ...

---

* The Polaris sales agreement between the US and British governments was signed in Washington on 6 April 1963.

Note on Documentary Sources

By far the most important and the largest published collection of official documents relevant to Anglo-American relations in the post-war quinquennium is the multi-volume *Foreign Relations of the United States*. Even this great collection is not exhaustive; the State Department records in the National Archives are a valuable adjunct.

The first volume of the British counterpart of the *FRUS*, *Documents on British Policy Overseas*, appeared only in 1984; and at the time of writing the series, though impressively produced, does not extend beyond the end of 1945. The principal British collection of primary sources therefore remains that of the Public Record Office. Of these British sources, the most important, for the purposes of this study, are the *Private Papers of Mr Ernest Bevin* (FO 800); the *Papers of the Prime Minister's Office* (PREM); the Cabinet minutes, conclusions, and memoranda (CAB 128 and 129); papers of the Cabinet Atomic Committee (CAB 130); papers of the Cabinet Defence Committee (CAB 131); and—a confusing but rewarding rabbit-warren—the *Foreign Office General Political Papers* (FO 371). There are now more publicly available papers relating to atomic issues of this period than is generally supposed, but frequent reference to the works of the official atomic historians on both sides of the Atlantic is still essential.

For 1945, in Part II some cross-bearings through the Soviet records of the trilateral conferences now published have been taken; and on the critical issue of the Oder–Neisse line through Polish records as well (*Dokumenty i materiały do stosunków polsko-radzieckich*). In general, in spite of what has been said (in the Preface) about the doubtful value of diaries, there are exceptions: for example, those of Edward Halifax and Henry Stimson. And, on both sides of the Atlantic, there is a wealth of material in the press and in the Congressional, Presidential, and Parliamentary records covering these years, on which the historian must draw without, however, allowing such references to swamp his narrative: not always an easy balance to maintain, since there are passages in the *Hansard* debates and in the press conferences recorded verbatim in the *Public Papers of the Presidents of the United States* of this period which are so vivid that they almost speak for themselves.

There is also an even vaster—and constantly growing—mass of secondary material. Some of this material—for example, the memoirs of the principal actors of the period and certain biographies and historical studies written since—is required reading; some is less so, partly because much of it was written before 1981. A cross-section will be found in the Select Bibliography.

There follows here a list of the abbreviations employed in the Notes for sources to which particularly frequent reference is made; and also a list of other documentary sources.

ABBREVIATIONS

Acheson, *Present at the Creation*
Dean Acheson, *Present at the Creation: My Years at the State Department* (Norton, New York, 1969).

Berlin Papers
Foreign Relations of the United States, Diplomatic Papers: The Conference of Berlin (The Potsdam Conference), 1945, 2 vols. (US Government Printing Office, Washington, DC, 1960).

Bullock, *Bevin*
Alan Bullock, *Ernest Bevin: Foreign Secretary, 1945–1951* (Heinemann and Norton, London and New York, 1983 and 1984).

CAB
Cabinet Papers, Minutes, Conclusions, and Confidential Annexes, in the Public Record Office at Kew, Richmond, Surrey.

Churchill
Winston S. Churchill, *The Second World War*, 6 vols. (Cassell and Houghton Mifflin, London and New York, 1948–54 and 1948–53).

CWP
United States relations with China, with special reference to the period 1944–1949, Department of State Publication 3573, Washington, DC, August 1949—generally known thereafter as 'The China White Paper'.

DBPO
Documents on British Policy Overseas, ed. Rohan Butler and others, 1st series, vol. i– , HMSO, London, 1984–

DSDF
General Records of the Department of State, Decimal Files, RG 59, National Archives, Washington, DC.

FO 371
Foreign Office General Political papers in the Public Record Office at Kew, Richmond, Surrey.

FO 800
The Private Papers of Mr Ernest Bevin, in the Public Record Office at Kew, Richmond, Surrey.

FRUS
Foreign Relations of the United States, Diplomatic Papers, 1945– (US Government Printing Office, Washington, DC, 1955–).

Gowing, *Independence and Deterrence*
Margaret M. Gowing, *Independence and Deterrence: Britain and Atomic Energy, 1945–52*, 2 vols. (Macmillan, London, 1974). Since only vol. i is cited in this book, all references are to this volume.

Hewlett and Anderson
Richard G. Hewlett and Oscar E. Anderson, *A History of the United States Atomic Energy Commission, vol. i: The New World, 1939/1946* (Pennsylvania University Press, University Park, Pa., 1962).

Hewlett and Duncan
Richard G. Hewlett and Francis Duncan, *A History of the United States Atomic Energy*

Commission, vol. ii: *Atomic Shield, 1947/1952* (Pennsylvania University Press, University Park, Pa., 1969).

PP: *Harry S. Truman*
   Public Papers of the Presidents of the United States: Harry S. Truman, 1945–53, 8 vols. (US Government Printing Office, Washington, DC, 1961–6).

PREM
   Papers of the Prime Minister's Office in the Public Record Office at Kew, Richmond, Surrey.

SSNMK
   Sovetskii Soyuz na mezhdunarodnykh konferentsiyach perioda Velikoi Otechestvennoi Voiny, 1941–1945gg, Sbornik dokumentov, 6 vols. (Izdatel'stvo politicheskoi literatury, Moscow, 1978–9).

T
   Treasury Papers in the Public Record Office at Kew, Richmond, Surrey.

Truman, *Year of Decisions*
   Harry S. Truman, *Memoirs, vol. i: Year of Decisions 1945* (Doubleday, New York, 1955).

Truman, *Trial and Hope*
   Harry S. Truman, *Memoirs, vol. ii: Years of Trial and Hope, 1946–53* (Doubleday, New York, 1956).

Yalta papers
   Foreign Relations of the United States, Diplomatic papers: The Conferences at Malta and Yalta, 1945 (US Government Printing Office, Washington, DC, 1955).

OTHER DOCUMENTARY SOURCES

Hansard, *Parliamentary Debates*, 1945–1950

Congressional Record, 1945–1950

White Papers
   Among the White Papers cited in this book, by far the richest in the information that it provides is the so-called *US China White paper*, listed as *CWP* in the abbreviations above. The principal British White Papers to which reference is made are the following:
United Nations Monetary and Financial Conference, Bretton Woods, New Hampshire, USA, July 1 to July 22, 1944, Final Act, Cmd. 6546 (HMSO, London, 1944, reprinted 1945).
Financial Agreement Between the Governments of the United States and the United Kingdom, December 6, 1945, Cmd. 6708 (HMSO, London, 1945).
Statistical Material presented during the Washington Negotiations, Cmd. 6705 (HMSO, London, 1945).
Economic Survey for 1947, Cmd. 7046 (HMSO, London, 1947).
Anglo-French Treaty of Alliance and Mutual Assistance, Dunkirk, 4 March 1947, Cmd. 7217 (HMSO, London, 1947).
The Treaty of Economic, Social and Cultural Collaboration and Collective Self-Defence, Brussels, 17 March 1948, Cmd. 7367 (HMSO, London, 1948).
North Atlantic Treaty, 4 April 1949, Treaty Series, No. 56 (1949), Cmd. 7789 (HMSO, London, 1949).
United Kingdom Balance of Payments, 1946 to 1950, Cmd. 8065 (HMSO, London, 1950).

Anglo-French Discussions Regarding French Proposals for the Western European Coal, Iron and Steel Industries, May–June 1950, Cmd. 7970 (HMSO, London, 1950).

United Kingdom Defence. Outline of Future Policy, April 1957, Cmnd. 124 (HMSO, London, 1957).

Agreement with the United States on Cooperation on the Uses of Atomic Energy for Mutual Defence Purposes, Washington, 3 July 1958, Cmnd. 537 (HMSO, London, 1958).

Conference on Security and Cooperation in Europe, Final Act, Cmnd. 6198 (HMSO, London, 1975, repr. 1979).

Notes

PREFACE

1. Andrew Shonfield, in the preface to his *Modern Capitalism* (Oxford University Press, Oxford, 1965).
2. Although the author did not enter the (British) Foreign Service until the end of 1946, while awaiting demobilization from the Army in 1945–6 he served in the British Section of the Political Division of the Allied Commission for Austria (in effect, as Third Secretary in the equivalent of the Vienna Chancery).
3. Certain British papers are still in the 'retained' category.
4. This observation is Anna Akhmatova's, quoted by Isaiah Berlin, *Personal Impressions* (The Hogarth Press, London, 1981), p. 156.
5. The definition of intelligence in *The Diaries of Sir Alexander Cadogan, 1938–45*, ed. David Dilks (Cassell, London, 1971), p. 21.
6. *Soviet Studies* is published by Longmans, for the University of Glasgow.

CHAPTER I

1. The comparison is striking, however. See, for example, Henry Kissinger, *A World Restored: Metternich, Castlereagh and the Problems of Peace, 1812–1822* (Gollancz and Houghton Mifflin, London and New York, 1973), Chapter 9; Harold Nicolson, *Peacemaking 1919* (Constable, London, 1943), pp. xvii and 25 (for the Vienna concepts); C. K. Webster, *The Congress of Vienna, 1814–1815* (Humphrey Milford, London, 1919), p. 101, which quotes Castlereagh's exchange with the Tsar; and C. M. Woodhouse, *Capodistria* (Oxford University Press, Oxford, 1973), pp. 120 ff.
2. In May 1945 Stalin said to Harry Hopkins: 'The Versailles Conference had been badly prepared and many mistakes had been made ...' *Berlin Papers*, vol. i, p. 31.
3. The introduction that Nicolson added to the 1943 edition of his *Peacemaking 1919* (originally published in 1933) listed twelve main lessons which the author, himself a participant in the Versailles negotiations, believed twenty-four years afterwards that the negotiators of the next peace treaties might learn from 1919. The second of these lessons is the one quoted here (pp. ix–x).
4. E. H. Carr, *Conditions of Peace* (Macmillan, London, 1942), p. 238.
5. The 'policeman' idea was Roosevelt's, dating from the Teheran Conference in 1943.
6. This was the phrase used by Churchill in his message to Stalin of 27 September 1944, when he was looking forward to a proclamation being issued to Japan by the United States, Great Britain, and the Soviet Union (Churchill, vol. vi, p. 187). In the event, however, the three governments who issued the proclamation of 26 July 1945 were those of the countries already at war with Japan—the United States, Great Britain, and China (not a Potsdam participant). Molotov

was simply sent a copy of the allied proclamation by Byrnes. The text is reproduced in ibid., pp. 556–7.

7. Walter Lippmann, *US War Aims* (Little, Brown, Boston, 1944), p. 132 (a quotation for which I am indebted to Adam Ulam). Willkie was the Republican Party candidate in the 1940 presidential election. Wallace was US Vice-President, 1940–5; then Secretary of Commerce until 1946.

8. Enoch Powell, 'Losing the Peace', in *Sunday Times*, 6 November 1983. The years to which his article referred were 1945–51.

9. In the end, the Japanese surrender, unlike the German, was not wholly unconditional: the monarchy was preserved, although the Emperor was placed under MacArthur's direction.

10. We cannot be certain, because the third atomic bomb had not been assembled at the moment when Japan surrendered.

11. See, respectively, *Atomic Energy: general account of the development of methods of using atomic energy for military purposes under the auspices of the United States Government, 1940–1945*, written by H. D. Smyth at the request of Major General L. R. Groves (US Government Printing Office, Washington, DC, 1945 (reprinted by HMSO, London)), pp. 135-6; and Hewlett and Anderson, pp. xi–xii. For the British attitude towards the Smyth report in August 1945, see ibid, p. 400.

12. *St Louis Post-Dispatch*, 6 March 1946, carried the text of the speech verbatim.

13. *Wall Street Journal*, 19 March 1946. In the issue of 8 March, a *Wall Street Journal* columnist had described Churchill's speech as 'brilliant', with a 'hard core of indisputable fact'. The reaction in London was initially one of consternation: see Bullock, *Bevin*, pp. 224–6.

14. *The Times*, 18 March 1946; and the US Department of State *Bulletin*, **14**, p. 483. In his speech at the Overseas Press Club on 28 February 1946, Byrnes had said: 'we will do nothing to break the world into *blocs* or spheres of influence' (*Keesing's Contemporary Archives*, **6** (1946–8), pp. 7826–9).

15. *Pravda*, 14 March 1946.

16. 'Does it exist?'—author's personal recollection.

17. Of these epithets the last is the sub-title of an article, 'America's Falklands War', in *The Economist*, 3 March 1984; and the first is that used in Arthur Campbell Turner, *The Unique Partnership, Britain and the United States* (Pegasus, New York, 1971).

18. *Foreign Affairs*, **54**: 4 (July 1976), p. 645.

19. To cite two recent examples one from each side of the Atlantic, the first sentence of Elizabeth Barker's *The British between the Super-Powers, 1945–50* (Macmillan, London, 1983), is: 'most people would agree that the years 1945–50 were crucial in the second half of the twentieth-century'. And in *Foreign Affairs*, **64**: 1 (Fall 1985) Adam Ulam has written in his article 'Troubled Coexistence': 'these were the years which set the pattern for the often tempestuous coexistence of the two superpowers, a pattern that has in many ways endured to this day'.

20. Yves Delahaye, *La frontière et le texte* (Payot, Paris, 1977), p. 10. The original French runs: ... 'les hommes considèrent encore trop souvant le monde soit comme un champ clos où s'affrontent des héros inspirés, soit comme le théâtre d'un désordre impossible à démêler'.

21. The titles of the first volume of Truman's memoirs and of Acheson's speak for themselves.

CHAPTER 2

1. The *New York Times*, 6 March 1946, also carried the text of the speech. For Churchill's definition of the 'special' relationship, see Chapter 1.

2. H. G. Nicholas, *Britain and The United States* (Chatto & Windus, London, 1963), p. 33.

3. For example, H. C. Allen, *Great Britain and the United States: a history of Anglo-American relations (1783–1952)* (Odhams Press, London, 1954), p. 27, looked back on a 'ripening of friendship' and a 'persistent, even steady progress from mistrust to cordiality'.

4. *Allies of a Kind* is both the title and the theme of Christopher Thorne's study of the Anglo-American relationship in the war against Japan. The Churchillian description of the Far East is quoted in the OUP paperback edition of the book (Hamish Hamilton and Oxford University Press, Oxford and New York, 1979), p. 117.

5. I am indebted to Arthur Campbell Turner, *The Unique Partnership, Britain and the United States* (Pegasus, New York, 1971), pp. 22–3 for drawing my attention to this estimate (53.6 per cent), which is the one suggested by Julian Huxley and A. C. Haddon in *We Europeans* (Cape, London, 1935) quoted by C. E. Carrington in *The British Overseas: Exploits of a Nation of Shopkeepers* (Cambridge University Press, Cambridge, 1950), p. 511. (Carrington himself suggested a 'British Isles' percentage of 61 per cent for 1930.)

6. Alastair Buchan's article, quoted in the previous chapter, drew my attention to Wilson's admission, made to Colonel House (see Patrick Devlin, *Too Proud to Fight: Woodrow Wilson's Neutrality* (Oxford University Press, Oxford, 1974), pp. 561–2). The comparison of the size of the press corps in the two capitals is also derived from this article, p. 651.

7. 'Common law allies' is the phrase coined by Robert Sherwood, *Roosevelt and Hopkins* (Harper, New York, 1948), p. 270. For the US commitment, given to Halifax by Roosevelt on 1 December 1941, see *The Diary of Lord Halifax*, 2 December 1941, p. 28, at the Borthwick Institute of Historical Research, University of York. (The entry of the previous day, on the same page, shows how uncertain Halifax then was that the commitment would be given.) Halifax reported his conversation with the President in Washington telegram No. 5519 to the FO: see the War Cabinet's discussion of 'Far Eastern Policy', 2 December 1941, WP(41) 296—CAB 66/20—and the Defence Committee (Operations) meeting on the following day, DO (41) 71st Meeting—CAB 69/2. Even then it was considered necessary to instruct Halifax to 'clinch the three points that the President had made'.

8. The few honourable abstainers from the ovation included Churchill and Eden. Harold Nicolson, who also remained in his seat, described this shameful scene in his *Diaries and Letters 1930–39*, ed. Nigel Nicolson (Collins, London, 1969), p. 364.

9. Keith Feiling, *The Life of Neville Chamberlain* (Macmillan, London, 1946), p. 325.

10. Alastair Cooke is one of the happiest examples of the westward flow from Cambridge University, as is his contemporary Joseph Harsch of the eastward flow to the same university.

11. Stanley Hornbeck, a former Rhodes scholar at Oxford, was Head of the State Department's Far Eastern Division during the war. An enthusiastic supporter of Chiang's China, he delivered his condescending opinions in convoluted prose, from the saddle of a particularly high moral horse; see Thorne, *Allies of a Kind, passim*. On the British side, the pompous example quoted in Chapter 3 n. 12 is one of many.

12. Ibid., p. 97, quoting *The Diary of Lord Halifax*, 7 March 1941. Nevertheless, after 'successfully living down some early gaffes' he presided 'in somewhat aloof and lazy fashion' (ibid., p. 119) over the Washington Embassy for five critical years. He was asked to stay on there even after the Labour Party came to power and did so for the better part of a year.

13. On the British side, Sir Ronald (Ian) Campbell, afterwards Ambassador in Cairo, was one of the honourable exceptions (as was Roger Makins) to the general norm of this paragraph. As early as 15 August 1941 Campbell was attempting, in a despatch to the Foreign Office, to answer the question how Britain could become a partner of the United States without being politically and economically swamped, especially since 'the British Commonwealth will be much exhausted by the war and deeply indebted to the United States, who are likely to be at the height of their mobilized strength'—FO 371/26151/A 7415/18/45.

14. John Balfour, *Not too correct an aureole* (Michael Russell, London, 1983), p. 105. Better known to historians of the war years as Archibald Clark Kerr, Inverchapel received his peerage after his service as Ambassador in Moscow. The bizarre story of his Russian valet, whom he brought with him from Moscow, was related by Frank Giles in the 1980 *Sunday Times Bedside Book*.

15. In 1938 (the last complete pre-war year) the United Kingdom imported £118m. from the United States and exported £21m. to the United States: B. R. Mitchell, *European Historical Statistics 1750–1970* (Macmillan, London, 1975), pp. 497 and 574.

16. Dean Acheson succeeded George Marshall as Secretary of State in January 1949. See his *Present at the Creation*, p. 323.

17. Churchill, vol. iii, p. 539.

18. Just under 2,000 written messages and telegrams in all. *Churchill and Roosevelt: the complete correspondence* has been edited in 3 vols. by Warren F. Kimball (Princeton University Press, Princeton, 1984).

19. Philip Larkin, *All What Jazz, A Record Diary 1961–1971* (Faber & Faber, London, revised edn., 1985), in particular the Introduction. Cf. his 'What Armstrong did' in *Required Writing* (Faber & Faber, London–Boston, 1983), pp. 314–5.

20. Isaiah Berlin, *Personal Impressions* (Hogarth Press, London, 1981), pp. 1 ff. Berlin's own description of this and other chapters in his book is 'writings that resemble what in the eighteenth century were called *éloges*—addresses commemorating the illustrious dead', p. vii.

21. Nevertheless, American casualties in the European and Pacific theatres of war proved to be almost exactly equal: 300,000 'military dead . . . and no civilians'— A. J. P. Taylor, *The Second World War* (Hamish Hamilton, London, 1975), p. 229.

22. Churchill was at his starkest in describing the surrender to Roosevelt at the time. In his memoirs he wrote of this disaster: 'I ought to have known. My advisers ought to have known and I ought to have been told, and I myself ought to have asked'—Churchill, vol iv, p. 43.

23. S. Woodburn Kirby [and others], *The War against Japan*, vol. v, *The Surrender of Japan* (HMSO, London, 1969), p. 391.

24. The Americans 'regard China as their theatre and barely tolerate our presence here at all'—the Head of the British Military Mission, Chungking, Major-General Hayes, to the Secretary of State for War, 20 July 1945: *DBPO*, vol. i, p. 788.

25. Kirby, *The Surrender of Japan*, p. 391.

26. Ibid., p. 425. Given this argument, against the background of a general shortage of shipping, it is not surprising that Slim's army was not a large one. The force

that would have invaded Malaya, had the war lasted longer, consisted of only seven divisions (reduced to six—see ibid., pp. 2 and 91).

27. In his report on 'The winning of the war in Europe and the Pacific', (Biennial Report of the Chief of Staff of the US Army, 1 July 1943 to 30 June 1945, to the Secretary of War), quoted in Thorne, *Allies of a Kind*, p. 134.

28. Ronald Lewin, *Ultra goes to War—The Secret Story* (Hutchinson, London, 1978) is one of the earliest accounts of the 'Ultra Secret'. Chapter 9 ('The American involvement'), pp. 234 ff, is headed by Marshall's description of Ultra—to Eisenhower—as of 'supreme importance'.

29. Field-Marshal Sir John Dill was Chief of the Imperial General Staff before being appointed Head of the British Joint Services Mission in Washington. His grave is at Arlington.

30. Coleridge's definition of poetic faith, *Biographia Literaria*, Ch. 12.

CHAPTER 3

1. For an account of Anglo-American malaise at the turn of 1944/5, see Robert M. Hathaway, *Ambiguous Partnership* (Columbia University Press, New York, 1981), pp. 89 ff.

2. The three quotations are from Churchill, vol. vi, p. 137; vol. iv, p. 344; and vol. vi, p. 417.

3. Isaiah Berlin, *Washington Despatches 1941–45*, ed. H. G. Nicholas (Weidenfeld & Nicolson, London, 1981), p. 595.

4. This quotation and those in the preceding sentence are derived from Robert J. Donovan's *Conflict and Crisis* (Norton, New York, 1977), p. xvii and p. 27 respectively. Unascribed information in this and the next three paragraphs is indebted to this source.

5. After Truman became President, his friendship with Jacobson was to prove of great political significance in relation to the Palestine issue. David Holloway drew my attention to the Soviet jibe, which in 1949 was reflected in the title of a Moscow play called 'The Mad Haberdasher': see Adam Ulam, *Dangerous Relations: the Soviet Union in World Politics, 1970–1982* (Oxford University Press, New York, 1984), p. 271.

6. The 'secret of Harry Truman'—Abe Fortas, quoted in William Hillman, *Mr President* (Faber, Straus and Young, New York, 1952), p. 115. (Cf. Jean Monnet, *Mémoires* (Fayard, Paris, 1976), p. 298.) Truman's reference to Plutarch is in his *Year of Decisions*, p. 119.

7. Donovan, *Conflict and Crisis*, p. xi. The countries to which Byrnes had travelled included Britain, which he had visited as a senator in 1937, through the Inter-Parliamentary Union. For the quarrel that developed between Truman and Byrnes at the turn of 1945/6, see Chapter 6.

8. For Taft's views, see, for example, Donovan, *Conflict and Crisis*, p. 260.

9. Vandenberg's speech is in *Congressional Record*, vol. 91, pt. 1, pp. 164–7. For Byrnes and Vandenberg, see Chapter 7.

10. Truman, *Year of Decisions*, pp. 98 and 195.

11. Shortly before his death, Roosevelt sent Baruch (whom he once described as 'that old Pooh Bah') on a visit to London: see Hathaway, *Ambiguous Partnership*, p. 137.

12. This paper—AN 1538/16/45, in FO 371/38523—to which further reference will be made later in this book, was not circulated outside the Foreign Office.

13. Stalin expected a Conservative Party majority of about 80: Churchill, vol. vi, p. 549.

14. Lessons described as 'a new realism' by Kenneth O. Morgan in his *Labour in Power 1945–51* (Clarendon Press, Oxford, 1984), p. 44.

15. This was the election of the 'give 'em hell, Harry' slogan. The mid-term election of 1946, which was a very different affair, was described by the Speaker, Sam Rayburn, as a 'damned beefsteak election' (quoted in Donovan, *Conflict and Crisis*, p. 235). It was Hugh Dalton who called 1946 Labour's *annus mirabilis*.

16. Kenneth Harris, *Attlee* (Weidenfeld & Nicolson, London, 1982), p. 248.

17. Patrick O'Donovan, writing in *The Observer*, 7 October 1951. I am indebted to Kenneth Harris for drawing my attention to this quotation.

18. Harold Wilson, in *A Prime Minister on Prime Ministers* (Weidenfeld & Nicolson and Michael Joseph, London, 1977), p. 297, described Attlee as 'tone deaf' in economic questions.

19. This was what many in Britain, including the author, believed in 1941. For evidence of the similar view then held in Washington, see the quotation from the State Department's letter to Winant in Thorne, *Allies of a Kind*, pp. 96–7 ('with Bevin as the possible head of government . . . Roosevelt sees this possibility . . .').

20. Lord Sherfield, 'Britain's nuclear story, 1945–52', review article in *Round Table*, **65** (1975), p. 199.

21. Dalton had expected the Foreign Office. Had he become Foreign Secretary, his xenophobia would surely have guaranteed failure. His tenure of the Treasury, which was not successful, was cut short by a pre-budget indiscretion in November 1947. Attlee said of him: 'Perfect ass. His trouble was he *would* talk'—Harris, *Attlee*, p. 353.

22. Ibid, p. 350. Sir Henry Campbell-Bannerman, a little known Prime Minister, who led the Liberal Party to victory in the general election of 1906.

23. Lord Pethick-Lawrence, George Hall, and Lord Addison were the three Secretaries of State in question. John Hynd was appointed Chancellor of the Duchy of Lancaster, without a seat in the Cabinet.

24. Bevin's prescient note about India (which also foresaw post-war developments in China), dictated on 21 June 1943, is quoted in Alan Bullock, *Life and Times of Ernest Bevin*, 3 vols. (Heinemann, London, 1967), vol. ii, pp. 278 ff.

25. For the 'take-over bid' and the 'violent passions' that it aroused, see Joe Garner, *The Commonwealth Office, 1925–68* (Heinemann, London, 1978), p. 301. Although much of this book is anecdotal, it offers a candid account (from the Commonwealth Office standpoint) of a characteristically British muddle, which ended only in 1968.

26. Wm. Roger Louis, *The British Empire in the Middle East, 1945–1951: Arab Nationalism, the United States and Postwar Imperialism* (Clarendon Press, Oxford, 1984), pp. 205 ff. describes the tussle over Cyprus between the Foreign Office and the Colonial Office.

27. Described as constituting 'the core of the specific war-time inheritance' in Peter Calvocoressi, *The British Experience 1945–1975* (Bodley Head, London, 1978), p. 23. The Education Minister in the coalition government was R. A. (Rab) Butler.

28. In preference to the 'machinery for long-term planning of the kind which became commonplace in Western Europe during the subsequent decade and a half', Andrew Shonfield, *Modern Capitalism* (Oxford University Press, Oxford 1965), pp. 89 ff.

29. For an analysis of different Labour Party views on foreign policy in 1945, see Bullock, *Bevin*, pp. 59 ff; and Morgan, *Labour in Power*, pp. 232 ff.

30. *The Economist*, 21 July 1945, p. 74.

31. George Orwell, *The English People* (Collins, London, 1947), p. 40.

32. Bevin was met by General Sir Hastings ('Pug') Ismay, Military Secretary to the Cabinet, who later became the first Secretary-General of NATO: see *Memoirs of General Lord Ismay* (Heinemann, London, 1960), p. 403.

33. He was a trade unionist throughout his political life until 1940, when Churchill invited him to join his wartime coalition government as Minister of Labour; this obliged him to stand for Parliament for the first time.

34. *Manchester Guardian*, 15 May 1948.

35. The view of his biographer (Bullock, *Bevin*, p. 193), in which I concur.

36. *FRUS, 1945*, vol. ii, p. 629. The record in *FRUS* is stated to be British, there being no American record of this meeting. (Since this chapter was drafted, this record has been published, in *DBPO*, vol. ii, p. 734. See also p. 868 in the same volume, 'Note of a conversation at the Kremlin', 24 December 1945.)

37. FO 800/478/MIS/45/14, Memorandum by the Secretary of State for Foreign Affairs.

38. Letter from Halifax to Bevin, 12 December 1945, FO 800/478/MIS/45/28.

39. CAB 131/2, DO (46) 40, 13 March 1946.

40. *DBPO*, vol. i, p. 188. Sargent's memorandum is in pp. 181 ff.

41. Elizabeth Barker, *The British between the Super-Powers, 1945–50* (Macmillan, London, 1983), p. 1, recalls Sir Orme Sargent on the day the war ended in Europe, standing 'on the steps of the Travellers' Club, watching the passing crowd with a kindly but pitying eye and an ironic smile'. Nicknamed 'Moley', he was one of the last members of the pre-war Foreign Office cadre, which (unlike the pre-war Diplomatic Service) did not have to serve abroad.

42. *DBPO*, vol. i, pp. 1189–90.

43. FO 371/44557, minute by Sir Orme Sargent, 1 October 1945.

44. *DBPO*, vol. i, pp. 181 ff.

CHAPTER 4

1. By the turn of the year 1944/5, 15 Army Group (perhaps the most polyglot allied force assembled during the war) had reached its lowest strength since the Italian mainland was first invaded, in September 1943. In August 1944 Alexander had been obliged to release seven divisions (for the invasion of southern France)—a Washington decision which Churchill had accepted only 'under a solemn protest'. Three divisions had been sent to Greece to support the Greek Government in the civil war; and the Canadian Corps was transferred to France after the Malta Conference. See Alan Shepperd, *The Italian Campaign* (Arthur Barker, London, 1968), pp. 281 ff, 320, 339 ff, and 341.

2. Marshall's estimate was that it might cost half a million American lives: Truman, *Year of Decisions*, p. 417.

3. The five objectives are quoted in full in Jan M. Ciechanowski, *The Warsaw Rising of 1944* (Cambridge University Press, Cambridge, 1974), pp. 217–18. His account of the rising, derived largely from Polish sources, remains the best. His figures of Polish casualties (p. 314) have therefore been followed, although estimates still vary. The most vivid account is the 300-page collection of photographs published in Jan Gużewski and Stanisław Kopf, *Dni powstania* (Pax Publishing House, Warsaw, 1957). For the Soviet part in this tragedy, see John Erickson, *The Road to Berlin* (Weidenfeld & Nicholson, London, 1983), pp. 269 ff.

4. The text of Stalin's telegram to Churchill and Roosevelt of 22 August 1944 is in Churchill, vol. vi, p. 120.

5. *Yalta Papers*, p. 225 and p. 31 respectively. References in the remainder of this

chapter not specifically attributed to other sources are derived from this volume of *FRUS*. The Soviet records of the Yalta Conference are in *SSNMK*, vol. iv.

6. See Churchill, vol. vi, pp. 197 ff, for the 'dire threats'.

7. All the other references in the last two sentences of this paragraph are derived from *Yalta Papers*: Leahy's letter (to Cordell Hull), p. 108; the briefing book paper, p. 1202; and the President's letter to the Polish Prime Minister, pp. 209–10.

8. *Yalta Papers*, pp. 266–7, has the State Department's official statement of 5 December 1944 on the unilateral British veto on Carlo Sforza as Prime Minister or Foreign Minister. This veto arose in part from Churchill's personal dislike of Sforza and in part from a tendency of the British Government, which lasted for several years after the war, to behave towards the Italian Government rather like an injured maiden aunt. The State Department's statement included a side-swipe at British intervention in Greece.

9. What he said on this occasion caused deep resentment among US commanders on the western front: Chester Wilmot, *The Struggle for Europe* (Collins, London, 1959), pp. 701 ff. Montgomery's own account of his performance ('I should ... have said nothing') is in Viscount Montgomery, *Memoirs of Field-Marshal Montgomery* (Collins, London, 1958), pp. 310 ff.

10. Memorandum for the President from Edward R. Stettinius, 2 January 1945, *DSDF*, 711.41/1–245, National Archives. *The Economist* article, entitled 'Noble Negatives', was dated 30 December 1944.

11. Let alone to fulfil 'the hope ... long cherished that you would pay your long-promised visit to Britain'—a hope for bilateral consultation, which by 19 November 1944 Churchill had to describe as 'destroyed' by Roosevelt's tripartite project: *Yalta Papers*, p. 17. Churchill's doggerel is in ibid., p. 26.

12. Ibid., p. 29.

13. 'The most violent disagreements and disputes of the entire war': Robert E. Sherwood in *The White House Papers of Harry L. Hopkins* (Eyre & Spottiswoode, London, 1949), vol. ii, p. 761.

14. *Yalta Papers*, pp. 39–40 has the remark to Hopkins; and pp. 32–3 has the exchange between Churchill and Roosevelt on the likely duration of the conference ('Even the Almighty took seven').

15. E.g. *The Times*, 13 February 1945, wrote of 'unbounded satisfaction', 'mutual confidence and unanimity of counsel', and 'the proposals for the settlement of the Polish question ... one of the greatest achievements of the conference'.

16. Sherwood, *White House Papers*, pp. 851 and 859.

17. *Yalta Papers*, pp. 711 ff. In the end the two republics were the Ukraine and Belorussia.

18. Ibid., p. 941. For confirmation of this view, see Truman, *Year of Decisions*, p. 272.

19. The text of the Yalta Agreement on the Far East of 11 February 1945 is contained in *CWP*, pp. 113 ff. Chapter 4 also offers a clear account of US-Soviet-Chinese negotiations in 1945.

20. Ibid., pp. 585 ff.

21. Adam Ulam's account of this renegotiation is in his *Expansion and Coexistence* (Secker & Warburg, London, 1968), pp. 492 ff. For Khrushchev's visit to Peking in 1955, see ibid, p. 555.

22. *CWP*, p. 116.

23. Ibid., p. 115.

24. Churchill, vol. vi, p. 342 and Anthony Eden, *The Memoirs of Sir Anthony Eden*, 3 vols., *The Reckoning* (Cassell, London, 1965), p. 513.

25. *Yalta Papers*, pp. 572–3.
26. Ibid, p. 899.
27. Ibid., p. 979.
28. For a different assessment, contrast Zbigniew Brzezinski's 'The Future of Yalta', *Foreign Affairs*, **63**: 2 (Winter 1984/5), pp. 277–302.
29. *Yalta Papers*, pp. 973–4.
30. 'Understanding between the Polish Committee of National Liberation and the Government of the USSR regarding the Polish-Soviet state frontier' (signed by Molotov and Osóbka-Morawski) Moscow, 27 July 1944—Polish text published in *Dokumenty i materiały do historii stosunków polsko-radzieckich*, vol. viii, Warsaw, 1974.
31. The description of the line in this document is identical with that of the Soviet 'Proposal of the Soviet Delegation regarding the western frontier of Poland' put forward at Potsdam on 20 July 1945: *SSNMK*, vol. iv, p. 459.
32. *The Memoirs of Wanda Wasiliewska* (published, in Polish, by Archivum Ruchu Roboticznego, Warsaw, 1982), vol. viii, p. 394 records a discussion of the frontier with Stalin 'in the late spring or at the beginning of the summer of 1943'. For a summary of Sazonow's proposals see Norman Davies, *God's Playground* (Clarendon Press, Oxford, 1981), vol. ii, p. 510.
33. Milovan Djilas, *Conversations with Stalin* (Rupert Hart-Davis, London, 1962), p. 105. For Churchill's private view, see—for example—John Colville, *The Fringes of Power: Downing Street diaries, 1939–1955* (Hodder & Stoughton, London, 1985), p. 555, entry dated 23 January 1945.
34. The literal translation of the Soviet record, in *SSNMK*, vol. iv, pp. 100 and 101.
35. 'The utmost tenacity and clarity of purpose on the Anglo-American side' would have been essential: Ulam, *Expansion and Coexistence*, p. 375—one of the earliest (and soundest) verdicts reached by an historian after studying the conference documents soon after they were first published.
36. Davies, *God's Playground*, p. 558, quotes Gomułka's statement in the Polish Communist Party's Plenum of 21–2 May 1945: 'the masses do not regard us as Polish communists at all, but just as the most despicable agents of the NKVD'.
37. *Yalta Papers*, pp. 971–2.
38. Ibid., pp. 93 ff., especially John Hickerson's memorandum of 8 January 1945. See also Lord Gladwyn, *The Memoirs of Lord Gladwyn* (Weidenfeld & Nicolson, London, 1972), p. 156: 'the general consensus [in London] was that nobody was committed very much' by the terms of the declaration.
39. Stimson's memorandum of 23 January 1945 to the President, is in *Yalta Papers*, p. 80. Kennan's phrase is in his *Memoirs 1935–1950* (Little, Brown, Boston, 1967), p. 212. That this was also Kennan's view at the time, with which Harriman 'could not agree', is clearly shown by W. Averell Harriman and Elie Abel, *Special Envoy to Churchill and Stalin, 1941–1946* (Random House, New York, 1975), p. 414.
40. *Yalta Papers*, Roosevelt to Churchill, 18 November 1944, p. 16.
41. Charles de Gaulle, *War Memoirs—Salvation 1944–46*, trans. Joyce Murchie and Hamish Erskine (Weidenfeld & Nicolson, London, 1960), p. 111.
42. Ibid., p. 115.

CHAPTER 5

1. *Yalta Papers*, pp. 974–5.
2. Churchill, vol. vi, p. 368.
3. Isaiah Berlin, *Personal Impressions* (Hogarth Press, London, 1981), p. 12.

4. *FRUS, 1945,* vol. i, pp. 121 ff. contains Charles Taussig's memorandum (dated 15 March 1945) of his conversation with the President, during which Roosevelt, for the first time, expressed his willingness for France to resume control of Indo-China, 'with the proviso that independence was the ultimate goal'. The final paragraph records Roosevelt's incipient change of mind over the future of Indo-China (and New Caledonia).

5. *Yalta Papers,* p. 383.

6. Churchill, vol. vi, p. 377.

7. Ibid., pp. 426–7.

8. 'My poor friend, what have you done? For ourselves, I no longer see any way out other than a fourth partition of Poland'—Robert Coulondre, *De Staline à Hitler: entre deux Ambassades, 1936–1939* (Hachette, Paris, 1950), pp. 165–7 (a quotation for which I am indebted to Jonathan Haslam).

9. Churchill, vol. vi, pp. 633–5, contains the complete text of Roosevelt's message to Stalin of 29 March 1945. The text of Churchill's message to Roosevelt of 13 March is on p. 374. Stalin's statement of the Soviet position is derived from his message to Churchill of 24 April, summarized in ibid., pp. 429–30.

10. Ibid., p. 398.

11. *Berlin Papers,* vol. i, pp. 716 ff. contains the briefing book paper on Poland, dated 29 June 1945.

12. Churchill, vol. vi, p. 506.

13. *Berlin Papers,* vol. i, pp. 24 ff. contains the record of the Hopkins Mission. Truman's press conference of 13 June 1945 is in *PP, Harry S. Truman, 1945,* pp. 118–27. His account of his first meeting with Molotov is in Truman, *Year of Decisions,* p. 82.

14. *Berlin Papers,* vol. i, pp. 733 ff.

15. For the Wolff negotiations, see Alan Shepperd, *The Italian Campaign 1943–45* (Arthur Barker, London, 1968), pp. 367 ff. The text of Eisenhower's telegram is in Churchill, vol. vi, pp. 401–2 and John Ehrman, *Grand Strategy* (HMSO, London, 1956), vol. vi, p. 132.

16. Charles Bohlen was the eyewitness; see Charles E. Bohlen, with Robert H. Phelps, *Witness to History* (Norton, New York, 1973), p. 209. The text of Roosevelt's final message to Stalin is in Churchill, vol. vi, p. 398.

17. On the question of Berlin, see Churchill, vol. vi, pp. 401–11 and 441–51. The importance of entering Berlin was stressed in his 'serious warning to Roosevelt', sent on 5 April (p. 446).

18. Churchill's message of 7 May to Eisenhower about Prague is in ibid., p. 442.

19. The President's refusal to delay the US withdrawal 'struck a knell in my breast'— Churchill, vol. vi, p. 525. Of the many messages on this subject sent by Churchill to Truman, perhaps the most dramatic was that of 12 May, beginning with the sentence: 'An iron curtain is being drawn down in their [the Russian] front' and ending with the words 'this issue of a settlement with Russia before our strength has gone seems to me to dwarf all others': *Berlin Papers,* vol. i, pp. 8–9.

20. *St Louis Post-Dispatch,* 6 March 1946. (For an account of the negotiations with Tito, which led to the withdrawal of Yugoslav troops from Trieste on 9 June 1945, see Truman, *Year of Decisions,* pp. 244 ff.)

21. One of the earliest indictments of the US decisions taken in April–May 1945, was Chester Wilmot's *The Struggle for Europe,* referred to in the previous chapter. Zhukov's directive and the Soviet controversy about it are described in John Erickson, *The Road to Berlin* (Weidenfeld & Nicolson, London, 1983), pp. 473 ff and pp. 741 ff.

22. Dwight D. Eisenhower, *Crusade in Europe* (Doubleday, New York, 1948), p. 396. For a recent assessment, see the revised edition of Stephen E. Ambrose's *Eisenhower and Berlin, 1945: the Decision to Halt at the Elbe* (Norton, New York, 1986).

23. *New York Herald Tribune*, 24 January 1945. For a cruder version of the same view, widely held in the United States at the time, see *Chicago Daily News*, 9 April 1945: 'we must have those naval and air bases', quoted in Thorne, *Allies of a Kind*, p. 664.

24. *Yalta Papers*, pp. 617 and 660.

25. Ibid., pp. 310 ff., contains an account of the Soviet proposal for a post-war loan. Henry Morgenthau actually raised the figure to US$10bn. in subsequent discussion within the Administration in Washington. Roosevelt's own view is recorded on 26 January as being one of 'keen interest'; and he believed that 'it should not be pressed further pending actual discussion between himself and Marshal Stalin and other Soviet officials'—ibid., p. 323. He did not discuss this with Stalin at Yalta, nor did Truman at Potsdam.

26. *Berlin Papers*, vol. i, p. 27.

27. In Stalin's own words: 'the question of a peace conference to settle the European War ... was ... knocking at the door'—ibid., p. 30.

28. The texts of the messages exchanged between the President and the Prime Minister, beginning with Churchill's message of 6 May (suggesting a meeting 'as soon as possible' of the three heads of government) are in *Berlin Papers*, vol. i, pp. 3 ff. The invitation to visit Britain was extended on 11 May (p. 5). Truman's message about 'ganging-up' was sent on the same date (p. 8).

29. Churchill, vol. vi, pp. 502–5, carries the full text of his 'Note on Mr Davies's Message', dated 17 May. Davies's report to the President, dated 12 June, is in *Berlin Papers*, vol. i, pp. 64 ff; his 1954 statement is recorded on p. 76 n. 22; Churchill's message to Truman of 31 May is on p. 89.

30. *Berlin Papers*, vol. i, p. 53. Churchill's telegrams of 11 and 12 May to Truman are on pp. 5 ff.

31. The key file in the UK Public Record Office on this phase of the subject is PREM 3/139, '1945-Explosives'. PREM 3/139/10 contains photocopies of the originals of the Quebec Agreement, the Declaration of Trust, and the Hyde Park Aide-Memoire. The texts of all three are reproduced as Appendices 1–3 in John Baylis, *Anglo-American Defence Relations, 1939–1984: the special relationship*, 2nd edn (Macmillan, London, 1984), pp. 130 ff; in Margaret Gowing, *Britain and Atomic Energy, 1939–1945* (Macmillan, London, 1964), as Appendices 4, 7, and 8; and also in Andrew Pierre, *Nuclear Politics* (Oxford University Press, Oxford, 1972).

32. On 18 July—PREM 3/139/9. The American copy of the Hyde Park Aide-Memoire was discovered years later in the files of Roosevelt's naval aide, who may have assumed that 'Tube Alloys', the British wartime codeword for the atomic bomb project, had something to do with torpedoes.

33. Gowing, *Britain and Atomic Energy*, p. 363.

34. Minute by Sir John Anderson to the Prime Minister, 29 June 1945—PREM 3/139/11A. Anderson was Chancellor of the Exchequer, but he had held special atomic responsibility from the outset of the project. Sir Henry Maitland Wilson succeeded Sir John Dill as Chief of the British Join Services Mission in Washington after the latter's death.

35. A partial text of the Combined Policy Committee's meeting of 4 July 1945 is in *Berlin Papers*, vol. i, p. 221.

36. Churchill, vol. vi, p. 553.

37. Hewlett and Anderson, pp. 350–2 and 360.

38. Minute by the Prime Minister to the Paymaster-General (better known as Professor Lindemann—'the Prof'), 27 May 1944—PREM 3/139/11A.

39. Ibid., telegram sent by Anderson to Wilson, following the Prime Minister's minute to Anderson of 21 May 1945.

40. Hewlett and Anderson, p. 372–3: 'A check with the British' is the subtitle of the relevant pages.

41. Ibid, p. 340. The summary of the Interim Committee's recommendations given in the following paragraph is derived mainly from pp. 352 ff.

42. Gowing, *Britain and Atomic Energy*, pp. 34 ff, gives a vivid account of Niels Bohr's part in this controversy, including his unhappy meetings with Churchill and Roosevelt. Churchill's extraordinary telegram, suggesting that Bohr 'ought to be confined or at any rate made to see that he is very near the verge of mortal crimes' (sent, from the Quebec Conference, to Cherwell on 20 September 1944) is in PREM 3/139/8A.

43. *Berlin Papers*, vol. i, loc. cit. in n. 35 to this chapter.

44. Cf. Churchill's conclusion, derived from his talk with Byrnes at Potsdam on 23 July, that 'the United States do not at the present time desire Russian participation in the war against Japan' (Churchill's minute of that date to Eden, *DBPO*, vol. i, p. 573). Whatever Byrnes may have thought, it was by then far too late to change course, at a moment when US-Soviet military plans for the Far Eastern war were being discussed in great detail, as the *Berlin Papers* clearly show. And Marshall made the telling point that the Soviet Union would now attack in the Far East, in its own interests, whether the United States wished it or not: see Hewlett and Anderson, vol. i, pp. 391–2.

45. Churchill, vol. vi, pp. 551 ff. The text of the proclamation eventually issued to Japan on 26 July is on pp. 556–7; also in *Berlin Papers*, vol. ii, pp. 1474 ff.

46. The main minute is by Cherwell, the subsequent ones by the Prime Minister's Private Secretary. Cherwell evidently believed that 'a few weeks' could elapse after the Potsdam Conference before 'concealment is no longer possible'—see *DBPO*, vol. i, p. 211.

47. In 1945 Vannevar Bush and James Conant correctly estimated that the Soviet Union could catch up in four years. Groves considered twenty years a much likelier figure—Hewlett and Anderson, p. 354. For the 'royal flush' calculation, see ibid., pp. 350–1.

48. The text of the Potsdam Protocol is in *Berlin Papers*, vol. ii, pp. 1478 ff. The Russian text is in *SSNMK*, vol. vi, pp. 459 ff. The English text is also in *DBPO*, vol. i, pp. 1262 ff.

49. *DBPO*, vol. i, p. 1143.

50. *Berlin Papers*, vol. ii, pp. 345 ff.

51. The first discussion of the Turkish and of the trusteeship issues took place in the plenary session of 22 July (ibid., pp. 252 ff.); for the Soviet request for a base in the Straits, see ibid., p. 303; the agreements reached on Iran and on Eastern Europe and Finland are in the Protocol, p. 1496 and pp. 1492 ff. respectively. The British record of the famous 'iron curtain' exchange between Churchill and Stalin is in *DBPO*, vol. i, p. 649; the Russian version, which omits the phrase 'iron curtain' but includes Stalin's retort ('fairy tales'), is in *SSNMK*, vol. vi, p. 183.

52. *Berlin Papers*, vol. ii, pp. 51, 84–5, 115, and 1465.

53. East Prussia was covered by Article V of the Potsdam Protocol, *DBPO*, vol. i, p. 1271 (VI in *Berlin Papers*, vol. ii, p. 1489).

54. *Berlin Papers*, vol. ii, pp. 332 ff. contains the 'Summary of Statements' made by members of the Polish Delegation to the meeting of the three Foreign Ministers,

24 July. Mikołajczyk, now deputy Prime Minister, in summing up for the Delegation maintained that 'if Polish claims were satisfied, Poland would lose 20 per cent of her territory, whereas Germany would lose 18 per cent'.

55. Ibid., pp. 204 ff. records the first plenary discussion of Poland. Truman's description of the Polish frontier dispute as an 'impasse' is on p. 215. Churchill's concluding remark is on p. 385. In the British version (*DBPO*, vol. i, p. 692) it is softer: 'it must be admitted that the conference would have failed'. The Russian version (*SSNMK*, vol. vi, p. 195) follows the British.

56. *Berlin Papers*, vol. ii, pp. 471 ff., 480 ff., and 1150 f.

57. Ibid., pp. 597 and 601.

58. Ibid., p. 517 records Bevin's 'All right, then' on German reparations (and p. 520 records Truman's 'then they were all agreed on the Polish question'). Bevin's remark is more gently recorded in *DBPO*, vol. i, p. 1078—'Mr Bevin then indicated that he would accept the percentages claimed by Premier Stalin'. The record of the British staff conference on 31 July is in *DBPO*, vol. 1, pp. 1052 ff; and that of the Anglo-Polish meeting on the same day is in ibid., pp. 1065 ff. The Polish President, Bolesław Bierut, can have been left in little doubt at this meeting that the chief British concern was what kind of statement could be made to the House of Commons on Bevin's return to London.

59. Churchill, vol. vi, p. 581.

60. *The Diaries of Sir Alexander Cadogan*, ed. David Dilks (Cassell, London, 1971), pp. 776–8.

61. The comment about Byrnes believing in the possibility of a 'deal' is in Bullock, *Bevin*, p. 24. Truman's remark is recorded in *Berlin Papers*, vol. ii, p. 367. The other references are in *DBPO*, vol. i, pp. 145, 957, and 947–8 (Cadogan's note of 28 July, given to Attlee and Bevin on their arrival in Berlin); and in *The Diaries of Sir Alexander Cadogan*, pp. 766–7.

62. The Russian text of the Soviet-German Treaty of 12 August is in *Pravda*, 13 August 1970; an English translation is in *Survival*, **12**: 10 (IISS, London, 1970); the text of the Helsinki Final Act is in HMSO, *Conference on Security and Cooperation in Europe Final Act*, Cmnd. 6198 (London, 1975 (repr. 1979)). For the 'device', see Robin Edmonds, *Soviet Foreign Policy—the Brezhnev Years* (Oxford University Press, Oxford, 1983), pp. 89 and 148.

63. *Berlin Papers*, vol. ii, p. 1484.

64. Sir David Waley's memorandum of 2 August, *DBPO*, vol. i, pp. 1257 ff., provides a clear record of Anglo-American differences on this subject; it also throws light on differences within the US delegation. (Waley was a member of the British delegation to the Reparations Commission). *The Diaries of Sir Alexander Cadogan*, p. 777, describes Byrnes on 30 July as having 'gone and submitted various proposals to Molotov which go a bit beyond what we want at the moment'.

65. *Berlin Papers*, vol. ii, pp. 1484–6.

66. France, already a member of the Allied Control Commission (see Chapter 4), was invited to join the Reparations Commission as a consequence of the Potsdam Conference (Attlee proposed this at the final plenary session: see *Berlin Papers*, vol. ii, pp. 569–70). For an exposition of the French view on the German question, as expressed by Bidault to Byrnes in Washington three weeks after Potsdam, see ibid., pp. 1557 ff. (this record is a translation from the French original). See also 'Memorandum by the French Delegation to the Council of Foreign Ministers'. 13 September 1945 (*DBPO*, vol. ii, p. 150, from which the quotation in this paragraph is derived).

67. Stettinius' memorandum is in *Berlin Papers*, vol. i, pp. 283 ff. Cadogan's talk with James Dunn on 14 July 1945 is recorded in ibid., p. 295.

68. At the plenary session on 17 July (*Berlin Papers*, vol. ii, 56 ff). At that stage Truman saw 'no objection to China being excluded from the Council'. For Molotov's earlier reservation, see *Berlin Papers*, vol. i, pp. 290–1.

69. The text of the Potsdam agreement on the Council of Foreign Ministers is in ibid., vol. ii, pp. 1478 ff. and in *DBPO*, vol. i, pp. 1263 ff.

70. Ibid., p. 1360.

71. Ibid., p. 47.

72. Churchill lunched alone with the President. The passage in Churchill, vol. vi, p. 554, is the first paragraph of his long 'summarised' note of the conversation, which covered a wide range of other topics as well. PREM 3/139/11A contains a photocopy of Copy No. 2 of Churchill's note of 18 July.

73. *Berlin Papers*, vol. ii, p. 225.

74. In the interval (ibid., pp. 1155 ff.) Stimson had written his 'reflections on the basic problems which confront us'—an unofficial paper, which he gave to Truman on 21 July. This paper reached the conclusion that until the Soviet internal system was effectively liberalized, 'we must go slow in any disclosures or agreeing to any Russian participation whatever and constantly explore the question how our head start in X [the atomic bomb] and the Russian desire to participate can be used to bring us nearer to the removal of the basic difficulty', which he defined as the fact that 'no permanently safe international relations can be established between two such fundamentally different national systems'.

75. Ibid., p. 243, and Churchill, vol. vi, pp. 552–3. There is no record, however.

76. Truman, *Year of Decisions*, pp. 420–1, gives the complete text of the order to Spaatz.

77. Ibid., p. 416, and Churchill, vol. vi, p. 580. There is a brief account in the introductory chapter to *SSNMK*, vol. vi, p. 15.

78. PREM 8/1101, minute of 26 September 1949 from Sir Henry Tizard to the Minister (of Defence), who showed it to the Prime Minister. Attlee wrote that there should be 'no action' on this minute. I am indebted to Margaret Gowing for drawing my attention to this remarkable document.

79. See Gregg Herken, *The Winning Weapon* (Knopf, New York, 1980), p. 303. The President's statement of 23 September 1949 is in *PP, Harry S. Truman, 1949*, p. 485.

80. The text is in *Berlin Papers*, vol. ii, pp. 1474 ff. and in Churchill, vol. vi, pp. 556–7.

81. Stimson's message is in *Berlin Papers*, vol. ii, pp. 1374–5.

82. Kenneth Harris, *Attlee* (Weidenfeld & Nicolson, London, 1982), p. 277. The letters (which are in *DBPO*, vol. i, pp. 1096–7 and 1119) relate to the statements to be issued in Washington and London after the first bomb had been dropped.

83. Harvey H. Bundy, 'Remembered Words', *The Atlantic*, March 1957, p. 57.

CHAPTER 6

1. For the Hopkins mission to Moscow, see Chapter 5. For the Manchurian campaign, see *Finale*, ed. Marshal M. V. Zakharov (Moscow, Progress, 1972), pp. 78–89; and *Leavenworth Papers*, No. 7, 'August Storm: the Soviet 1945 strategic offensive in Manchuria', by LTC David M. Glantz (Combat Studies Institute, US Army Command and General Staff College, Fort Leavenworth, Kansas, 1983): I am indebted to John Erickson for both these sources.

2. This phrase was first used publicly by Truman on 27 October 1945 (*New York Times*, 28 October 1945), but it was Attlee who suggested to him, in a telegram on

8 August, that the two heads of government 'should without delay make a joint declaration of our intentions to utilise the existence of this great power . . . as trustees for humanity . . .' *FRUS, 1945,* p. 37.

3. Those venturing into revisionist territory for the first time may find the 'US Foreign Policy' section of the bibliography in Bullock, *Bevin,* pp. 865–6, a useful guide; it marks the works of revisionist historians with an asterisk.

4. When Bourke Hickenlooper, Chairman of the Senate Atomic Energy Commission, was at last shown the figures on US atomic bomb production early in 1947 by Truman, he was 'visibly shaken'—Truman, *Trial and Hope,* p. 339.

5. The text of Stimson's memorandum, dated 11 September, is in *FRUS, 1945,* vol. ii, pp. 41 ff. In spite of his emphasis on the need for a specifically US proposal to the Soviet Union, he expressly stated, at the outset of this memorandum, that 'Britain in effect already has the status of a partner with us in the development of this weapon'.

6. Hewlett and Anderson, p. 419; and Acheson, *Present at the Creation,* pp. 124–5.

7. David Holloway, *The Soviet Union and the Arms Race* (Yale University Press, New Haven and London, 1983), p. 18 (*uran* is Russian for uranium). The same author also drew my attention to the Soviet sources referred to in notes 8 and 10 of this chapter.

8. A. I. Ioirysh, I. D. Morokhov, and S. K. Ivanov, *A-Bomba* (Nauka, Moscow, 1980), pp. 377 and 390–1.

9. Zhukov's account is in G. K. Zhukov, *Vospominaniya i razmyshleniya* (Novosti, Moscow, 1969), p. 732; see also Yu. V. Sirintsev, *I. V. Kurchatov i yadernaya energetika* (Nauka, Moscow, 1980), pp. 10–11. Shtemenko's account is in S. M. Shtemenko, *General'nyi shtab v gody voiny,* 2 vols. (Voennoe izdatel'stvo, Moscow, 1981), vol. i, p. 425. Shtemenko was deputy to General Antonov, Soviet Chief of Staff.

10. Stalin's remark is quoted in A. Lavrent'yeva, 'Stroiteli novogo mira', *V mire knig,* **9** (Moscow, 1970), pp. 4–5. The British telegram is Moscow telegram No. 5192 to the FO, in CAB 130/3 ('at a blow the balance which had seemed set and steady was rudely shaken'); see also *DBPO,* vol. ii, pp. 650–2.

11. *28aya godovshchina velikoi oktyabr'skoi sotsialisticheskoi revolyutsii* (Gospodizdat, Moscow, 1946), pp. 28–9.

12. *The Journals of David E. Lilienthal,* vol. ii (Harper, New York, 1964), entry dated 16 January 1946, p. 10.

13. Although Churchill mentioned no names in this passage of his speech ('It would nevertheless be wrong and imprudent to entrust the secret knowledge or experience of the atomic bomb . . . to the world organisation, which is still in its infancy'), it is clear what he had in mind as the target of his attack—*St Louis Post-Dispatch,* 6 March 1946.

14. Zhukov, loc. cit. in n. 9. For a contemporary Soviet view, see *SSNMK,* vol. vi, p. 14.

15. Stalin told Harriman that 'Soviet scientists had been working on the problem but had not been able to solve it' (Truman, *Year of Decisions,* p. 426). For Stalin's view of Truman, see *Khrushchev Remembers,* trans. and ed. Strobe Talbott (Little, Brown, Boston, 1970), p. 221. (Khrushchev's own view is expressed on p. 361.) The official American historian's verdict on Truman's action on 24 July is recorded in Hewlett and Anderson, p. 394.

16. *The Economist,* 26 September 1945. The American record of this conference is in *FRUS, 1945,* vol. ii, pp. 112–555. The British record is in CAB 133/15; see also *DBPO,* vol. ii, pp. 100 ff.

17.   Quoted in Gregg Herken, *The Winning Weapon* (Knopf, New York, 1980), p. 53.
18.   Milovan Djilas, *Conversations with Stalin* (Rupert Hart-Davis, London, 1969), p. 52.
19.   *FRUS, 1945*, vol. ii, p. 61. Both Bevin and Byrnes appear to have believed at the time that the ultimate Soviet objective was the uranium deposits of the Congo.
20.   Cf. Bullock, *Bevin*, pp. 133 and 136.
21.   Djilas, *Conversations with Stalin*, p. 164.
22.   For Bevin's view, see GEN 75, 4th meeting, 11 October 1945 (CAB 130/2). The British official was the No. 2 at the Cabinet Office, Norman Brook, in a letter to Ronald Campbell, 6 October 1945 (*DBPO*, vol. ii, pp. 497–8).
23.   The text of Byrnes' telegram to Moscow of 23 November is in *FRUS, 1945*, vol. ii, p. 578. (He sent Bevin a message via Winant forty-eight hours later, although this did not reach Bevin until 26 November.) The telegram from the British Embassy in Moscow is No. 5086 to the FO, 24 December 1945, in *DBPO*, vol. ii, Chapter 3.
24.   Ibid., pp. 581–96 records the whole episode, from Winant's telegram of 26 November up to the press release of 7 December 1945. The British record of the teletype conversation of 27 November is in FO 800/446, reproduced in *DBPO*, vol. ii, pp. 639–41.
25.   E.g., *The Economist*, 5 January 1946, p. 3; and the *Observer*, 30 December 1945.
26.   The UK delegation's suggested terms of reference, which included several amendments made at Byrnes' suggestion, is in *FRUS, 1945*, vol. ii, pp. 771–2.
27.   Wm. Roger Louis, *The British Empire in the Middle East, 1945–1951: Arab Nationalism, the United States, and Post-War Imperialism* (Clarendon Press, Oxford, 1984), pp. 53 ff., provides an admirably clear account. For an example of the 'turning-point' view, see Bullock, *Bevin*, p. 238. In the end Soviet troops left Northern Iran in May 1946. Under the terms of the 1942 Anglo-Soviet agreement with the Iranian Government, all troops should have been withdrawn from Iran six months after the end of the war—a date confirmed by Molotov at the London meeting of the Council of Foreign Ministers.
28.   The text is in *FRUS, 1945*, vol. ii, pp. 815 ff., and in *DBPO*, vol. ii, pp. 905–13. The British record of the Moscow Conference is in CAB 133/82.
29.   The draft resolution, to be proposed by the five members of the Security Council and Canada, is in *FRUS, 1945*, vol. ii, pp. 823–4 and *DBPO*, vol. ii, pp. 912–13.
30.   Having taken Dulles with him instead of Vandenberg to the London conference in September, Byrnes then took no one from Congress to Moscow in December. The President was left to find out the terms of the Moscow communiqué for himself. The strange sequence of events is related in Donovan, *Conflict and Crisis* (Norton, New York, 1977), pp. 156–62. Although Byrnes remained in office for another year, on 16 April 1946 he submitted his resignation privately to Truman, who decided to replace him with Marshall. The likelihood of Marshall's appointment was leaked to the press on 4 March: see ibid., pp. 192–3.
31.   *PP, Harry S. Truman, 1946*, pp. 9–10.
32.   William Hillman, *Mr President* (Farrar, Straus, & Young, New York, 1952), pp. 21–3; and Truman, *Year of Decisions*, pp. 551–2.
33.   Truman's invitation was accepted by Churchill on 8 November 1945: see Donovan, *Conflict and Crisis*, p. 19.

CHAPTER 7

1.   The Quebec Agreement was not published by the British Government until 1954 (Cmd 9123). For the text of all three agreements, see Chapter 5 n. 31. The Anglo-

American muddle which these agreements resolved has been magisterially described by Margaret Gowing in her *Britain and Atomic Energy, 1939–1945* (Macmillan, London, 1964).

2. Lord Sherfield (previously Sir Roger Makins), 'Britain's Nuclear Story, 1945–52: politics and technology' review article published in *Round Table*, **65** (1975), p. 194.

3. For the strict interpretation, see ibid.; and it was so argued in Washington. See, however, Acheson's comment, quoted below. Churchill's interpretation comes from the telegram that he sent home at the time—Octagon to Admiralty, GUNFIRE 293—in PREM 3/139/8A.

4. For the text of the President's message and for his press conference, see *PP, Harry S. Truman, 1945*, pp. 362–6 and 381–8.

5. 'The Atomic Bomb', memorandum by the Prime Minister, CAB 130/3, GEN 75/1, 28 August 1945. This document, and some of the others cited in this chapter, have—since the time of writing—been published in *DBPO*, vol. ii, Chapter 2.

6. CAB 130/2, GEN 75, 2nd meeting, 29 August 1945.

7. Attlee's letter of 25 September to Truman forms Annex I to the 'Note by the Prime Minister', 26 September 1945, in PREM 8/117, GEN 75/3. Annex II contains Churchill's (undated) message to Attlee. The repetition in this message of the word 'not', which was standard procedure in British military telegraphese of the Second World War, suggests that it was sent through military channels. Churchill spent much of September in a villa on Lake Como, which Field-Marshal Alexander had put at his disposal: see Mary Soames, *Clementine Churchill* (Cassell, London, 1979), pp. 392 ff.

8. CAB 130/2, GEN 75, 4th and 6th meetings, 11 and 18 October 1945.

9. In the words of Margaret Gowing, *Independence and Deterrence*, p. 63, Britain's 'atomic relations with the United States were important to her not only because of the information she needed for her domestic programme but because they must be a key to the strength and balance of Anglo-American relations as a whole'.

10. The original text of the Tripartite Declaration is in PREM 8/117, the fourth paper in the file.

11. Hewlett and Anderson, p. 465.

12. Ibid., p. 468.

13. The text of the Groves–Anderson memorandum is in *DBPO*, vol. ii, pp. 630–2. See also Gowing, *Independence and Deterrence*, Appendix 5, pp. 85–6; and John Baylis, *Anglo-American Defence Relations, 1939–1984: the Special Relationship*, 2nd edn. (Macmillan, London, 1984), pp. 140–1.

14. Hewlett and Anderson, p. 469. Bush had been appointed Director of the Office of Scientific Research and Development in 1941: he was a key figure in the American atomic energy programme.

15. 415 *H.C.Debs.*, col. 1333, 7 November 1945.

16. Quoted in Gregg Herken, *The Winning Weapon* (Knopf, New York, 1980), p. 351.

17. CAB 129/4 contains the memorandum (CP (45)272) which the Prime Minister asked his colleagues to approve at this meeting; the minutes are in CAB 128/4 CM (45) 51st Conclusions, 8 November 1945.

18. Herken, *The Winning Weapon*, p. 351.

19. Hewlett and Anderson, pp. 469–70.

20. At that time Bevin also took the view that the UN Atomic Energy Commission, under discussion in Moscow, should 'be made responsible to the Security Council'—Moscow telegram No. 14, WORTHY to FO, 17 December 1945, 'for Prime Minister from Foreign Secretary', GEN 75/18, in CAB 130/3.

21. GEN 75, 12th meeting, 20 March 1946, in CAB 130/2.

22. Quoted in Gowing, *Independence and Deterrence*, p. 164.

23. CAB 130/2 contains the records of the decisions reached by the GEN 75 Committee at its 8th and 15th meetings, held on these dates. The record of the GEN 163 meeting held on 8 January 1947 is in CAB 130/16, Confidential Annex, Minute 1. GEN 163 was an even smaller committee than GEN 75 (set up in August 1945), which it succeeded in January 1947. A Ministerial Standing Committee on atomic energy was not set up until February 1947. For an account of the British governmental machinery established by the Labour Government for atomic matters, see Gowing, *Independence and Deterrence*, Chapter 2.

24. For some examples, see Andrew Pierre, *Nuclear Politics: The British experience with an Independent Strategic Force, 1939–1970* (Oxford University Press, Oxford, 1972), p. 73. The British press, starved of atomic information from domestic sources, relied largely on material gleaned from the United States.

25. Gowing, *Independence and Deterrence*, p. 56; and Sherfield, 'Britain's nuclear story', p. 199.

26. The remark ascribed to Byrnes is in David E. Lilienthal, *Journals*, vol. ii, *The Atomic Energy Years* (Harper & Row, New York, 1964), p. 59. At the beginning of 1946 Lilienthal had served as head of the panel of advisers to the Acheson Committee on the international control of atomic energy.

27. Sir James Chadwick, discoverer of the neutron, was the senior British scientist in the United States and adviser to the British representative on the CPC during the war. The personal relationship that he established with Groves was of crucial importance both during and immediately after the war.

28. See Chapter 2 n. 13 and Chapter 7 n. 42. As early as 24 September 1945 Makins had minuted: 'It is important to act quickly before the American views have crystallised and while the machinery of the CPC and of the Trust is still in active operation'—*DBPO*, vol. ii, p. 548.

29. The original is in PREM 8/117, the second paper in the file.

30. Sherfield, 'Britain's nuclear story', pp. 194–5, points out that even when Anglo-American atomic collaboration broke down in general, it continued in 'two important fields—raw materials and intelligence, where the machinery of collaboration continued without interruption'.

31. Truman, *Trial and Hope*, pp. 12–14; and Gowing, *Independence and Deterrence*, p. 101.

32. Text in Gowing, *Independence and Deterrence*, Appendix 7, pp. 126 ff.

33. CAB 130/3, GEN 75/28, 27 February 1945, has as its annex Washington telegram ANCAM 536 reporting this 'confused' meeting.

34. Hewlett and Anderson, p. 514. The text of the bill as originally presented is in pp. 714 ff.

35. Acheson, *Present at the Creation*.

26. Quoted in Herken, *The Winning Weapon*, p. 137.

37. *The Times*, 18 February 1952; and Gowing, *Independence and Deterrence*, pp. 107 ff.

38. Gowing, ibid., p. 121 gives an account of Acheson's briefing. See also Acheson, *Present at the Creation*, pp. 167–8.

39. The State Department official was the Special Atomic Assistant in the State Department, Gordon Arneson, quoted by Herken, *The Winning Weapon*, p. 147 n.

40. Sherfield, 'Britain's nuclear story', p. 196.

41. Gowing, *Independence and Deterrence*, p. 114 and pp. 338 ff. John Cockcroft, who (like Chadwick) was a Nobel Prize winner, was recalled from the Chalk River project to become the first Director of the UK atomic research establishment at Harwell in January 1946. He, Christopher Hinton, and William Penney were the

three outstanding men who were later, at working level, to succeed in bringing the British atomic project to fruition.

42. From March 1947. Thereafter Sir Roger Makins (afterwards Lord Sherfield) became successively Ambassador in Washington, Joint Permanent Secretary to the Treasury, and Chairman of the UK Atomic Energy Authority.

43. For the 'long telegram', see George Kennan, *Memoirs 1925–1950* (Little, Brown, Boston, 1967), Appendix C, pp. 295 ff. The 'X Article', published in *Foreign Affairs*, **25**: 4 (July 1947) defined 'containment' as a 'long term, patient but firm and vigilant containment of Russian expansionist tendencies' (pp. 575–6) and (pp. 580 ff) 'a policy of firm containment designed to confront the Russians with unalterable counter-force at every point where they show signs of encroaching upon the interests of a peaceful and stable world'.

44. *Foreign Affairs*, **25**: 4 (July 1947), pp. 583 ff.

45. Quoted in Gowing, *Independence and Deterrence*, p. 265.

46. *FRUS, 1950*, vol. vii, p. 1463.

47. Hewlett and Duncan, vol. ii, pp. 275 ff.

48. Full text in Gowing, *Independence and Deterrence*, Appendix 9, pp. 266 ff. and in Baylis, *Anglo-American Defence Relations, 1939–1984*, Appendix 6, pp. 142 ff.

49. The Vice-Chief of the Air Staff, Air Marshal Sir William Dickson, (later Chief of the Defence Staff). For his view, and the Chiefs of Staff's counter-arguments, see Gowing, *Independence and Deterrence*, p. 251.

50. That the information was valuable is the view of Sherfield, 'Britain's Nuclear Story', quoting Cockcroft. But, for example, permission to visit the US production plants at Hanford continued to be withheld.

51. David E. Lilienthal, *Journals*, vol. ii, *The Atomic Energy Years*, p. 455.

52. The US record of the CPC meetings is in *FRUS, 1949*, vol. i, pp. 520–662. See also Gowing, *Independence and Deterrence*, pp. 283 ff.

53. Acheson, *Present at the Creation*, pp. 316–20; and Gowing, *Independence and Deterrence*, pp. 277 ff.

54. Acheson, *Present at the Creation*, p. 321.

55. Hewlett and Duncan, pp. 312–14; Gowing, *Independence and Deterrence*, p. 333; and Herken, *The Winning Weapon*, p. 397.

56. 450 *H.C.Debs.*, col. 2117, 12 May 1948.

57. All four of the British 1944 documents quoted in this paragraph are to be found in PREM 3/139/11A.

58. The Acheson-Lilienthal plan is described in Hewlett and Anderson, vol. ii, pp. 538 ff.; the quotation is the title of the relevant chapter.

59. Gowing, *Independence and Deterrence*, pp. 254 and 264.

60. Sherfield, 'Britain's nuclear story', p. 195.

61. Lilienthal, *Journals*, vol. ii, p. 548: '... if the alternative were full partnership *or* continuing the limited area-by-area arrangement under the Modus Vivendi, my personal view would be to drop the whole thing and each go it alone'.

62. The eventual outcome was 'one of the most successfully executed programmes in British scientific and technical history'—Gowing, *Independence and Deterrence*, p. 57.

63. By the Anglo-American *Agreement for Cooperation on the Uses of Atomic Energy for Mutual Defence Purposes*, 3 July 1958: see Chapter 20.

CHAPTER 8

1. Although such a loan is interest-free, IDA (the 'soft window' agency of the World Bank) levies a service charge of 0.75 per cent and a charge of 0.50 per cent on

undisbursed credits: source—*1983 Review of Development Cooperation: Effects and Policies*, Development Assistance Committee, OECD.

2. A 'conjunction of power and idealism in economic affairs ... unprecedented in American history': Richard N. Gardner, *Sterling–Dollar Diplomacy in Current Perspective* (Columbia University Press, New York, 1980), p. 13.

3. US Congress, House Committee on Ways and Means, *Extension of Reciprocal Trade Agreements Act*, 76th Congress, first session (1940), vol. i, p. 38.

4. For examples of this hyperbole and denial, see Gardner, *Sterling–Dollar Diplomacy*, pp. 136 ff. Admittedly some of the arguments that had to be overcome in Congress were ludicrous. Taft, for example, said that US participation in the IMF would mean 'pouring money down a rat hole'—91 *Congressional Record*, 7573, 16 July 1945.

5. *United Nations Monetary and Financial Conference, Bretton Woods, New Hampshire, USA, July 1 to July 22, 1944, Final Act*, Cmd 6546 (HMSO, London, 1944 (repr. 1945)), p. 38, Article xx(e).

6. Whereas 'the Americans claimed that the Bretton Woods institutions would meet Britain's post-war needs and that no additional measures would be required', the British saw ratification of the Bretton Woods agreements as 'contingent upon bold new measures of transitional aid': Gardner, *Sterling–Dollar Diplomacy*, p. 143, which gives a clear account of the two countries' contrasting states of opinion.

7. Adam Ulam, *Expansion and Coexistence* (Secker & Warburg, London, 1968), p. 410; and Hannes Adomeit, *Soviet Risk-Taking and Crisis Behaviour* (Allen & Unwin, London, 1982), p. 115.

8. US$25bn. were envisaged under the Keynes scheme; see R. F. Harrod, *Life of John Maynard Keynes* (Macmillan, London, 1951), pp. 526 ff.

9. *Bretton Woods Final Act*, pp. 40, 46, 47, 49, and 65.

10. *The Times*, 2 July 1945.

11. CP(45)112, 'Appreciation by Lord Keynes of "Our Overseas Financial Prospects" ', 13 August, circulated by the Chancellor of the Exchequer on 14 August 1945—CAB 129/1.

12. The text of the White House press release of 21 August 1945 is in *PP, Harry S. Truman, 1945*, p. 232, footnote.

13. 410 *H.C.Debs.* cols. 955–8, 29 August 1945.

14. For a discussion of the so-called Stage II, see Gardner, *Sterling–Dollar Diplomacy*, pp. 180–2 and 185–7; and Robert M. Hathaway, *Ambiguous Partnership* (Colombia University Press, New York, 1981), pp. 72 ff.

15. *DBPO*, vol. i, pp. 350–1, Memorandum from Truman for the Prime Minister, 17 July 1945, and telegram from Sir John Anderson to Anthony Eden, for the Prime Minister, 18 July 1945.

16. Sterling indebtedness at the end of the war totalled £3,355m.: *Statistical Material presented during the Washington Negotiations*, Cmd. 6705 (HMSO, London, 1945), p. 11.

17. Later used by Gardner, to head the Introduction to his *Sterling–Dollar Diplomacy*.

18. Harrod, *Keynes*, pp. 597, 604, 612, and 627; and Gardner, *Sterling–Dollar Diplomacy*, p. 201.

19. *Bretton Woods Final Act*, pp. 16 and 33; and *Financial Agreement between the Governments of the United States and the United Kingdom*, Cmd 6708 (HMSO, London, 1945), Articles 7 and 8.

20. Harrod, *Keynes*, pp. 606–8.

21. *Financial Agreement*, Section 10.

22. Under the terms of the 'Joint statement regarding settlement for Lend-Lease,

reciprocal aid, surplus war property and claims', attached to the Financial Agreement.

23. CP(49)179, 24 August 1949, Note by the President of the Board of Trade, Harold Wilson, covering a Treasury memorandum, 23 August 1949, 'Sterling balances of the sterling area'—CAB 129/36/71933.

24. 441 *H.C.Debs.*, col. 1670, 7 August 1947.

25. The sterling area remained 'the focal point' of British overseas financial policy for many years after the end of the war: Andrew Shonfield, *British Economic Policy since the War* (Penguin Books, Harmondsworth, 1959), p. 150.

26. For examples, see Gardner, *Sterling–Dollar Diplomacy*, pp. 246–7.

27. 92 *Congressional Record* 4080, 22 April 1946.

28. 417 *H.C.Debs.*, col. 439, 12 December 1945.

29. 138 *H.L.Debs.*, cols. 790–4, 18 December 1945.

30. Source: *UK Balance of Payments*, Central Statistical Office Pink Book (HMSO, London, August 1985).

31. *The Economist*, 15 December 1945, p. 850.

32. CAB 129/1, cited in n. 11 to this chapter.

CHAPTER 9

1. *The Federal Reserve Bulletin*, quoted by Richard Gardner, *Sterling–Dollar Diplomacy in Current Perspective* (Columbia University Press, New York, 1980), p. 318.

2. *Economic Survey for 1947*, Cmd. 7046 (HMSO, London, 1947); and *United Kingdom Balance of Payments 1946 to 1950*, Cmd. 8065 (HMSO, London, 1950). Years later, revised statistical methods have produced different, less cataclysmic, figures for the UK post-war balance of payments deficits; but in this chapter it has seemed best to follow the figures on which judgements were based on both sides of the Atlantic at the time. On this whole subject, see Alec Cairncross, *British Economic Policy, 1945–51* (Methuen, London, 1985).

3. *The Times*, 22 August 1947. The British negotiator of this agreement was Sir Wilfrid Eady, Second Secretary at the Treasury (a post which he had held since 1942).

4. For an account of the deliberations of the divided Cabinet and its subsequent 'near panic', see Kenneth O. Morgan, *Labour in Power, 1945–1951* (Clarendon Press, Oxford, 1984), pp. 114 ff.

5. *FRUS, 1947*, vol. iii, pp. 275–6 and 293: both quotations from the US record of the series of Anglo-American meetings held in London in June 1947, further discussed in Chapter 13. To be fair to Dalton, he did describe the 'timetable' of the 1945 Financial Agreement as 'so wrong'. But he never mentioned convertibility once in his diary during 1947: see Morgan, *Labour in Power*, p. 348.

6. The Walcheren expedition, 1809.

7. *The Economist*'s leader was dated 16 August 1947. Douglas' telegram, to Lovett, dated 13 September 1947, was sent in reply to a Personal telegram from Lovett, No. 3805 to London, 2 September 1947: *DSDF*, 711.41/9–1347 and 711.41/9–1247, respectively.

8. *Daily Mail*, 23 July, quoting Reuters, and *The Times*, 5 August 1947.

9. *The Economist*, 23 August 1947, pp. 305–6.

10. So strong were feelings about Section 9 in London that on 18 September the US Embassy issued a categorical statement that it 'was never intended to constitute a straitjacket on British trade', *Financial Times*, 19 September 1947.

11. By 31 December 1947 British reserves had fallen to £958m.: less than the figure at the end of the war.

12. *Economic Survey*, Cmd. 7647 (HMSO, London, 1949), p. 511.

13. Cabinet Economic Policy Committee (49) 22nd meeting, 24 June 1949, in CAB 134/220, which contains the minutes of the EPC meetings of June–July 1949. Quotations in this paragraph are derived from the same source.

14. *The Economist*, 7 May 1949, p. 825.

15. Gaitskell, who was not yet in the Cabinet, became the key figure in this revolt. The crisis and the revolt are described in Morgan, *Labour in Power*, pp. 380 ff.

16. FO telegram No. 3015 to New York, 21 September 1949. An unctuous paragraph at the end reads: 'it has not passed without comment that sterling, in spite of its critics in America, is still so much the world's currency that its movement has taken nearly every other in a similar direction'—PREM 8/973.

17. Ibid, Washington telegram No. 4555 to FO, 20 September 1949, which also contains the Acheson quotation in the preceding sentence. A copy of this telegram, which was addressed personally to the Cabinet Secretary and the Permanent Under-Secretary at the Foreign Office, was sent to 10 Downing Street.

18. Minute on 'Gold and Dollar Reserves—31 December 1949', 4 January 1950, in T229/231. Hall, afterwards Lord Roberthall, was then Director of the Cabinet Office Economic Section.

19. The text of the communique (which made no mention of devaluation) issued on 12 September 1949 is in PREM 8/973, Washington telegram No. 4324 to FO of that date.

20. 'The most shattering moment of truth for Britain in the early post-war years': Margaret Gowing, *Independence and Deterrence*, p. 4.

21. *FRUS, 1949*, vol. i, pp. 520 ff.

22. Ibid., p. 514.

23. Scott Newton, 'The 1949 sterling crisis and British policy towards European integration', *Review of International Studies* **11**, (1985), p. 180. For what happened in the 'aftermath', see Chapters 15 and 17.

24. I am indebted for this quotation to R. B. Manderson-Jones, *The Special Relationship: Anglo-American relations and Western European unity, 1947–1956* (Crane, Russak, New York, 1972), p. 61.

25. 'Paper prepared in the Department of State' (in preparation for the bilateral Anglo-American talks held in London from 26 April to 1 May 1950), 19 April 1950: *FRUS, 1950*, vol. iii, p. 878.

26. These strongly worded judgements have been quoted deliberately because they represent a view that cannot be regarded as biased against the Attlee Government: that of Morgan, *Labour in Power*, pp. 388–9.

27. In 1949 there had indeed been some relaxation of rationing, food and otherwise; but the point, made by Andrew Shonfield in *British Economic Policy since the War*, rev. edn. (Penguin Books, Harmondsworth, 1959), p. 169, is that in 1950 'no one seriously thought' of launching an attack on the whole system of food rationing, which was not finally abolished until 1954.

28. Makins informed Petsche, the French Finance Minister, at the British Embassy in Washington (FO 800/465, Washington telegram to FO No. 4487, 17 September 1949). In the words of Makins' telegram, Petsche described the devaluation as '*une décision brutale*, which he had hoped might now have been avoided'. (The second round of Washington talks, in which the French had taken part, took place from 13–17 September.) The subsequent reaction of French officials was described by Makins as 'one of complete consternation. Alphand in particular

became somewhat hysterical'. Hervé Alphand was at that time Head of the French Delegation to the OEEC.

CHAPTER 10

1. Sir Reader Bullard was Ambassador in Tehran until his retirement in 1946. The way in which his Soviet colleague (A. A. Smirnov, afterwards Deputy Foreign Minister) spoke of him to the author in Moscow in 1970 was the more remarkable in that Bullard was under no illusions about the Soviet system.
2. The description—undated, but probably 1963—of Sir Walter Smart, Oriental Minister in Cairo at the end of the war, is Julian Amery's in *Walter Smart by Some of His Friends*, copy no. 105 (Chichester Press), p. 1. There is a copy in the Middle East Centre at St Antony's, Oxford.
3. Annex to COS (45) 175th meeting, 12 July 1945. *DBPO*, vol. i, p. 208.
4. Memorandum by the Chiefs of Staff, DO(46)47, 2 April 1946, p. 5—CAB 131/2.
5. 'A small but most irritating part', Wm. Roger Louis, *The British Empire in the Middle East, 1945–1951: Arab Nationalism, the United States, and Postwar Imperialism* (Clarendon Press, Oxford, 1984), p. 4.
6. *FRUS, 1945*, vol. viii, pp. 2–3. 'His reassurance,' Roosevelt added, 'concerned his own future policy as Chief Executive of the US Government'. See also p. 702.
7. *DBPO*, vol. i, pp. 25 and 108 (Churchill's minute and the Chiefs of Staff's comments on it).
8. 426 *H.C.Debs.*, col. 1253, 1 August 1946.
9. Truman's memorandum of 24 July 1945 is in *DPBO*, vol. i, pp. 1043–4, which also records the minutes written on this memorandum by Foreign Office officials. His letter of 31 August 1945 to Attlee is in Truman, *Years of Trial and Hope*, pp. 138–9.
10. The figure of 100,000 'acquired an almost symbolic significance' as time went by: Louis, *British Empire in the Middle East*, pp. 386 ff.
11. *DBPO*, vol. i, p. 1042.
12. 415 *H.C.Debs.*, col. 1934, 13 November 1945.
13. Bullock, *Bevin*, pp. 182–3; and for Bevin's press conference and the Zionist reaction to it, see pp. 181–2.
14. Personal recollection of the author.
15. Sargent's letter of 26 November 1945 to W. E. Houstoun-Boswall, then Minister at Sofia, quoted by Halifax in his letter of 12 December 1945 to Bevin, in FO 800/478/Misc.
16. The British Middle East Office, established in Cairo after the war, was one of Bevin's many disappointments in the area. Papers in the Foreign Office's 'Nile Waters' file were numerous and treated very seriously in the late 1940s: the author's personal recollection.
17. Bevin to Campbell, 21 June 1946, in FO 800/457/EG46/32. The parlous condition of Egypt at this time was well documented for the RIIA by Charles Issawi in his *Egypt: an Economic and Social Analysis* (Oxford University Press, Oxford, 1947).
18. Although the reasons for this dislike were largely irrational, it was a real factor in British domestic politics at the time. See the correspondence about it quoted by Louis, *British Empire in the Middle East*, p. 120.
19. Minute to the Prime Minister by Emmanuel Shinwell, Secretary of State for War, 9 December 1949 in PREM 8/1230. The same minute, which Attlee

welcomed, also stated that there was 'no other suitable location' for the main British base in the Middle East.

20. Memorandum by the Prime Minister and the Minister of Defence, DO(46)27, 2 March 1946, in CAB 131/2.

21. This strategic concept, codenamed Pincher, is quoted in Gregg Herken, *The Winning Weapon* (Knopf, New York, 1980), pp. 219 ff. Air bases for atomic forces in Britain, Egypt, and India were also assumed in the 1947 version of Pincher, codenamed Broiler (ibid., pp. 227 ff.).

22. Memorandum by the Secretary of State for Foreign Affairs, DO(46)40, 13 March 1946, in CAB 131/2. (For the ideological argument, see p. 28.)

23. Ibid.—reports by the Chiefs of Staff, 'Location of Middle East Forces', DO(46)48, 2 April 1946, and 'Egyptian Treaty Revision', DO(46)56, 15 April 1946.

24. The extent of the fuss made by Killearn over his transfer is revealed in FO 800/457/Eg 46, papers 5 ff. Campbell had served before in Cairo, where he was widely respected.

25. 422 *H.C.Debs.*, cols. 883–6, 7 May 1946.

26. See, for example, the Permanent Under-Secretary's Committee's 1949 Memorandum on 'The Near East', in FO 371/76384: '. . . the strategic key to this area is Egypt, to which there is no practical alternative as a main base . . . Cyrenaica and Transjordan can afford adjuncts but not a substitute'.

27. The text of the Bevin–Sidky 'Anglo-Egyptian Treaty of Mutual Assistance' is in FO 800/457/EG 46/32 (the Sudan Protocol was Annex E—paper 118 in this file). The description of Sidky as 'the most astute operator of them all' was Sir Laurence Grafftey-Smith's, quoted by Louis, *British Empire in the Middle East*, p. 235.

28. Sidky having been succeeded as Egyptian Prime Minister by Nokrashi Pasha (assassinated in December 1948), the Wafd Party, which had cooperated with the British during the war, was returned to power after the elections of January 1950.

29. Acheson, *Present at the Creation*, p. 169. For Henderson, Sumner Welles, and George Wadsworth (examples, each in his own way, of the first American school of thought about the Middle East) see Louis, *British Empire in the Middle East*, pp. 37–9 and 163–5.

30. See the references *passim* to all three in Robert Donovan, *Conflict and Crisis* (Norton, New York, 1977), but especially in Chapter 34 and 39 (also in relation to Weizmann).

31. This description of Weizmann is Isaiah Berlin's, *Personal Impressions* (The Hogarth Press, London, 1981), pp. 57–8.

32. See ibid., and also Bullock, *Bevin*, p. 173 (respectively, 'the British Government . . . wounded him as no one else ever could', and 'he would never forgive' this affront).

33. Quoted in R. H. S. Crossman, *Palestine Mission* (Hamish Hamilton, London, 1947), p. 163.

34. *PP, Harry S. Truman, 1946*, pp. 218–19.

35. 422 *H.C.Debs.*, col. 197, 1 May 1946.

36. It is a moot point which of Bevin's public utterances about Palestine did him most harm, but on any reckoning this one must come high on the list: 'I hope I will not be misunderstood in America if I say that this [100,000] was proposed with the purest of motives. They did not want too many Jews in New York'.

37. 'A classic Colonial Office solution of dyarchy', Louis, *British Empire in the Middle East*, p. 432. The use of Morrison's name was due to the fact that it was he who announced the terms of the plan in Parliament.

38. *FRUS, 1946*, vol. vii, Truman to Attlee, 12 August 1946, p. 682.

39. Acheson, *Present at the Creation*, p. 176. The text of Truman's statement is in *PP, Harry S. Truman, 1946*, pp. 442–4.
40. In PREM 8/627/5, Part 5, ending with the sentence: 'I shall await with interest to learn what were the imperative reasons which compelled this precipitancy'.
41. Truman, *Trial and Hope* (Doubleday, New York, 1956), p. 154.
42. Including the High Commissioner in Palestine, General Sir Alan Cunningham, and—initially—Brigadier I. N. Clayton, the principal Arabist at the British Middle East Office in Cairo: see the despatch of 20 September 1946 from the High Commissioner to the Secretary of State for the Colonies, in PREM 8/627, Part 5.
43. CAB 128/11, CM (47), 6th Conclusions, Minute 3, Confidential Annex, 15 January 1947.
44. See, for example, David Horowitz, *State in the Making* (Knopf, New York, 1952), p. 143.
45. Donovan, *Conflict and Crisis*, pp. 329–31.
46. Quoted from a 'Memorandum on the essential solution of the Palestine problem', by Bernard Burrows, Head of the Eastern Department in the Foreign Office, 10 June 1948, in FO 371/68566. This passage in the memorandum in fact attributed the prediction to the Arabs. But Bevin himself spoke more than once of Israel's becoming another China.
47. CM (47), 93rd Conclusions, 4 December 1947, in CAB 127/10.
48. *FRUS, 1947*, vol. v, pp. 488–626 contains the US record. (The 'General Statement by the American Group' is on pp. 582 ff.) The British record is in DO 800/476/ME/47/21.
49. Minute by Michael Wright, 20 January 1948, in FO 371/68041.
50. Donovan, *Conflict and Crisis*, pp. 371 ff. offers a vivid account. Truman's statement announcing US recognition of the State of Israel is in *PP, Harry S. Truman, 1948* p. 258.
51. *FRUS, 1948*, vol. v, Part 2, p. 993 (published 1976).
52. Minute by Wright, 15 May 1948 (after a talk with the Foreign Secretary that morning), in FO 371/68665/E6758.
53. *FRUS, 1948*, vol. v, Part 2, Douglas to Lovett, 22 May 1948, p. 1031.
54. For the 'abscess', see minute by Wright, 23 June 1948, in FO 371/68650/E8121. The second quotation is the conclusion of Burrows' memorandum of 10 June 1949 quoted above, in FO 371/68566.
55. '. . . a bold move to recapture the British political initiative in the Middle East. His goal was a system of defensive alliances to be concluded with each of the important Arab states. The first steps would be taken with Iraq and Jordan and would eventually encompass Saudi Arabia as well. On the basis of successful treaties with those countries Bevin hoped to bring even Egypt back into the British fold. The underlying rationale was that the Arab states would now look to the British for leadership because the British had avoided a pro-Zionist line . . .' Louis, *The British Empire in the Middle East*, p. 200.
56. As reported in paragraph 2 of Washington telegram No. 247 to the Foreign office, 13 January 1949 (FO 371/75334/E614).
57. Memorandum by Bevin, 'Middle East Policy', CP(49)188, 25 August 1949, in CAB 129/36, Part 2.
58. *FRUS, 1948*, vol. v, Part 2, pp. 1512 ff; and Truman, *Trial and Hope*, pp. 168–9. Truman's statement of 24 October 1949 on Israeli boundaries is in *PP, Harry S. Truman, 1948*, p. 843.
59. To which, however, Hugh Dow's minutes make a refreshing contrast: for example, in FO 371/75054/E2478 and FO 371/91184. Sir Hugh Dow was Consul-General in Jerusalem.

60. Bevin's warning to Marshall on 27 October 1948 is recorded in *FRUS, 1948*, vol. v, Part 2, p. 1521. Louis, *British Empire in the Middle East*, pp. 564–5, discusses how near Anglo-Israeli conflict came in January 1949, when four RAF reconnaissance aircraft were shot down by Israeli forces.

61. Minute by Hector McNeil to the Secretary of State, 14 January 1949, in FO 371/65337/E1881/G.

62. Bullock, *Bevin*, p. 652—a judgement fully supported by a Top Secret letter which Bevin wrote to the Ambassador in Washington three weeks later: Bevin to Franks, 3 February 1949, E1932 G, also in FO 371/75337.

63. Acheson, *Present at the Creation*, p. 396. The British Government effectively disowned the declaration in 1956. For its text, see 'Joint Declaration with the United Kingdom and France on the Arab States and Israel', 25 May 1950, in *PP, Harry S. Truman, 1950*, pp. 147–8.

64. 'A British responsibility in case of a hot war, at least during the first two years of such a war': General Joseph Collins, expressing this view on behalf of the US Joint Chiefs of Staff at the 'US–UK Political Military Conversations, Washington, 23–26 October 1950': *FRUS, 1950*, vol. iii, p. 1693.

65. This was Franks' summary of the British view, at the talks in Washington referred to in the previous note.

66. ... 'of a kind that permanently and deeply alter many human lives', Berlin, *Personal Impressions*, p. 33.

67. See Donovan, *Conflict and Crisis*, p. 323 (on Lewis), and Halifax's letter to Churchill of 3 August 1945, in the microfilm of Halifax Papers in Churchill College Library, Cambridge (reel II, 410), for the description of Bevin.

68. Cf. Donovan, ibid., p. 387.

69. A point rammed home by the British team in the papers submitted at the Pentagon Talks in October 1947 (see *FRUS, 1947*, vol. v, pp. 580–1) and three years later, in the Political-Military Conversations held in Washington (see notes 64 and 65 to this chapter).

70. The words of Robert Lovett, US Under Secretary, reported by Franks to Bevin in the telegram quoted above (FO 371/75334/E614).

CHAPTER 11

1. *FRUS, 1942*, vol. i, Mohandas K. Gandhi to President Roosevelt, 1 July 1942, p. 677.

2. In the words of Acheson's letter to the President of 30 July 1949, transmitting the 'China White Paper', American 'friendship for that country [China] has always been intensified by the religious, philanthropic and cultural ties which have united the two peoples ...' *CWP*, pp. iii–iv.

3. For a description and an assessment of this remarkable lobby, see Robert Donovan, *Tumultuous Years, 1949–52* (Norton, New York, 1982), Chapters 2, 6, and 7. (Henry Luce was himself the son of a missionary to China.)

4. *New York Times*, 23 January and *Christian Science Monitor*, 25 July 1942.

5. E.g., *The Times*, on China's 'resplendent destiny', 12 January 1943.

6. For examples of Clark Kerr's overestimation of Chiang Kai-shek and his misunderstanding of Mao, see Christopher Thorne, *Allies of a Kind* (Hamish Hamilton and Oxford University Press, New York and Oxford, 1979), pp. 67, 68,

and 185. Cripps also went overboard both for Chiang and for his wife (pp. 193–4).

7. *CWP*, p. 129.

8. The Long March had reduced 90,000 men to 20,000: Stanley Karnow, *Mao and China* (Macmillan, London, 1973), pp. 44–5. Karnow estimates their post-war military strength as half a million (p. 50). Edgar Snow estimates CPC membership in 1936 in the north-west as 'no more than 40,000': *Journey to the Beginning* (Victor Gollancz, London, 1959), p. 177.

9. M. I. Sladovskii, *Kitai i Angliya* (Nauka, Moscow, 1980), p. 201.

10. Francis Valeo, *The China White Paper*, a summary of the State Department volume, which was prepared for the Library of Congress Legislative Reference Service, Public Affairs Bulletin No. 77 (October 1949), p. 39.

11. *CWP*, p. 314.

12. *FRUS, Yalta Papers*, p. 771. Byrnes, however, was still saying much the same thing at the Moscow Meeting of Foreign Ministers in December 1946. See *DBPO*, vol. ii, p. 850.

13. Truman, *Year of Decisions*, pp. 102 ff. The same paper also said that the return of Hong Kong to China was something to be 'welcomed'.

14. For examples of British views, see Thorne, *Allies of a Kind*, p. 556 and 561–2; also the even earlier—and remarkably accurate—forecast in a minute by George ('Gerry') Young, then in the Far Eastern Department of the FO, which Thorne quotes on p. 321.

15. Truman's own account is in his *Year of Decisions*, pp. 446 ff.

16. For British ideas, such as they were, see Thorne, *Allies of a Kind*, pp. 556 ff.

17. Foreign investments of this kind are notoriously hard to assess accurately, but for a Chinese estimate, see Wu Chengming, *Imperialist Investments in Old China* (People's Publishing Press, Peking, 1956).

18. *CWP*, p. x.

19. For a sterner verdict, see William W. Stueck Jr., *The Road to Confrontation: American policy toward China and Korea, 1947–50* (The University of North Carolina Press, Chapel Hill, 1981), p. 123: 'Both temperamentally and intellectually, Acheson was poorly suited to dealing in an astute manner with the Communists'.

20. Robert Donovan, *Conflict and Crisis* (Norton, New York, 1977), p. 104.

21. Marshall's instructions were virtually impossible of fulfilment. Even if he could not secure 'the necessary action by the Generalissimo', which he thought 'reasonable and desirable, it would still be necessary for the US Government, through me [Marshall] to continue to back the Nationalist Government of China through the Generalissimo': *FRUS, 1945*, vol. vii, p. 770, Marshall's memorandum of 14 December 1945, recording his final conversation with Truman and Acheson before leaving for China. See also Truman, *Years of Trial and Hope*, pp. 67 ff.

22. Stueck, *The Road to Confrontation*, p. 58.

23. Annual review for 1948, enclosed in Washington despatch No. 694 to the Foreign Office of 1949: FO 371/74159. For an account of the bureaucratic dissensions in Washington during these years, see Stueck, *The Road to Confrontation*, e.g. pp. 109–10.

24. Acheson's letter of transmittal to the President is on p. xiv of the *CWP*.

25. *Washington Post*, 12 September 1949.

26. See Rusk's televised interview of 22 October 1961 (Department of State *Bulletin*, **45** (1961), pp. 802–4).

27. The author's personal recollection of a long discussion with Senator Jackson over lunch in the Senate, in August 1961.
28. *Pravda*, 15 February 1950. For an assessment of the treaty and of Sino-Soviet relations in the preceding years, see Adam Ulam, *Expansion and Coexistence* (Secker & Warburg, London, 1968), pp. 470 ff.
29. Memorandum prepared by the Foreign Office Permanent Under-Secretary's Committee, 30 August 1949 (FO 371/76385). The paper, however, reached the conclusion that the British aim in South-East Asia should be 'some form of regional association . . . in partnership with the Atlantic powers'.
30. Edwin Pauley's report, 12 November 1946, on this subject is in *CWP*, pp. 598 ff.
31. Esler ('Bill') Dening, a much respected official (who had been Mountbatten's Political Adviser at South-East Asia Command during the war), later became Ambassador to Japan. Had 1950 turned out differently, he would have been Britain's first Ambassador to the Chinese People's Republic.
32. Dening to Sir Cecil Syers, 18 March 1949, in FO 371/76023/F4468, enclosing the draft brief on 'South East Asia and the Far East'.
33. Ibid., second and third paragraphs of the brief: 'We must face the prospect that the whole of China will come under eventual Communist domination' and 'it is by no means certain that the Republic of South Korea will, in the long run, be able to prevail against Communist pressure from the north'.
34. Minute by Dening to the Foreign Secretary, 23 March 1949: ibid.
35. FO 800/448, 'Secretary of State's Conversation and meetings during his visit to Washington . . .', Chapter 9, 30 March 1949. See also *FRUS, 1949*, vol. viii, pp. 1135–7 and 1204–8, and vol. vii, pp. 1215–20. For a different view, see Ritchie Ovendale, *The English-Speaking Alliance: Britain, the United States, the Dominions and the Cold War, 1945–1951* (Allen & Unwin, London, 1985), Chapter 6.
36. Malcolm MacDonald, Singapore despatch No. 16 to the Foreign Office, 23 March 1949: FO 371/76633/F4545.
37. *The Times*, 20 May 1949.
38. *FRUS, 1949*, vol. ix, pp. 225–6, memorandum of conversation by Jacob D. Beam, pp. 1140–1. See also the record of a meeting between Acheson and Bevin at the State Department, 13 September 1949: FO 800/462/FE/49/21.
39. Memorandum on China by the Secretary of State for Foreign Affairs, CP (49) 180, 23 August 1949—CAB 129/36, Part 2. Cf. John L. Gaddis' remark in 'Korea in US politics', *The Origins of the Cold War in Asia, International Symposium (1975)*, ed. Yonosuke Nagai and Akira Iriye (Columbia University Press and University of Tokyo Press, New York and Tokyo, 1977), p. 292: 'men who had set out to exploit a Sino-Soviet split wound up instead encouraging Sino-Soviet unity'.
40. *FRUS, 1949*, vol. ix, pp. 225–6. For the run-up to the British decision to recognize, see the account in Ovendale, *The English-Speaking Alliance*, pp. 193–200.
41. For the 'dust settling', see Acheson, *Present at the Creation*, p. 306. This decision was approved by Truman at a meeting of the National Security Council held on 30 December 1949: see Donovan, *Tumultuous Years*, pp. 88 and 416.
42. Thorne's summing up of US policy towards China during the Second World War (in *Allies of a Kind*, p. 574) applies with equal force to the policy that followed it: 'the Great Republic has known better moments in its foreign relations . . . those elements of genuine well-meaning and concern that, alongside hypocrisy, condescension and self-seeking, formed a part of the United States' response to China, deserved something more than this'.
43. By the time that British and Chinese forces found themselves fighting on opposite

sides in Korea, at the end of 1950, Sino-British relations were still entangled in argument between the two governments. They remained in a diplomatic limbo until 1954. For the episode of Dening's application for a Chinese visa, see FO 371/ 84519. I am indebted to Mr James Tang for drawing my attention to this interesting file in the Public Record Office.

CHAPTER 12

1. US records of the fourth meeting of the CFM are in *FRUS, 1947*, vol. ii, pp. 138–471. British records are in FO 371/64176–64207.
2. In his Memorandum of 8 November 1945, 'The Foreign Situation': see Chapter 3.
3. For the background to this treaty, see Duff Cooper, *Old Men Forget* (Hart-Davis, London, 1954), pp. 359 ff. Its signature had been long delayed; a draft treaty was included among the briefs taken to Potsdam by the British delegation: see *DBPO*, vol. i, pp. 248 ff.
4. In Article 11 of the *Anglo-French Treaty of Alliance and Mutual Aid*, Dunkirk, 4 March 1947, Cmd. 7217 (HMSO, London, 1947).
5. CM(47)15, 3 February 1974, in CAB 128/9.
6. Page 4 of Sir Frank Roberts' record of 'Anglo-French Conversations', 17 December 1947, in FO 371/67674/Z 11010 (G).
7. Gowing, *Independence and Deterrence*, vol. i, p. 91, based on DO(47)44—in CAB 131/4, which is, however, a folio still withheld from the Public Record Office.
8. In the French case this took the form of a resolution by the National Assembly opposing German rearmament and German accession to the North Atlantic Pact. For British views at this juncture, see Bullock, *Bevin*, pp. 739–40.
9. The last talk in the series took place between the British Ambassador, Sir Maurice Peterson, and Yakov Malik, in the Soviet Foreign Ministry, on 28 May 1947: FO 800/502. For an account of these negotiations, see Elizabeth Barker, *The British between the Super-Powers, 1945–50* (Macmillan, London, 1983), pp. 71 ff.
10. *FRUS, 1948*, vol. ii, pp. 999–1006. See also Walter Bedell Smith, *My Three Years in Moscow* (Lippincott, Philadelphia, 1950), p. 244.
11. *The Times*, 20 September 1946.
12. Ibid., 15 May 1947. See also Michael Charlton, *The Price of Victory* (British Broadcasting Corporation, London, 1983), pp. 38–40.
13. Jean Monnet, *Mémoires* (Fayard, Paris, 1976), frontispiece ('We are not making coalitions of States, we are uniting men') and p. 350.
14. In the course of Bevin's meeting with Clayton at 10 Downing Street on 24 June 1947: *FRUS, 1947*, vol. iii, p. 273.
15. The text of this broadcast is in RIIA *Documents on International Affairs, 1947–48*, ed. Margaret Carlyle (London, 1952), pp. 471–81. Before his departure from Moscow Marshall had left his Soviet hosts in no doubt about his impatience to get on with the task of European reconstruction, in the speech that he made at the farewell banquet at the Kremlin, to which their reaction was frosty: I am indebted to Michael Cullis for this personal recollection.
16. 'The United States and England might be willing to give up reparations; the Soviet Union could not . . . ten billion dollars of reparations were very popular in the Soviet Union'. The US record (a copy of which was given to the British) is in *FRUS, 1947*, vol. ii, pp. 337–414. The British copy is in FO 800/502/71933. The quotation above, about 'skirmishes', is derived from the same source: an important document in the history of the cold war.

17. Evgenii Varga's phrase, in his article published in *Foreign Affairs* (cited in Chapter 7 n. 44) four months after the presidential message. In the shift of Congressional opinion, Vandenberg played the key role from the outset: see Acheson, *Present at the Creation*, p. 219.

18. *FRUS, 1947*, vol. v, pp. 32 ff. contains the texts of the two aides-mémoire, one on Greece and the other on Turkey.

19. Francis Williams, *A Prime Minister Remembers*, (Heinemann, London, 1961), p. 172.

20. Margaret Truman, *Harry S. Truman* (Morrow, New York, 1973), p. 343.

21. The text of Truman's Special Message to Congress on Greece and Turkey is in *PP, Harry S. Truman, 1947*, pp. 176–9.

22. Milovan Djilas, *Conversations with Stalin* (Hart-Davis, London, 1962), p. 164.

23. C. M. Woodhouse, *The Struggle for Greece* (Hart-Davis, McGibbon, London, 1976), p. 149.

24. For the voting, which angered Bevin (a 'stab in the back'), see Kenneth O. Morgan, *Labour in Power, 1945–1951* (Clarendon Press, Oxford, 1984), pp. 63–4.

25. FO 800/475/ME/46/22.

26. New York telegram No. 2295 to FO, for Minister of State from Foreign Secretary, 5 December 1946: GRE/46/40 in FO 800/468.

27. The Memorandum by the Prime Minister, 5 January 1947, entitled 'Near Eastern Policy', is in FO 800/502/71933.

28. Bevin's reply to Attlee is in FO 800/476/ME/47/4, 9 January 1947. According to Montgomery, at his prompting, the Chiefs of Staff let Attlee know privately that they would resign if he persisted in his view of the Middle East: Viscount Montgomery, *Memoirs of Field-Marshal Montgomery* (Collins, London, 1958), p. 436. Attlee's memorandum of 5 January 1947 was indeed 'the most radical criticism Bevin had to face from inside the Government during his five and a half years as Foreign Secretary', Bullock, *Bevin*, pp. 348–54.

29. FO 800/475/ME/46/25, minute by Nicholas Henderson, 28 December 1946 (recording the talk at Chequers); FO 371/67032, minute to the Prime Minister from the Chancellor of the Exchequer, 11 February 1947 (with Bevin's manuscript concurrence); and minute by the Secretary of State, 18 February (recording his talk with the Chancellor of that date).

30. *FRUS, 1947*, vol. i, pp. 750–1.

31. Marshall's view, according to *Forrestal Diaries*, ed. Walter Millis and E. S. Duffield (Viking Press, New York, 1951). For an account of the crisis in Washington written not long after the event, see Joseph Jones, *Fifteen Weeks* (Viking Press, New York, 1955).

32. Acheson, *Present at the Creation*, p. 219, 'These congressmen had no conception of what challenged them, it was my task to bring it home' (at the first meeting held at the White House, on 26 February 1947, to discuss the issue).

33. Senate Committee on Foreign Relations, *Legislative Origins of the Truman Doctrine*, p. 95, quoted in Robert Donovan, *Conflict and Crisis* (Norton, New York, 1977), p. 286.

34. For example, George Kennan and Gladwyn Jebb. For Jebb's later account, see Lord Gladwyn, *The Memoirs of Lord Gladwyn* (Weidenfeld and Nicolson, London, 1972), pp. 199–201.

35. Bullock, *Bevin*, p. 371, citing Dalton's *Memoirs*.

36. The opportunity to make 'his most decisive personal contribution as Foreign Secretary to the history of his times': Bullock, *Bevin*, pp. 404–5.

CHAPTER 13

1. *The Times*, 6 June 1947. The edition of the following day did much better.
2. Bullock, *Bevin*, p. 404.
3. Hugh Dalton, Diary, 24 February 1947 (written just before Bevin left for Moscow), quoted in ibid., p. 362.
4. FO telegram No. 999 to Paris, 9 June 1947, in FO 800/465. The same file contains Bevin's subsequent telegram, sent to Paris on 14 June.
5. *FRUS, 1947*, vol. iii, p. 239, which is also the source of the quotation that follows in the next paragraph.
6. In Monnet's apt description, 'L'affaire fut menée entre cinq ou six personnes dans un total secret et avec une rapidité fulgurante' ('The business was dealt with among five or six people in complete secrecy and with lightning speed'): Jean Monnet, *Mémoires* (Fayard, Paris, 1967), p. 315.
7. *FRUS, 1947*, vol. v, pp. 94–5.
8. US Department of State *Bulletin*, 18 May 1947, p. 991.
9. *FRUS, 1947*, vol. iii, p. 233.
10. Ibid., pp. 230–2. Quotations that follow from Clayton's memorandum of 27 May (though it may have been written earlier) are from this source.
11. Ibid., pp. 223–30. Quotations from Kennan's paper that follow are from this source.
12. The underlinings in these quotations from the memorandum are Clayton's.
13. *FRUS, 1947*, vol. iii, p. 226.
14. Ibid., p. 221, Memorandum by the Director of the Policy Planning Staff, 16 May 1947.
15. Kenneth Royall's remark that he 'felt free to boost German industrial production without consulting the French' brought the Marshall Plan negotiations to a temporary halt: see Bullock, *Bevin*, p. 432, to which I am also indebted for the description of the agreement between the two US Departments.
16. *FRUS, 1947*, vol. iii, p. 239.
17. The title (perhaps ironical) of a scathing article by K. Morozov, published in *Pravda Ukrainy*, 11 June 1947, quoted in a translation telegraphed to the State Department by the US Embassy, Moscow at the time: *FRUS, 1947*, vol. iii, pp. 294–5.
18. The conference 'ended with sufficient appearance of success to keep hope alive': Bullock, *Bevin*, p. 460.
19. In mid-1947 the strength of the British Army, excluding colonial troops and locally enlisted personnel abroad, was 845,805 (source: *Strength Return of the British Army*, June 1947, War Office publication)—a strength comparable with that of the US Army at the time, which was 989,664: (source: *Strength of the Army, STM-30*, Department of the Army, Washington, DC, 1 July 1949). I am indebted for the latter figure to statistics supplied by Lieutenant Colonel Robert Frank, Infantry Chief, Research and Analysis Division, Department of the Army Center of Military History.
20. All the quotations from these five meetings that follow are derived from the summaries prepared by the American side, copies of which were sent by the US Ambassador on 1 July 1947 to the Foreign Secretary and to British officals: *FRUS, 1947*, vol. iii, pp. 268 ff.
21. Sir Edward Bridges, who was also Treasury Permanent Secretary; Sir John Henry Woods spoke for the Board of Trade.
22. *FRUS, 1947*, vol. iii, pp. 284–8. The chief British reservation was that the US 'continental approach' to European problems presented 'very special difficulties'

for Britain, which was 'not merely a European country but an international trader'.

23. By Adam Ulam, *Expansion and Coexistence* (Secker & Warburg, London, 1968), p. 434.

24. *FRUS, 1947*, vol. iii, pp. 319–20 has the text of Klement Gottwald's telegram of 9 July 1947, describing two meetings with Stalin in Moscow, which looks authentic. On p. 305 of the same volume is Couve de Murville's account of the Paris meeting on 1 July, quoted by the US Ambassador in Paris in his telegram to Washington, 2 July 1947.

25. Ibid., p. 303, Bevin, quoted by the US Ambassador in Paris, in his telegram to Washington, 1 July 1947.

26. Ibid., p. 307, has the text of the US Ambassador in London's telegram to Washington, 3 July 1947. Cf. Bevin's remark to Dixon quoted in Bullock, *Bevin*, p. 422.

27. The International Communist Information Bureau.

28. Andrei Zhdanov, defender of Leningrad during the Second World War, was a member of the Politburo, best remembered inside the Soviet Union for his post-war cultural repression, known as the *Zhdanovshchina*. He died in mysterious circumstances in August 1948. The text of his speech at the inaugural meeting of the Cominform, held at Wielniza Góra in Poland on 22–3 September 1947, was carried in the first issue of the Cominform's journal, *For a lasting peace, for a people's democracy*.

29. Ulam, *Expansion and Coexistence*, p. 461.

30. The American record is in *FRUS, 1947*. ii, pp. 676–830; and the British is in FO 871/65341 and 64645–6.

31. See, for example, the record of Marshall's meeting with Bevin at the US Ambassador's residence on 4 December 1947, *FRUS, 1947*, vol. ii, pp. 750–3; and FO 800/466/GER/47/47.

32. *FRUS, 1947*, vol. ii, pp. 694 and 698, memoranda of conversations by Jacob D. Beam, 30 October and 4 November 1947.

33. Ibid., pp. 822–7 (a British record) and 827–9 (a US record).

34. FO 800/466/GER/47/49. 'Anglo-American Conversations', 17 December 1947; and *FRUS, 1947*, vol. ii, pp. 815–22 (a British record of Bevin's meeting with Marshall). Frank Roberts' record of Bevin's talk with Bidault is in FO 371/67674/ Z 11010/G, 'Anglo-French Conversations', 17 December 1947. Bevin also talked to the Canadian High Commissioner.

CHAPTER 14

1. *Berlin Papers*, vol. i, pp. 256 ff.

2. Quoted by John Baylis in *International Affairs*, **60**: 4 (Autumn 1984), 'Britain, the Brussels Pact and the Continental commitment', p. 619. On the 1944 discussion of Western Europe in Whitehall, see also his article.

3. The western divisions were outnumbered by at least three to one. How many Soviet divisions were then capable of offensive operations in Europe is still a matter of debate: perhaps about thirty (certainly nothing remotely approaching the figure of 175 with which the Red Army was credited by some western estimates at the time). Khrushchev (*Pravda*, 15 January, 1960) gave the figure of 2,874,000 as the level to which the total Soviet armed forces had been reduced by 1948—not greatly in excess of the combined Anglo-American total at the time—although he may have exaggerated the extent of Soviet demobilization. For two

views of the figures, see Samuel F. Wells, Jr., 'Sounding the Tocsin: NSC 68 and the Soviet Threat' in *International Security*, **4**: 2 (Fall 1979), pp. 152–3, and Hannes Adomeit, *Soviet Risk-Taking and Crisis Behaviour* (Allen & Unwin, London, 1982), pp. 138–41.

4. See Nicholas Henderson, *The Birth of NATO* (Weidenfeld & Nicolson, London, 1982), pp. 9 and 24–25, for examples.

5. DO(50), 5th Meeting, 23 March 1950 (CAB 131/9), which is further discussed in Chapter 15. See also Elizabeth Barker, *The British between the Super-Powers, 1945–1950* (Macmillan, London, 1983), pp. 114–15.

6. 446 *H.C.Debs.*, col. 383–409, 22 January 1948, based on Bevin's Cabinet Memorandum, 'The First Aim of British Foreign Policy', CP(48)6, 4 January 1948 (CAB 129/25).

7. Attlee to Bevin, 28 January 1948, quoted in Kenneth Harris, *Attlee* (Weidenfeld & Nicolson, London, 1982), p. 601; *The Economist*, 31 January 1948; and Marshall to Inverchapel, 20 January 1948, *FRUS, 1948*, vol. iii, pp. 8–9.

8. Jonathan Haslam, 'E. H. Carr and the History of Soviet Russia' in *The Historical Journal* **26**: 4 (1983), p. 1021.

9. This phrase was used in Prague telegram No. 19 (Saving) to FO, 27 February 1948, annexed to Bevin's report to the Cabinet, CP(48)71, 3 March 1948 (in CAB 129/25).

10. See the correspondence with the British Ambassador in Prague quoted in Bullock, *Bevin*, p. 525.

11. The epithet used by George Kennan to describe both the Prague take-over and the Berlin blockade: 'The United States and the Soviet Union, 1917–1976', *Foreign Affairs*, **54**: 4 (July 1976), pp. 683–4.

12. *FRUS, 1948*, vol. iii, pp. 32–3. The British record of this conversation, on 26 February 1948, records Bevin as ascribing the same view to Churchill: Frank Roberts' minute of 26 February 1948 (in FO 800/502).

13. CP(48)72, 3 March 1948 (in CAB 129/25).

14. *FRUS, 1948*, vol. iii, p. 478.

15. Lovett to Inverchapel, 2 February 1948, ibid., p. 18.

16. Ibid., Douglas to Secretary of State, 26 February 1948, 18 and pp. 32–3. Montgomery's own assessment of the success of his personal effort, made during a visit to Washington in September 1946, to initiate Anglo-US-Canadian defence talks, has to be taken with a grain of salt: see Viscount Montgomery, *The Memoirs of Field-Marshal the Viscount Montgomery* (Collins, London, 1958), pp. 440–3.

17. *The Treaty of Economic, Social and Cultural Collaboration and Collective Self Defence*, 17 March 1948, Cmd. 7599 (HMSO, London, 1948). For a discussion of the problems involved in the negotiations leading up to the Brussels Pact, see John Baylis, *Anglo-American Defence Relations, 1939–1984: the special relationship*, 2nd edn. (Macmillan, London, 1984).

18. Montgomery, *Memoirs*, pp. 504–7.

19. *North Atlantic Treaty, 4 April 1949*, Treaty Series, No. 56 (1949), Cmd. 7789 (HMSO, London, 1949). A comparison of the full texts of the draft of 24 December 1948 and of the treaty signed on 4 April 1949 may be found in Appendices A and B of Henderson, *Birth of NATO*, pp. 119 ff.

20. CP(50)220, 6 October 1950, in CAB 129/42.

21. Henderson, *Birth of NATO*, p. x.

22. For an account of the controversy that accompanied the passage of this bill through Congress, see Acheson, *Present at the Creation*, pp. 307–13.

23. It was none the less regarded in London as sufficiently important to be reproduced in full as an appendix (II) to Bevin's memorandum DO(48)64, 20

September 1948, reporting to the Cabinet Defence Committee on the 'Washington Exploratory Conversations on Security'—FO 800/468/GER/49/5.

24. Minute by the Foreign Secretary to the Prime Minister, 31 December 1948, referred to the Cabinet Defence Committee DO(48)88 (in CAB 131/6).

25. Henderson, *Birth of NATO*, p. 56.

26. 'If any of the High Contracting Parties shall be the object of an armed attack in Europe, the other High Contracting Parties will, in accordance with the provisions of Article 51 of the Charter of the United Nations, afford the Party so attacked all the military and other aid and assistance in their power'; see n. 17 to this chapter.

27. Acheson, *Present at the Creation*, pp. 280–3.

28. Bullock, *Bevin*, p. 672.

29. Henderson, *Birth of NATO*, p. 57.

30. *FRUS, 1948*, vol. iii, pp. 72–5; summarized in Bullock, *Bevin*, p. 542.

31. The full text of these recommendations forms the annexes to CP(48)143, 4 June 1948. In CAB 129/270. *FRUS, 1948*, vol. ii, pp. 1–702 contains the US records of 'The London Conference on Germany' and 'The Implementation of the London Conference'.

32. See Adomeit, *Soviet Risk-Taking*, pp. 94–7; and—for the haziness of the 1945 agreement—ibid., pp. 82–3 and 105.

33. Lewis Douglas, quoted by Bullock, in *Bevin*, p. 587.

34. He had, on Marshall's instructions. On 11 May 1948 Bevin summoned Douglas (who was unbriefed) and asked him 'what deduction was one to draw from this rather extraordinary action?': FO despatch No. 688 to Washington, 11 May 1948, in FO 800/502.

35. Douglas' telegram No. 3625 to the State Department, 11 August 1948, *DSDF*, 711.41/8–1148, National Archives (a telegram written in the first person singular).

36. Robert Donovan, *Conflict and Crisis* (Norton, New York, 1977), pp. 423–5, makes it clear that Truman's political motive was purely domestic.

37. *FRUS, 1948*, vol. iii, pp. 951 and 995–1099, 'The meetings of Representatives of the United States, the United Kingdom, France, and the Soviet Union in Moscow'.

38. For the concession, see Adomeit, *Soviet Risk-Taking*, pp. 99–100. For the Anglo-American difference of view in late August 1948, see the exchange of minutes between Bevin and Attlee on 24 August, in FO 800/502/SU/48/29 (General Sir Brian Robertson was Clay's British opposite number in Germany).

39. Acheson, *Present at the Creation*, pp. 270 and 273. The meeting, fruitless so far as Germany was concerned, succeeded in making substantial progress towards an Austrian State Treaty, although in the event final agreement on this treaty was not reached until 1955. On the whole subject, see Michael Cullis, 'The Austrian Treaty Settlement', British International Studies Association *Review of International Studies*, **7**, (1981), pp. 159–64.

40. Oliver S. Franks, *Britain and the Tide of World Affairs*, BBC Reith Lectures 1954 (Oxford University Press, London, 1955), p. 5.

41. See *FRUS, 1948*, vol. iii, pp. 926–8; also Gregg Herken, *The Winning Weapon*, pp. 256–8 and 383. There is a similar doubt about who first suggested the Berlin airlift; for the British claim, see Bullock, *Bevin*, p. 567.

42. Agreement had in fact been given by the Germany Committee of the Cabinet. The formal record of Bevin's talk with Douglas is in FO 800/467/GER/48/33, FO despatch No. 924 to Washington, 28 June 1948.

43. Hewlett and Duncan, vol. ii, pp. 521–1.

44. See Andrew Pierre, *Nuclear Politics* (Oxford University Press, Oxford, 1972), p. 79.

45. The main British records are FO 800/467/GER/48/33, 38, 39, and 41; and Bevin's report referred to in note 46.

46. Memorandum by the Foreign Secretary to the Defence Committee of the Cabinet, DO(48)59, 'United States Heavy Bombers', 10 September 1948 (in CAB 131/6).

47. See FO 800/456/DEF/51/1 ff.

CHAPTER 15

1. Acheson, *Present at the Creation*, p. 290.

2. Ibid., p. 387.

3. See Chapter 13; the references for the quotations in this paragraph will be found in notes 10–11 and 20–22 to that chapter.

4. See n. 39 to this chapter.

5. Although only half a dozen votes were cast against the Treaty in the House of Commons, there were 112 abstentions—a number comparable with the 'stab in the back' vote that had angered Bevin in November 1946. The wish for an alternative, socialist foreign policy remained alive in the Labour Party and was to receive a stimulus from the events of 1950–1.

6. Acheson, *Present at the Creation*, pp. 286–90.

7. Schuman's German policy described by André François-Poncet (speaking under instructions to Kennan in Frankfurt), 21 March 1949: *FRUS, 1949*, vol. iii, pp. 113–14. Sir Ashley Clarke (then Minister at the British Embassy in Paris) recalls it being said of Schuman at the time that he regarded the Germans as 'nos voisins de campagne' ('our neighbours in the country').

8. Jean Laloy, to whom I am indebted for his recollection of this lapidary remark, which may perhaps be translated: 'The Germans were refused everything when they should have been given something and they were given everything when they should have been refused everything: I, for my part, would like to do something different'.

9. Bullock, *Bevin*, p. 667, and Acheson, *Present at the Creation*, p. 271.

10. Acheson, *Present at the Creation*, Chapter 33 contains his account of the 1949 meeting of the CFM.

11. Acheson's account of the November 1949 German negotiations is in ibid., pp. 337 ff. The British records are in FO 800/467/GER/49/5.

12. Marshall to Robert Murphy, 6 March 1948, in *FRUS, 1948*, vol. iii, p. 389.

13. Robert Lovett to the US Ambassadors in London and Paris, 8 April 1948: ibid., p. 417.

14. Lord Franks in conversation with Michael Charlton: Charlton, *The Price of Victory* (British Broadcasting Corporation, 1983), pp. 72–3.

15. Ibid., p. 75.

16. See Bullock, *Bevin*, pp. 103, 318, and 358; and Elizabeth Barker, *The British between the Super-Powers, 1945–50* (Macmillan, London, 1983), pp. 95–6.

17. CP(48)6, 4 January 1948 (in CAB 129/5).

18. *FRUS, 1948*, vol. iii, pp. 4–6.

19. Paul-Henri Spaak, quoted in Charlton, *The Price of Victory*, p. 54.

20. The description in despatch No. 320 to the Foreign Office, from the British Ambassador at The Hague, Sir Philip Nichols, 19 May 1948—FO 371/73095. For Churchill's Zurich speech, see Chapter 12.

21. Harold Macmillan, *Tides of Fortune 1945–55* (Macmillan, London, 1969), pp. 163–4. Churchill had appointed Macmillan head of the London committee of the European Movement.

22. See Sir Roderick Barclay (Bevin's Principal Private Secretary, 1949–51), in conversation with Charlton, *Price of Victory*, p. 78.

23. See, for example, Kennan's report of 13 September 1949, quoted on p. 110.

24. CP(48)75, 6 March 1948, 'European Economic Cooperation' (in CAB 129/25).

25. CM(48)20, 8 March 1948 (in CAB 128/12).

26. Richard ('Otto') Clarke to Roger Makins, 1 March 1948, FO/371/71808, which contains a copy of the February memorandum, sent by Healey to Hector McNeil, Minister of State.

27. FO 371/67674, Personal letter of 16 October 1947 from the British Ambassador in Paris to the Secretary of State. See also Duff Cooper, *Old Men Forget* (Hart-Davis, London, 1954), pp. 379–81.

28. Record of conversation between the Foreign Secretary and the French Minister of National Defence, at the French Ministry of War, 21 October 1948 (in FO 371/72979). FO 371/79214 contains the record of the talk between Bevin and Schuman at the Foreign Office, 14 January 1949.

29. Bullock, *Bevin*, p. 586. This conviction, however, must be distinguished from the thesis, recently expressed by Ritchie Ovendale in his *The English-Speaking Alliance: Britian, the United States, the Dominions, and the Cold War, 1945–1951* (Allen and Unwin, London, 1985), that Bevin's policy was essentially 'built on a Commonwealth–USA basis'. Not only were the continental Europeans a vital component of the Atlantic defence community. The language of the passage on which Ovendale has heavily relied as support for his thesis (Patrick Dean's minute of 23 August 1950, recording the Foreign Secretary's talk with Spofford and Holmes in FO 800/517/US/50/35), though striking, owes a great deal to the date of the conversation and to the particular circumstances that gave rise to it.

30. Report by the Chiefs of Staff on the 'United Kingdom Contribution to the Defence of Western Europe', DO(50)20, 20 March 1950, circulated by the Minister of Defence to the Cabinet Defence Committee(CAB 131/9), the source of all the quotations in this paragraph, the preceding paragraph, and the following one, with the exception of the Defence Committe's decision of 10 January 1949, which is recorded in DO(49) 2nd meeting, Defence Committee (in CAB 131/8).

31. The records of the time abound in forceful expressions of Makins' views, but the memorandum that he wrote for the Prime Minister on the Schuman Plan in 1950 is a good summary, which he himself selected in talking to Charlton (*Price of Victory*, p. 122) thirty years afterwards.

32. As FO 800/517 makes clear, the language was Makins', converted into a minute from Bevin to Attlee, 12 January 1951, The 'deliberations' referred to are those summarized in the Permanent Under-Secretary's Committee's memorandum, 'Third World Power or Western Consolidation', approved by Bevin on 27 March 1949—FO 371/76384.

33. DO(50) 5th meeting, Defence Committee, 23 March 1950 (CAB 138/8).

34. Said to Charlton: *Price of Victory*, p. 307.

35. Sir Edmund Hall-Patch, quoted in ibid., pp. 71–2. I am indebted to Michael Charlton for drawing my attention to this prescient minute.

36. See Kenneth O. Morgan, *Labour in Power, 1945–51* (Clarendon Press, Oxford, 1984), pp. 391–8.

37. CP(49)204, Memorandum by the Foreign Secretary, 'Council of Europe', 24 October 1949: CAB 129/37, Part 1.

38. CP(49)203, Memorandum by the Foreign Secretary and the Chancellor of the Exchequer, 'Proposals for the Economic Unification of Europe', 25 October 1949, also in CAB 129/37, Part 1. Both these memoranda were discussed by the Cabinet on 27 October 1949: CM (49) 62nd Conclusions, in the same file. For the Washington communiqué, see Chapter 12.

39. *FRUS, 1949*, vol. iv, pp. 470–1.

40. Charles Bohlen, Minister at the US Embassy in Paris, ibid., p. 493. Subsequent quotations from this conference are derived from the record in the same source (pp. 472–96).

41. Paris telegram No. 4422 to Washington, 22 October 1949, sent by George Perkins, Assistant Under Secretary for European Affairs (who had taken the chair at the conference); ibid., pp. 342–4.

42. *New York Times*, 1 November 1949.

43. FO 800/448, 'meetings of the Foreign Ministers of the Western Powers, Paris, 9th–10th November 1949—Germany'.

44. 469 *H.C.Debs.*, col. 2214, 17 November 1949.

45. Acheson, *Present at the Creation*, p. 387.

CHAPTER 16

1. Of £229m.

2. *The Economist*, 15 July 1950.

3. Acheson, *Present at the Creation*, p. 384: 'found him in distressing shape ... he was taking sedative drugs that made him doze off, sometimes quite soundly, during the discussion'.

4. Attlee finally announced Bevin's resignation from the Foreign Office and his appointment as Lord Privy Seal on 9 March 1951, his seventieth birthday. Bevin died five weeks later.

5. Kenneth Younger, from 28 February 1950 (Hector McNeil became Secretary of State for Scotland).

6. See Chapter 12. For a discussion of Attlee's views in the earlier post-war years, see Raymond Smith and John Zametica, 'The Cold Warrior: Clement Attlee reconsidered, 1945–7', in *International Affairs*, **61**: 2 (Spring 1985). The authors of this article are at a loss to explain why in January 1947 Attlee gave way (p. 251). It should perhaps be added that Attlee's ideas on military matters tended to be erratic: see, for example, his 1942 'Memorandum on British Military failures', quoted in Kenneth Harris, *Attlee* (Weidenfeld and Nicolson, London, 1982), pp. 585–8. Attlee was then under the impression that a British Army division numbered 40,000 men and that 'in this war artillery has so far played a minor part'.

7. The first quotation forms part of the message sent by Attlee to Truman on 6 July 1950, proposing joint Anglo-American military planning in a political context: see Harris, *Attlee*, p. 254. (Truman accepted Attlee's proposal on 10 July: see PREM 8/1405/Part 1.) 'Cold Warrior' is the title of Chapter 18 of Harris' *Attlee*. The second quotation is from ibid., p. 456.

8. *FRUS, 1950*, vol. iii, pp. 1657–68 and 1686 ff.

9. CM53(50), in CAB 128/18. National service was extended from eighteen months to two years in September 1950.

10. See Makins' minute of 8 December 1950, 'Record of Conversation in Mr Harriman's house': FO 800/517/US/50/56. Attlee, Franks, and Plowden were present on the British side. 'Harriman renewed forcibly his plea for an increased

United Kingdom defence effort over the £3,600 million . . . Acheson . . . appeared to agree with Mr Harriman's analysis'. Immediately after his return to London, Attlee acted in Cabinet: see CAB 128/18.

11. 484 *H.C.Debs.*, col. 728–40, 15 February 1951.
12. This description is that of Kenneth O. Morgan, *Labour in Power* (Clarendon Press, Oxford, 1984), p. 435. His account of the 1951 financial crisis is in pp. 477–9.
13. 487 *H.C.Debs.*, col. 34–43, 25 April 1951.
14. Harris, *Attlee*, pp. 447–8: 'Attlee himself would have left office standing far higher in reputation if the Conservatives had secured those few extra seats in February 1950'.
15. A former State Department official (a member of the US delegation at Yalta and Secretary-General of the UN founding conference in San Francisco in 1945), he was sentenced to five years' imprisonment in January 1950 for lying to a grand jury when accused of spying for the Soviet Union.
16. Marshall, who resigned from the State Department after an operation at the end of 1948, was appointed Defense Secretary in September 1950.
17. *Executive Sessions of the Senate Foreign Relations Committee (Historical Services)*, vol. 2, Eighty-first Congress, First and Second Sessions, 1949–50, pp. 4–37.
18. 'The very gifts of mind and manner that won . . . favour in London and Paris' for Acheson were to 'make him vulnerable to vengeful Republicans looking for a target', Robert Donovan, *Tumultuous Years* (Norton, New York, 1982), p. 34.
19. Acheson, *Present at the Creation*, p. 360. Characteristically, he entitled this chapter of his memoirs 'The attack of the primitives begins'.
20. Donovan, *Tumultuous Years*, p. 135.
21. Quoted in ibid., pp. 162–3.
22. Truman's handwritten note, dated 16 April 1950, is quoted in full in ibid., p. 171.
23. He did, however, manage to travel to Brussels to attend the meeting of the North Atlantic Council a fortnight later—his last international conference.
24. Ottawa telegram No. 1287 to the CRO, from the Prime Minister for the Foreign Secretary, 10 December 1950: FO 800/517/US/50/57.
25. *FRUS, 1950*, vol. iii, pp. 1698–1706.
26. The phrase 'political dynamite' was used by Truman at one point during this conference, which is discussed in Chapter 18.

CHAPTER 17

1. Acheson, *Present at the Creation*, p. 321.
2. Not that the British Government (as opposed to Churchill, who was already advocating a 'parley at the summit') would have been likely to welcome such a dialogue at the time; and seven years later the first British hydrogen bomb was exploded near Christmas Island.
3. For a discussion of this scepticism in the early years of the decade, see Oliver Franks, *Britain and the Tide of World Affairs*, BBC Reith Lectures, 1954 (Oxford University Press, Oxford, 1955), pp. 6 ff.
4. Ottawa telegram No. 1287 to the CRO, 10 December 1950, reporting Attlee's assessment ('for Foreign Secretary from Prime Minister') of his Washington summit meeting: FO 800/517/US/50/57.
5. Ibid.—minute by the Foreign Secretary to the Prime Minister, 12 January 1951.
6. W. T. R. Fox, *The Super-Powers—their responsibility for peace* (Yale Institute of International Studies, New York 1944), pp. 20–1.
7. This was a favourite phrase in the planning documents of the period. It will be

found both in the (American) NSC 68, quoted below, and in the (British) 'Threat to Civilisation', quoted in Chapter 14.

8. His speech—'Crisis in China—An examination of United States Policy', was delivered from 'a page or two of notes': Acheson, *Present at the Creation*, pp. 354–7.

9. Rusk's statement is quoted in Robert Donovan, *Tumultuous Years* (Norton, New York, 1982), p. 182. (Rusk was then Assistant Secretary for Far Eastern Affairs). Acheson's is on p. 176. For Truman's statement see *PP, Harry S. Truman, 1950*, p. 285.

10. Personal recollection of the author, who was Resident Clerk on duty at the Foreign Office for the whole of that weekend. The British Minister remained in Seoul, where he was captured, with his staff.

11. The development of Anglo-American relations in this field is discussed by Ritchie Ovendale, 'Britain, the United States, and the Cold War in the South-East Asia, 1949–50', in *International Affairs*, **58**: 3 (Summer 1982); since followed by Chapter 6 of his *English-Speaking Alliance: Britain, the United States, the Dominions and the Cold War, 1945–1951* (Allen and Unwin, London, 1985). See also *FRUS, 1949*, vol. viii, pp. 1215–20 and 1950, vol. vi, pp. 791–2. Donovan (*Tumultuous Years*, Chapter 13) rightly identifies *French* policy, in Indo-China, as the determining factor in the formation of the new US policy towards South-East Asia in the early months of 1950.

12. Acheson, *Present at the Creation*, p. 399. The whole of Chapter 43 ('Balanced and Collective Forces for Europe') is relevant.

13. See Chapter 15. Just before the Council met, the British Cabinet had also taken the formal decision that 'for the original conception of Western Union we must now begin to substitute the 'wider conception of the Atlantic Community'—CM(50)29, 8 May 1950, in CAB 128/17.

14. *PP, Harry S. Truman, 1950*, p. 138; and *FRUS, 1950*, vol. i, p. 542.

15. President of Harvard; and (as Chairman of the National Defense Research Committee) one of the principal figures in launching the wartime Manhattan Project; he had also been a member of the President's Interim Atomic Committee in 1945.

16. *FRUS, 1949*, vol. i, pp. 570–3, and Hewlett and Duncan, pp. 383–5.

17. *FRUS, 1950*, vol. i, pp. 22–40, 'International Control of Atomic Energy', 20 January 1950, submitted by Kennan as a personal paper because Nitze was 'not entirely in agreement with the substance'.

18. For the press, see the *Washington Post* leading article, 1 February 1950; and for the advice of the Joint Chiefs of Staff, see *FRUS, 1949*, vol. i, pp. 595–6.

19. Accounts of the process of this decision may be found in Hewlett and Duncan, Chapter 9; Donovan, *Tumultuous Years*, Chapter 14; and Gregg Herken, *The Winning Weapon* (Knopf, New York, 1980), Chapter 15.

20. *FRUS, 1950*, vol. i, pp. 141–2.

21. Acheson, *Present at the Creation*, p. 374.

22. The full text of the report is in *FRUS, 1950*, vol. i, pp. 235 ff.; this quotation is from p. 260; its conclusions are on pp. 287–8. For a detailed assessment, see Samuel F. Wells, Jr., 'Sounding the Tocsin: NSC 68 and the Soviet Threat', *International Security*, **4**: 2 (Fall, 1979), pp. 116 ff.

23. A further recommendation, of immense long-term significance, was for the 'intensification of . . . measures and operations by covert means in the fields of economic warfare and political and psychological warfare . . .'.

24. *FRUS, 1950*, vol. i, p. 400.

25. Wells, 'Sounding the Tocsin', p. 139.

26. *FRUS, 1950*, vol. i, pp. 221–5.

27. 'All but an outcast in Acheson's own party', Donovan, *Tumultuous Years*, p. 165.

28. 482 *H.C.Debs.*, col. 1456, 14 December 1950.

29. CM 34(50), 2 June 1950—CAB 128/17.

30. The most vivid account (which does not, however, quote documentary sources) is that of Michael Charlton, *The Price of Victory* (British Broadcasting Corporation, London, 1983), Chapter 4. At the time the press in Britain, France, and the United States gave a blow-by-blow account of what was happening in all the capitals concerned: the notes exchanged between the British and French governments in May–June 1950 were summarized in—for example—the *Manchester Guardian* and *Le Monde*, 14 June 1950.

31. Lord Sherfield (Sir Roger Makins), in conversation with Michael Charlton, *The Price of Victory*, p. 93. The reference is presumably to the Anglo-American bilateral talks, held in London from 26 April to 1 May 1950. The US record, however, gives a somewhat different impression: see *FRUS, 1950*, vol. iii, pp. 887–9.

32. CP(50)92 and (50)80, 8 May 1950—in CAB 129/39.

33. See Memorandum, 27 February 1950, 'US–British relations', by Henry R. Labouisse, Jr. to Mr. Perkins, and Memorandum of Conversation in the State Department, 7 March 1950: *DSDF*, 611.41/2—2750 and 611.41/3—750, respectively, National Archives.

34. Acheson, *Present at the Creation*, p. 358. This paper is discussed further in Chapter 21.

35. Ibid., pp. 382 and 385.

36. Ibid., p. 385.

37. Lincoln Gordon, in conversation with Charlton, *Price of Victory*, p. 97.

38. Monnet, *Mémoires*, pp. 329–32; and Charlton, *Price of Victory*, pp. 82–8. In conversation with the author, Lord Franks (who knew Monnet well) has said that he was not surprised by the fact that Monnet produced the plan, but by its timing.

39. Described, with masterly understatement, by his Principal Private Secretary, talking to Charlton, *Price of Victory*, p. 91.

40. See the President's news conference, May 18, 1950: *PP, Harry S. Truman, 1950*, p. 418.

41. 'That is our breakthrough': Charlton, *Price of Victory*, p. 105, quoting Sir Con O'Neill. Cf. Monnet, *Mémoires*, pp. 364–7.

42. FO 371/85342/CE 2338—an illuminating file for the British historian.

43. Notably, the view of the British Ambassador in Paris, Sir Oliver Harvey, on 20 May 1950 was that 'this represents a turning point in European and indeed in world affairs . . . We really have the ball at our feet'—ibid., CE 2342. He used this phrase in a personal letter to the Permanent Under-Secretary, asking him to be sure to read his despatch No. 324 of the previous day, which he had marked 'Most Immediate' (a classification normally used only for telegrams).

44. *The Times*, 14 June 1950. For the 'Brown Paper' (entitled *European Unity*), see 476 *H.C.Debs.*, col. 2149, 27 June 1950. The White Paper was entitled *Anglo-French Discussions regarding French Proposals for the Western European Coal, Iron and Steel Industries, May–June 1950*, Cmd. 7970 (HMSO, London, 1950).

45. *Daily Telegraph*, 16 June 1950.

46. *Christian Science Monitor*, 13 June 1950.

47. *The Times*, 14 June 1950.

48. *New York Times*, 21 June 1950, and *Le Figaro*, same date: *nous ne pouvons pas concevoir l'Europe sans elle (l'Angleterre)* ('we cannot imagine Europe without her (England).)

49. CM 38(50), 22 June 1950—CAB 128/17, which also agreed on the wording of a governmental amendment to counter Churchill's motion.

50. 476 *H.C.Debs.*, cols. 2104 and 2171, 27 June 1950.

51. Ibid., cols. 2139–59. The adjective 'devastating' is Harold Wilson's, in conversation with Charlton, *Price of Victory*, p. 110. (See also *The Times*, among other British newspapers, of 28 June 1950.)

52. Churchill said this in French in the House of Commons: 476 *H.C.Debs.*, col. 2152. The expression is best left in the original language, but may be freely translated 'People who do not take part are always in the wrong'.

53. Ibid., cols. 2159–60.

CHAPTER 18

1. Robert Donovan, *Tumultuous Years* (Norton, New York, 1982), p. 196.

2. Quoted by Robert O'Neill, *Australia in the Korean War, 1950–53*, (2 vols.) vol. i, *Strategy and Diplomacy* (The Australian War Memorial and the Australian Government Publishing Servics, Canberra 1981), p. 14.

3. Ibid., p. 15.

4. *PP: Harry S. Truman, 1950*, pp. 527–36.

5. CM (50) 39th Conclusions, minute 4, 27 June 1950—PREM 8, 1405, Part I. See also Dalton's diaries, quoted in Bullock, *Bevin*, p. 791.

6. An important, though not necessarily reliable, strand in the evidence that has come to light since the Korean War is Nikita S. Khrushchev, *Khrushchev Remembers*, trans. and ed. by Strobe Talbott (Little, Brown, Boston, 1970), pp. 367–73. For a recent discussion of the evidence, see O'Neill, *Australia in the Korean War*, vol. i, pp. 15 ff.

7. For example, see the remarks about 'International Communism' and 'the Communist' made by the Australian Prime Minister, Sir Robert Menzies, in his broadcast 'Defence Call to the Nation', 20–25 September 1950, quoted in O'Neill, ibid., p. 103; Menzies' statement is also one of the earliest expositions of the 'proxy' theory.

8. *FRUS, 1950*, vol. vii, pp. 155–6.

9. H. G. Nicholas, *The United Nations as a Political Institution* (Oxford University Press, Oxford, 1962), p. 51.

10. The news of the fall of Seoul reached Washington in the early hours of 28 June.

11. *PP, Harry S. Truman, 1950*, p. 492.

12. MacArthur had returned to Tokyo after making a personal reconnaissance in Korea.

13. Truman, *Trial and Hope*, p. 343, and *FRUS, 1950*, vol. vii, p. 263 (the Joint Chiefs of Staff to the Commander in Chief, Far East, 30 June 1950).

14. *PP, Harry S. Truman, 1950*, p. 492. For the Joint Chiefs of Staff's earlier view on Formosa, see *FRUS, 1949*, vol. vii, NSC report to the President on 'The Position of the United States with respect to Asia', 30 December 1949; and for the President's public assurance, given a week later, that 'the United States has no predatory designs on Formosa, or any other Chinese territory', see *PP, Harry S. Truman, 1950*, p. 11.

15. *New York Times*, 29 June 1950. See also the Department of State *Bulletin*, 4 September 1950, p. 396.

16. Andrew Shonfield, *British Economic Policy since the War*, rev. ed. (Penguin Books, Harmondsworth, 1959), p. 57. The GNP statistics are derived from R. N. Rosencrance, *Defence of the Realm: British Strategy in the Nuclear Epoch* (Columbia University Press, New York, 1968), pp. 295–6.

17. DO (50) 12th Meeting, 6 July 1950—CAB 131/8.

18. Franks to Attlee, 15 July and Washington telegram No. 2036, 25 July 1950—PREM 8/1405, Part I (a file which also contains copies of the minutes of the Defence Committee and Cabinet relevant to the Korean decision).

19. DO (50) 15th Meeting, 24 July 1950—CAB 131/8.

20. CM 50 50th meeting, 25 July 1950—CAB 128/19.

21. *FRUS, 1950*, vol. vii, pp. 347–52. This message was in fact received by Attlee, who sent it on to Bevin on 11 July: see PREM 8/1405, Part I.

22. Dean Acheson, *Present at the Creation* p. 418, in a section of Chapter 45 entitled 'Anglo-Indian Peace Initiatives'.

23. CM 44 (50)—CAB 128/18.

24. FO despatch No. 989 to Washington, 29 July, 1950 in FO 800/462/FE/50/29.

25. Washington telegram No. 2361 to the FO, 31 August 1950—FO 800/517/US/50/40.

26. *FRUS, 1950*, vol. vii, p. 839. PREM 8/1405, Part III contains the text of New Delhi telegram No. 2805 to the Commonwealth Relations Office in London, reporting two messages of 3 October from the Indian Ambassador in Peking. The same file contains Tokyo telegram No. 1371 to the FO. The telegram No. 1010 sent to the FO 'from the Secretary of State on board the Queen Mary' is in FO 800/462/FE/50/39.

27. For a recent analysis, see Donovan, *Tumultuous Years*, Chapters 26 and 27.

28. Department of State Press Release, 17 October 1950, *FRUS, 1950*, vol. iii, pp. 1427–8.

29. Truman, *Trial and Hope*, pp. 331–48.

30. *PP, Harry S. Truman, 1950*, p. 626.

31. Acheson, *Present at the Creation*, p. 440.

32. CM 59 (50), 15 September 1950—CAB 128/18.

33. The US records of the meetings held in New York, 12–20 September 1950, are in *FRUS, 1950*, vol. iii, pp. 1108–301; and the British records are in FO 800/449/CONF/58/47. For Bevin's later change of mind, see Bullock, *Bevin*, pp. 827–8. I am also indebted to the same source (p. 807) for the quotation in the preceding sentence.

34. Acheson, *Present at the Creation*, p. 486. Charles Spofford was Acheson's Deputy in the North Atlantic Council.

35. The US record is in *FRUS, 1950*, vol. iii, pp. 531–604. The British record is in FO 800/449/CONF/50. For the subsequent 'Great Debate' in Washington, see Acheson, *Present at the Creation*, Chapter 51. Truman's letter to Eisenhower, on his appointment as Supreme Commander, is in *PP, Harry S. Truman, 1950*, pp. 754–5.

36. O'Neill, *Australia in the Korean War*, vol. i, p. 130.

37. *FRUS, 1950*, vol. vii, pp. 1237–8.

38. *PP, Harry S. Truman, 1950*, p. 724–8 (the President's press conference), p. 727, note (the subsequent White House statement), and pp. 741 ff. (the President's televised broadcast).

39. *Izvestiya*, 5 November 1950.

40. Quoted in Donovan, *Tumultuous Years*, p. 369.

41. Kenneth Harris, *Attlee* (Weidenfeld & Nicolson, London, 1981), p. 463. The Cabinet's decision that the Prime Minister should go to Washington is CM(50), 80th Conclusions, 30 November 1950—PREM 8/1405, part 4.

42. E.g. by Kenneth Younger, in the interview transcript quoted by Kenneth O. Morgan, *Labour in Power* (Clarendon Press, Oxford, 1984), p. 429; and Herbert Morrison, quoted by Harris, *Attlee*, p. 466.

43. 'No records of Mr. Attlee's conversations in Washington (in 1945) have been traced in the (British) archives': *DBPO*, vol. ii, p. xiii.

44. The US records of these talks are in *FRUS, 1950*, vol. iii, pp. 1706–88; the British are in FO 800/445/COM/50/17.

45. FO telegram No. 5434 to Washington, 4 December 1950, 'for Prime Minister'—a personal telegram from Bevin, who said that he had 'had little chance to get views of Cabinet owing to length of agenda on coal'—FO 371/84107. The record of the Cabinet meeting is CM 50(81), item 2, 4 December 1950—CAB 128/18.

46. FO 800/517/US/50/57. Attlee's statement to Parliament is in 482 *H.C.Debs.*, col. 1356, 14 December 1950.

47. See Gowing, *Independence and Deterrence*, vol. i, p. 313.

48. The agreed Final Communiqué, 8 December 1950, is in *FRUS, 1950*, vol. iii, pp. 1783–7. The manner of the US 'unachieving' is recorded in *FRUS, 1950*, vol. vii, 'Memorandum for the Record', by R. Gordon Arneson, 16 January 1953, pp. 1462–5. The British record of Truman's assurance is in paragraph 18 of Annex 5, in FO 800/445/COM/50/17.

49. See Gowing, *Independence and Deterrence*, pp. 314–18 and John Baylis, *Anglo-American Defence Relations, 1939–1984: The Special Relationship*, 2nd edn. (Macmillan, London, 1984), pp. 40–1—'a matter for joint decision by the two governments in the light of the circumstances prevailing at the time'. The gist of this formula (the so-called Attlee–Churchill Understandings) was quoted in Parliament in December 1957.

50. FO 800/517/US/51/50. The same file contains Bevin's reply about this to Attlee, which was drafted by Makins.

51. FO 800/517/US/50/570. The to-ing and fro-ing that followed in January 1951, over the UN resolution condemning China as the aggressor in Korea (on which the British Cabinet wavered), offers little support for this view of Anglo-American 'equality'. For an account, see Ritchie Ovendale, *The English-Speaking Alliance: Britain, the United States, the Dominions and the Cold War, 1945–1951* (Allen and Unwin, London, 1985), pp. 227 ff.

52. Acheson, *Present at the Creation*, pp. 484–5. For the less acerbic view that he took in the last year of his life, see Harris, *Attlee*, pp. 463–4.

CHAPTER 19

1. Harold Macmillan was a Classical Exhibitioner at Balliol College, Oxford. It is true that the Romans admired Greek art and literature and that the Romans and the Greeks shared a common mythology. But there the analogy surely ends. Not only did they speak different languages. The Romans acquired their eastern provinces by defeating the Greeks in war; and they went on to pillage the immense riches of the Greek world on a vast scale.

2. Robert Murphy, *Diplomat among Warriors* (Doubleday, New York, 1964), p. 164.

3. Harold Macmillan, *Riding the Storm, 1956–59* (Macmillan, London, 1971), Appendix 1, has the text of the plan.

4. 'The Anglo-American Schism' is the title of Chapter 4 of Macmillan's *Riding the Storm, 1956–59.*

5. The absence of official documents open to the public beyond 1954, at the time of writing.

6. For a discussion of this important point, see, for example, Stephen E. Ambrose, *Eisenhower: the President* (Simon & Schuster, New York, 1984), Chapter 15.

7. 'With the election of Eisenhower and the prospect of possibly four years or more of power or influence wielded by these two men ... it is hard to believe that all will not be well': H. C. Allen, *Great Britain and the United States: A History of Anglo-American Relations (1783–1952)* (Odhams Press, London, 1954), p. 983.

8. Dwight D. Eisenhower, *The White House Years*, 2 vols., *Mandate for Change, 1953–1956* (Doubleday, New York, 1963), p. 97.

9. Harold Macmillan, *Tides of Fortune, 1945–1955* (Macmillan, London, 1969), p. 523.

10. Eisenhower, *Mandate for Change* (Doubleday, New York, 1963), pp. 249–50.

11. D. Cameron Watt, *Succeeding John Bull, America in Britain's Place, 1900–1975* (Cambridge University Press, Cambridge, 1984), p. 127. On Churchill's mental capacity during his third Administration, see John Colville, *The Fringes of Power: Downing Street Diaries 1939–1955* (Hodder & Stoughton, London, 1955), p. 634.

12. Written in order to persuade the Prime Minister to reverse the British Cabinet's earlier decision not to send ground or air forces to Korea: see Chapter 18 n. 18.

13. Attlee's message to Truman of 6 July 1950 is in FO telegram No. 3070 to Washington—PREM 8/1405, Part I. Truman's positive reply was delivered by the US Ambassador on 10 July.

14. Watt, *Succeeding John Bull*, p. 22.

15. Gowing, *Independence and Deterrence*, Chapter 12.

16. This description is Raymond Aron's in his 'Historical Sketch of the Great Debate', *France defeats the EDC*, ed. Daniel Lerner and Raymond Aron (Praeger, New York, 1957), p. 10. Acheson's words at Bad Godesberg are quoted in *Present at the Creation*, p. 647.

17. Described in Anthony Eden, *The Memoirs of Sir Anthony Eden*, 3 vols., *Full Circle* (Cassell, London, 1960), Chapter 7.

18. Russell Bretherton, the Board of Trade Under-Secretary who 'represented' the United Kingdom in Brussels during the negotiations of the Six between the Messina and Venice Conferences, quoted by Michael Charlton, *The Price of Victory* (BBC, London, 1983), p. 200.

19. Eisenhower, *Mandate for Change*, p. 335.

20. 515 *H.C.Debs.*, col. 897, 11 May 1953.

21. Colville, *Fringes of Power*, pp. 697 ff. provides a vivid account of the genesis of the 'Molotov telegram' and its repercussions in the Cabinet. Malenkov was then Soviet Prime Minister (he was ousted by Khrushchev six months afterwards). In the form in which it was finally sent to Moscow, on 26 July 1954, the telegram was 'not over-enthusiastically' received by the Soviet Government.

22. Eden, *Full Circle*, Chapter 6.

23. Eisenhower, *Mandate for Change*, pp. 335 and 349.

24. The Anglo-Egyptian Treaty of 1936 and the 1899 agreement between the two countries establishing the Anglo-Egyptian Condominium over the Sudan.

25. Acheson, *Present at the Creation*, p. 564. Ibid, pp. 562–7, gives a fair summary of the proposal.

26. A scepticism that was not confined to Acheson. See, for example, the personal letter written by the British Ambassador in Cairo, Sir Ralph Stevenson, to Sir

James Bowker, Assistant Under-Secretary responsible for the Middle East in the Foreign Office, 14 August 1951: FO 371/90136/179 G.

27. Wm. Roger Louis, *The British Empire in the Middle East, 1945–1951: Arab Nationalism, the United States, and Postwar Imperialism* (Clarendon Press, Oxford, 1984), p. 646. Aramco's decision was taken for reasons of American taxation.

28. A well documented account is provided by ibid., Part V, Chapter 3.

29. CM 60(51), 27 September 1951—CAB 128/20.

30. The text of the US press release of 19 July 1956, announcing this decision, is in Dwight D. Eisenhower, *The White House Years, Waging Peace, 1956–61* (Double-day, New York, 1965), Appendix 4, p. 663.

31. That the indications of Israeli intentions were plentiful is evident from ibid., pp. 56–73, particularly p. 70.

32. Macmillan, *Riding the Storm*, p. 165; and Eden, *Full Circle* pp. 554–6. Bulganin's message to Eden asked him 'what kind of position would Britain be in if she had been attacked by stronger powers with all kinds of modern offensive weapons at their disposal ... such as rockets'—*Pravda*, 6 November 1956. See also *Istoriya vneshnei politiki SSSR*, ed. A. A. Gromyko and B. N. Ponomarev, 4th edn. (Nauka, Moscow, 1981), vol. ii, pp. 259–60.

33. Macmillan's own account (he was Chancellor of the Exchequer at the time) is in his *Riding the Storm*, pp. 163–79. Even he wrote of 'humiliations almost vindic-tively inflicted upon us at the instance of the United States Government'.

34. Elizabeth Monroe, *Britain's Moment in the Middle East, 1914–58* (Chatto & Windus, London, 1981), p. 209, cited by Wm. Roger Louis, in 'American anti-colonialism and the British Empire', *International Affairs*, **61**: 3 (Summer 1985).

35. Armand de Richelieu, *Testament politique*, part 2 (Imprimerie Le Breton, 1864), p. 41: 'Bien que ce soit un dire commun, que quiconque a la force, a d'ordinaire raison, il est vrai toutefois que de deux Puissances inégales, jointes par un traité, la plus grande risque d'être plus abandonnée que l'autre'.

36. Eisenhower, *Waging Peace*, p. 99.

37. George Kennan, 'The United States and the Soviet Union, 1917–1976', *Foreign Affairs*, 54: 4 (July 1976), pp. 684–6.

CHAPTER 20

1. He resigned on health grounds. Created Earl of Avon, he left public life altogether, although he lived for another twenty years. The Latin epitaph may be roughly rendered: 'No one would have doubted his ability to rule, had he never ruled'; Tacitus, *Histories*, 1, xlix.

2. Harold Macmillan, *Riding the Storm*, p. 185. Since he did not accept his present title (Earl Stockton) until nearly thirty years afterwards, it has seemed best to refer to him throughout as Macmillan.

3. Shorter Oxford English Dictionary (Oxford University Press, Oxford, 1968), p. 1433 (attributed to Thomas Burton, a member of the Cromwellian parliaments from 1656–9—I am indebted for this information to the Editor, the SOED).

4. Harold Macmillan, *Riding the Storm, 1956–59* (Macmillan, London, 1971), p. 195.

5. Ibid., p. 240.

6. An extract from a letter from the Prime Minister to the Queen, 24 December 1961, reproduced in Harold Macmillan, *At the End of the Day, 1961–1963* (Macmillan, London, 1973), pp. 147–8.

7. Ormsby Gore was a close friend of the Kennedys: see Macmillan, *Pointing the*

Way, 1959–1961 (Macmillan, London, 1972), p. 339. In addition, although John Kennedy and Harold Macmillan met for the first time in 1961, there was a connection between Macmillan himself and the Kennedy family (his wife's nephew, killed in the war, had married one of John Kennedy's sisters).

8. One of the most fruitful areas of this interpenetration was that of strategic studies. From the European anxiety to establish an intellectually rigorous transatlantic dialogue on strategic (above all, nuclear) questions—an anxiety shared by the Ford Foundation—grew the Institute for Strategic Studies, founded in London in November 1958. By the end of its first year this (British) Institute was being advised by an International Council. Its first Director was Alastair Buchan. It has since developed into the modern International Institute for Strategic Studies.

9. D. Cameron Watt, *Succeeding John Bull: America in Britain's place, 1900–1975* (Cambridge University Press, Cambridge, 1985), p. 13. (It will, however, be evident that the author does not share all the conclusions drawn by Watt from this concept.)

10. Selwyn Lloyd, Eden's Foreign Secretary, had offered his resignation after the Suez failure. Macmillan kept him on at the Foreign Office, thus ensuring that the reins of foreign policy would be held in his own hands at 10 Downing Street.

11. Dwight D. Eisenhower, *The White House Years, 1956–1961, Waging Peace* (Doubleday, New York, 1965), p. 122.

12. Ibid., p. 124.

13. Department of State *Bulletin*, 'The Evolution of Foreign Policy', xxx: 761 (25 January 1954).

14. *United Kingdom Defence. Outline of Future Policy*, April 1957, Cmnd.124 (HMSO, London, 1957).

15. Macmillan, *Riding the Storm*, p. 263. By giving priority to a strategy of nuclear deterrence, the way was opened to the abolition of conscription. For a study of the development of Soviet nuclear doctrine, see Stephen M. Meyer, 'Soviet Theatre Nuclear Forces, Part I, Development of Doctrine and Objectives', *Adelphi Papers*, No. 187 (IISS, London, 1983/4).

16. The text of the 'Declaration of Common Purpose' may be found at Appendix 3 in ibid., pp. 756–9. Macmillan's remark is in ibid., p. 323. Eisenhower's is in *Waging Peace*, p. 219.

17. See Andrew Pierre, *Nuclear Politics: The British Experience with an Independent Strategic Force, 1939–1970* (Oxford University Press, Oxford, 1972), p. 141, which also quotes the precise definition of 'substantial' progress.

18. *Agreement for Cooperation on the Uses of Atomic Energy for Mutual Defence Purposes*, Cmnd. 537 (HMSO, London, 1958), further amplified by the amendment signed in Washington on 7 May 1959, Cmd. 859 (HMSO, London, 1959).

19. Macmillan, *Pointing the Way*, pp. 250–8 describes the background to this decision; the quotation is on p. 258.

20. As has been well argued by Pierre, *Nuclear Politics*, pp. 319–21.

21. 'Which no other ally enjoyed in day-to-day discussions of nuclear targeting and war planning', ibid., p. 144.

22. See, for example, Macmillan, *Riding the Storm*, p. 460.

23. Quoted in Pierre, *Nuclear Politics*, p. 178; both quotations are from February 1958.

24. 'Dangerous, expensive, prone to obsolescence and lacking in credibility as a deterrent' (quoted in Pierre, *Nuclear Politics*, p. 208). At the December 1962 meeting of the North Atlantic Council, moreover, McNamara had proposed the formation of a 'multilateral nuclear force'.

25. His own account is in Harold Macmillan, *At the End of the Day*, pp. 341–4.

26. Discussed further in the next chapter.
27. George Ball, *The Past has another Pattern: Memoirs* (Norton, New York, 1982), p. 268.
28. By 1960, the (seven) German divisions in Allied Forces Central Europe already outnumbered the British and French divisions combined: *The Communist Bloc and the Free World: the Military Balance, 1960* (IISS, London, November 1960).
29. Cf. Henry Kissinger's description of Metternich's *'tour de force* of a solitary figure', in *A World Restored* (Gollancz, London, 1973), p. 322.
30. His own account is in Macmillan, *Riding the Storm*, Chapter 18. For Eisenhower's views, see his *Waging Peace*, pp. 342–60 and pp. 401–3.
31. Charles de Gaulle, *Discours et Messages, 1962–65* (Plon, Paris, 1970): *1963*, p. 69. The translation is the author's.
32. For example, the phrase 'massive retaliation' in Dulles' famous speech of 12 January 1954 and his 'brink of war' article in *Life* magazine.
33. The Soviet Government recognized Cuba as a 'socialist' country in June 1961, but Cuba was not formally admitted to the communist bloc for another twelve months. For an assessment of the Soviet conduct of the Cuban missile crisis, see Robin Edmonds, *Soviet Foreign Policy: the Brezhnev Years* (Oxford University Press, Oxford, 1983), Chapter 3.
34. In Harold Macmillan, *At the End of the Day* Chapter 7.
35. One of the earliest detailed analyses of this crisis—Graham Allison's *Essence of Decision: Explaining the Cuban Missile Crisis* (Little, Brown, Boston, 1971)—has retained its high quality, despite the fact that this book appeared before the full text of the ten letters exchanged between Kennedy and Khrushchev in October 1962 were published: see the Department of State *Bulletin*, November 1973, **69**: 1975.
36. ExCom was the name of the small group set up by Kennedy to advise him during the Cuban missile crisis.
37. A description of Churchill cited by E. T. Williams in his notice of Churchill in the *Dictionary of National Biography Supplement, 1961–70* (Oxford University Press, Oxford, 1981), p. 214. It could equally well be said of Macmillan.

CHAPTER 21

1. Alastair Buchan, 'Greeks and Romans (Mothers and Daughters)', *Foreign Affairs*, **54**: 4 (July 1976), p. 645.
2. Planned at first as the year in which the transatlantic relationship would be rationalized and brought up to date, 1973 ended by becoming the most traumatic year that the North Atlantic Alliance had ever known.
3. For example, Ronald Reagan's 'anti-communist crusade', proclaimed during his tour of European capitals in 1981; and his speech describing Moscow as the 'focus of evil' in 1983. See *The Economist*, 19 March 1983, pp. 43 ff; and (for European concern about US nuclear concepts during the first Reagan Administration) Michael Howard's *Weapons and Peace*, Annual Memorial Lecture, 13 January 1983 (David Davies Memorial Institute of International Studies, London, 1983), pp. 12–13.
4. Text in *The Times*, 22 November 1985, p. 8.
5. The expression used by Cyrus Vance, in his speech on US foreign policy delivered at Harvard on 5 June 1980; the text is in the *New York Times*, 6 June 1980.
6. To be exact, of liquid hydrocarbons, well over half of which were exported in the first half of 1985 (source: *The Petroleum Economist*, September 1985, p. 322).
7. Text in *The Listener*, 13 May 1982.

8. See Henry Kissinger, *Years of Upheaval* (Weidenfeld & Nicolson and Michael Joseph, London, 1982), pp. 274 ff. (For the text of US-Soviet agreement see *The Times*, 23 June 1973.) And in 1976, when Kissinger entered the political arena of Southern Africa for the first time, the draft that he used for his negotiations with Ian Smith was British: see *The Listener*, 13 May 1982, p. 16.

9. On a different, though potentially serious level, the inequality between the resources devoted to Soviet and Eastern European studies in each country is beginning to render extremely difficult the maintenance of a sustained Anglo-American academic dialogue in this critical field.

10. A notable instance in 1985 was the deaths caused by British soccer fans in Brussels.

11. For an account, see *The Economist*, 12 November 1983, 'America and the Falklands: Case Study of the Behaviour of an Ally'.

12. Acheson, *Present at the Creation*, pp. 387–8. (For the occasion, see Chapter 17.) The Policy Statement of 7 June 1950, which seems to have been given a wide circulation to US Missions abroad, is to be found in *DSDF* 611.41/6–750, National Archives.

13. Notably, the paper never mentions the Schuman Plan and it talks of a 'disarmed' Germany—thus clearly proving that it must have been written before the events of May–June 1950.

14. 'The Essentials of an American Policy', 21 March 1944: AN 1538/16/45—FO 371/38523. Cf. the phrase 'inexperienced colossus', used seven years later in Bevin's memorandum to Attlee (quoted in Chapter 17).

15. Dated 24 August 1949, this memorandum was one of the series of 'long-term' papers prepared by the Foreign Office Permanent Under-Secretary's Committee in that year: FO 371/76384 and 76385.

16. See Oliver S. Franks, *Britain and the Tide of World Affairs*, The BBC Reith Lectures 1954 (Oxford University Press, Oxford, 1955), p. 12. Lord Franks has confirmed the fact of Churchill's admonition to him ('never forget the three circles'), in conversation with the author.

17. In both of which he served as an official within the wartime allied bureaucracy, working very closely with American and British colleagues.

18. Or, as Acheson called it, 'the first wrong choice': *Present at the Creation*, p. 387.

19. The phrase is Max Beloff's, in *A Century of Conflict, 1850–1950: Essays for A. J. P. Taylor*, ed. Martin Gilbert (Hamish Hamilton, London, 1966), p. 154.

20. For example, it was only natural to find 'going over the ground' on a particular issue, in English with an American official in Washington or abroad, a more agreeable experience than arguing the toss in a foreign language with a difficult interlocutor.

21. Sir Duncan Wilson.

22. Bullock, *Bevin*, p. 416.

23. See the final paragraphs of Chapter 9.

24. 138 *H.L.Debs*, cols. 790–4, 18 December 1945.

25. *Foreign Affairs*, **54**: 4 (July 1976), p. 656.

26. Chevaline is the codename of the refit (costing about £1bn.) of the British submarine nuclear deterrent force, designed to enable it to penetrate the Galosh anti-ballistic missile defences of Moscow. It was kept secret from most members of British governments of both parties until 1980.

27. Andrew Pierre, *Nuclear Politics* (Oxford University Press, Oxford, 1972), pp. 320–3. Cf. David Robertson, 'Could Britain Fight a Nuclear War?', Whitehall Paper No. 3 (Royal United Services Institute for Defence Studies, London, 1985).

28. *The Economist*, 3 November 1984, p. 34.

29. *The Financial Times*, 'Trident, the political costs of an addiction', 12 November 1984.

30. However, for an analysis of the problems of the transatlantic alliance that reaches a more encouraging conclusion than the author's, see Gregory F. Treverton, *Making the Alliance Work: the United States and Western Europe* (Macmillan, London, 1985).

31. On the intelligence relationship, see David Reynolds, 'America, Britain and the International Order', *International Affairs*, **62**: 1, Winter 1985/6 (RIIA, London, 1985), pp. 11 and 16.

32. 'The main condition for (defence) consensus in the 1980s is in fact that we should all grow up' was the wise conclusion of Michael Howard's 'Defence, Consensus, and Reassurance in the Defence of Western Europe', *Adelphi Paper*, No. 184 (IISS, London, 1983).

33. The policy is still not entirely clear, but—seen from a US perspective—it looks unattractive as it stands at present: a non-nuclear Britain, from which US strategic bombers, cruise missiles and the Holy Loch submarine base would have been removed.

34. As François Mitterrand did, on 20 January 1983.

35. In the final chapter of the author's *Soviet Foreign Policy: The Brezhnev Years*, 'An Option for Europe' (Oxford University Press, Oxford and New York, 1983).

36. *The Economist*, 26 April 1986, leading article, p. 13.

Select Bibliography

History, in A. J. P. Taylor's words, 'gets thicker as it approaches recent times'.* The literature relating to the years covered by this book, already vast, is growing with every year that passes. For this reason, the books listed in the bibliography that follows have, with a few exceptions, been confined to those quoted from or referred to in this book.

BOOKS

ACHESON, Dean, *Present at the Creation: My Years at the State Department* (Norton, New York, 1969).

ADOMEIT, Hannes, *Soviet Risk-Taking and Crisis Behaviour* (Allen & Unwin, London, 1982).

ALLEN, H. C., *Great Britain and the United States: a history of Anglo-American relations (1783–1952)* (Odhams Press, London, 1954).

ALLISON, Graham, *Essence of Decision: Explaining the Cuban Missile Crisis* (Little, Brown, Boston, 1971).

AMBROSE, Stephen E., *Eisenhower: the President* (Allen & Unwin and Simon & Schuster, London and New York, 1984).

Ambrose, Stephen E., *Eisenhower and Berlin, 1945: the Decision to Halt at the Elbe*, rev. edn. (Norton, New York, 1986).

BALFOUR, John, *Not too correct an aureole* (Michael Russell, London, 1983).

BALL, George W., *The Past has another Pattern: Memoirs* (Norton, New York, 1982).

BARKER, Elizabeth, *The British between the Super-Powers, 1945–1950* (Macmillan and University of Toronto Press, London and Toronto, 1983).

BAYLIS, John, *Anglo-American Defence Relations, 1939–1984: the Special Relationship*, 2nd edn. (Macmillan, London, 1984): *Anglo-American Defense Relations: 1939–1979; the Special Relationship* (St Martin's Press, New York, 1981).

BEDELL SMITH, Walter, *My Three Years in Moscow* (Lippincott, Philadelphia, 1950).

BERLIN, Isaiah, *Personal Impressions*, ed. Henry Hardy (The Hogarth Press and Viking, London and New York, 1981).

BERLIN, Isaiah, *Washington Despatches, 1941–45*, ed. H. G. Nicholas (Weidenfeld & Nicholson, London, 1981).

BOHLEN, Charles E., with PHELPS, Robert H., *Witness to History* (Norton, New York, 1973).

BULLOCK, Alan, *The Life and Times of Ernest Bevin*, 3 vols. (Heinemann, London, 1967–83); vol. iii is *Ernest Bevin: Foreign Secretary, 1945–1951* (Heinemann and Norton, London and New York, 1983 and 1984).

CADOGAN, Alexander, *The Diaries of Sir Alexander Cadogan*, ed. David Dilks (Cassell, London, 1971).

* *English History 1914–1945* (Clarendon Press, Oxford, 1965), Bibliography, p. 602.

CAIRNCROSS, Alec, *British Economic Policy, 1945–51* (Methuen, London, 1985).

CALVOCORESSI, Peter, *The British Experience 1945–1975* (Bodley Head, London, 1978).

CARR, Edward H., *Conditions of Peace* (Macmillan, London, 1942).

CARRINGTON, C. E., *The British Overseas: Exploits of a Nation of Shopkeepers* (Cambridge University Press, Cambridge, 1950).

CHARLTON, Michael, *The Price of Victory* (British Broadcasting Corporation, London, 1983).

CHURCHILL, Winston S., *The Second World War*, 6 vols. (Cassell and Houghton, Mifflin, London and New York, 1948–54 and 1948–53).

CIECHANOWSKI, Jan M., *The Warsaw Rising of 1944* (Cambridge University Press, Cambridge and New York, 1974).

COHEN, Michael, *Palestine and the Great Powers, 1945–48* (Princeton University Press, Princeton, 1982).

COLVILLE, John, *The Fringes of Power, Downing Street Diaries, 1939–1955* (Hodder & Stoughton, London, 1985).

COOPER, Duff, *Old Men Forget* (Hart-Davis, London, 1954).

COULONDRE, Robert, *De Staline à Hitler: entre deux Ambassades, 1936–1939* (Hachette, Paris, 1950).

CROSSMAN, Richard H. S., *Palestine Mission* (Hamish Hamilton, London, 1947).

DAVIES, Norman, *God's Playground: A History of Poland*, 2 vols. (Clarendon Press and Columbia University Press, Oxford and New York, 1981 and 1982).

DE GAULLE, Charles, *Discours et messages* (Plon, Paris, 1970), 5 vols. (vol. iv: *Pour l'effort, 1962–1965*).

DE GAULLE, Charles, *War Memoirs—Salvation 1944–46*, trans. Joyce Murchie and Hamish Erskine (Weidenfeld & Nicolson, London, 1960): *The Complete War Memoirs of Charles de Gaulle: 1940–46*, trans. Richard Howard (Da Capo Press, New York, 1984).

DELAHAYE, Yves, *La frontière et le texte* (Payot, Paris, 1977).

DEVLIN, Patrick, *Too Proud to Fight: Woodrow Wilson's Neutrality* (Oxford University Press, Oxford and New York, 1974 and 1975).

DIVINE, Robert A., *Eisenhower and the Cold War* (Oxford University Press, New York, 1981).

DJILAS, Milovan, *Conversations with Stalin*, trans. Michael B. Petrovich (Hart-Davis, London, 1962 and Harcourt, Brace, Jovanovich, Inc., New York, 1963).

DONOVAN, Robert J., *Conflict and Crisis: the Presidency of Harry S. Truman, 1945–1948* (Norton, New York, 1977).

DONOVAN, Robert J., *Tumultuous Years: the Presidency of Harry S. Truman, 1949–1953* (Norton, New York, 1982).

EDEN, Anthony, *The Memoirs of Sir Anthony Eden*, 3 vols. (Cassell, London, 1960–65).

EDMONDS, Robin, *Soviet Foreign Policy: the Brezhnev Years* (Oxford University Press, Oxford and New York, 1983).

EHRMAN, John, *Grand Strategy*, vols. v and vi of the *Official War History* (HMSO, London, 1956).

EISENHOWER, Dwight D., *Crusade in Europe* (Doubleday, New York, 1948).

EISENHOWER, Dwight D., *The White House Years*, 2 vols. (Doubleday, New York, 1963–5).

ERICKSON, John, *The Road to Berlin* (Weidenfeld & Nicolson, London, 1983).

FEILING, Keith, *The Life of Neville Chamberlain* (Macmillan and Shoe String Press, London and Hamden, Conn., 1946 and 1970).

FORRESTAL, James, *Forrestal Diaries*, ed. Walter Millis and E. S. Duffield (Viking Press, New York, 1951).

FOX, William T. R., *The Super-Powers, the United States, Britain and the Soviet Union—their responsibility for peace*, Yale University Institute of International Studies (Harcourt, Brace, New York, 1944).

FRANKS, Oliver S., *Britain and the Tide of World Affairs*, BBC Reith Lectures 1954 (Oxford University Press, London, 1955).

FREEDMAN, Lawrence, *Britain and Nuclear Weapons* (Macmillan, London, 1981).

GARDNER, Richard N., *Sterling–Dollar Diplomacy in Current Perspective* (Columbia University Press, New York, 1980).

GARNER, Joe, *The Commonwealth Office, 1925–68* (Heinemann and Heinemann Educational Books, London and Exeter, N.H., 1978).

GLADWYN, Lord, *The Memoirs of Lord Gladwyn* (Weidenfeld & Nicolson, London, 1972).

GOWING, Margaret, *Britain and Atomic Energy, 1939–1945* (Macmillan, London, 1964).

GOWING, Margaret, *Independence and Deterrence: Britain and Atomic Energy, 1945–1952*, 2 vols. (Macmillan, London, 1974; St Martin's Press, New York, 1975).

GROMYKO, A. A. and PONOMAREV, B. N. (eds.), *Istoriya Vneshnei Politiki SSSR, 1917–1975*, 2 vols. (Nauka, Moscow, 1981).

GUŻEWSKI, Jan and KOPF, Stanisław, *Dni Powstania* (Pax Publishing House, Warsaw, 1957).

HARRIMAN, W. Averell and ABEL, Elie, *Special Envoy to Churchill and Stalin, 1941–1946* (Random House, New York, 1975).

HARRIS, Kenneth, *Attlee* (Weidenfeld & Nicolson and Norton, London and New York, 1982 and 1983).

HARROD, Roy F., *Life of John Maynard Keynes* (Macmillan and Augustus M. Kelley, London and Fairfield, N.J., 1951 and 1952 (2nd edn.)).

HATHAWAY, Robert M., *Ambiguous Partnership* (Columbia University Press, New York, 1981).

HENDERSON, Nicholas, *The Birth of NATO* (Weidenfeld & Nicolson and Westview Press, London and Boulder, Col., 1982 and 1983).

HERKEN, Gregg, *The Winning Weapon* (Knopf, New York, 1980).

HEWLETT, Richard, and ANDERSON, Oscar, *A History of the United States Atomic Energy Commission*, vol. i: *The New World, 1939/1946* (Pennsylvania State University Press, University Park, Pa., 1962).

HEWLETT, Richard, and DUNCAN, Francis, *A History of the United States Atomic Energy Commission*, vol. ii: *Atomic Shield, 1947–1952* (Pennsylvania University Press, University Park, Pa., 1969).

HILLMAN, William, *Mr President* (Faber, Straus & Young, New York, 1952).

HOLLOWAY, David, *The Soviet Union and the Arms Race* (Yale University Press, New Haven and London, 1983).

HOROWITZ, David, *State in the Making* (Knopf, New York, 1953).

HUXLEY, Julian and HADDON, A. C., *We Europeans* (Cape, London, 1935: Kraus rep., Millwood, N.J.).

IOIRYSH, A. I., MOROKHOV, I. D., and IVANOV, S. A., *A-Bomba* (Nauka, Moscow, 1980).

ISMAY, Lord, *The Memoirs of General Lord Ismay* (Heinemann and Greenwood Press, London and Westport, Conn., 1960 and 1974).

ISSAWI, Charles, *Egypt: an economic and social analysis* (Oxford University Press, Oxford, 1947).

JONES, Joseph M., *The Fifteen Weeks, February 21–June 5, 1949* (Viking, New York, 1955).

KARNOW, Stanley, *Mao and China: From Revolution to Revolution* (Macmillan and Viking, London and New York, 1973 and 1972).

KENNAN, George F., *Memoirs 1925–1950* (Little, Brown, Boston, 1967).

KHRUSHCHEV, Nikita, *Khrushchev Remembers*, 2 vols., trans. and ed. Strobe Talbott (Little, Brown, Boston, 1970).

KIMBALL, Warren F. (ed.), *Churchill and Roosevelt: the complete correspondence*, 3 vols. (Princeton University Press, Princeton, 1984).

KIRBY, S. Woodburn [and others], *The War against Japan* (HMSO, London, 1969).

KISSINGER, Henry, *A World Restored: Metternich, Castlereagh and the Problems of Peace, 1812–1822* (Gollancz and Houghton Mifflin, London and New York, 1973).

LARKIN, Philip, *Required Writing* (Faber & Faber and Farrar, Straus & Giroux, London and New York, 1983 and 1984).

LARKIN, Philip, *All What Jazz, A Record Diary, 1961–1971* (Faber & Faber, London, 1985).

LEWIN, Ronald, *Ultra goes to War: the Secret Story* (Hutchinson and Pocket Books, London and New York, 1978 and 1981).

LILIENTHAL, David E., *The Journals of David E. Lilienthal*, 2 vols. (Harper and Row, New York, 1964).

LIPPMANN, Walter, *U.S. War Aims* (Little, Brown, Boston, 1944).

LOUIS, Wm. Roger, *The British Empire in the Middle East, 1945–51: Arab Nationalism, The United States, and Postwar Imperialism* (Clarendon Press, Oxford, 1984).

LOUIS, Wm. Roger and BULL, Hedley, eds. *The Special Relationship: Anglo-American Relations since 1945* (Clarendon Press, Oxford, 1986).

MACDONALD, Ian S. (ed.), *Anglo-American Relations since the Second World War* (St Martin's Press, New York, 1974).

MACMILLAN, Harold, *Memoirs*, 6 vols. (Macmillan, London, 1966–73).

MANDERSON-JONES, R. B., *The Special Relationship: Anglo-American relations and Western European Unity, 1947–1956* (Crane, Russak, New York, 1972).

MITCHELL, B. R., *European Historical Statistics, 1750–1970* (Macmillan and Facts on File, London and New York, 1975 and 1980).

MONNET, Jean, *Mémoires* (Fayard, Paris, 1967); trans. Richard Mayne (Doubleday, Garden City, 1978).

MONROE, Elizabeth, *Britain's Moment in the Middle East, 1914–56* (Chatto & Windus and Johns Hopkins Univ. Press, London and Baltimore, Maryland, 1981).

MONTGOMERY, Viscount, *The Memoirs of Field-Marshal the Viscount Montgomery* (Collins, London, 1958).

MORGAN, Kenneth O., *Labour in Power: 1945–1951* (Clarendon Press and Oxford University Press, Oxford and New York, 1984).

MURPHY, Robert, *Diplomat among Warriors* (Doubleday, New York, 1964).

NAGAI, Yonosuke and IRIYE, Akira (eds.), *The Origins of the Cold War in Asia* (Columbia University Press and University of Tokyo Press, New York and Tokyo, 1977).

NICHOLAS, H. G. *The United Nations as a Political Institution* (Oxford University Press, Oxford and New York, 1962 and 1975).

NICHOLAS, H. G. *Britain and the United States* (Chatto & Windus, London, 1963).

NICOLSON, Harold, *Diaries and Letters 1930–39*, ed. Nigel Nicolson (Collins Fontana Books, London, 1969).

NICOLSON, Harold, *Peacemaking 1919* (Constable and Peter Smith, London and Magnolia, Mass., 1943 and 1984).

O'NEILL, Robert, *Australia in the Korean War, 1950–53*, 2 vols. (The Australian War Memorial and the Australian Government Publishing Services, Canberra, 1981–5).

ORWELL, George, *The English People* (Collins and Haskell House, London and New York, 1947 and 1984).

OVENDALE, Ritchie, *The English-Speaking Alliance: Britain, the United States, the Dominions and the Cold War, 1945–1951* (Allen & Unwin, London, 1985).

PIERRE, Andrew, *Nuclear Politics: the British Experience with an Independent Deterrent, 1939–1970* (Oxford University Press, Oxford, 1972).

POLLARD, Robert A., *Economic Security and the Origins of the Cold War, 1945–1950* (Columbia University Press, New York, 1985).

RICHELIEU, Armand de, *Testament Politique*, part 2 (Imprimerie Le Breton, 1864).

ROSENCRANCE, R. N., *Defence of the Realm: British Strategy in the Nuclear Epoch* (Columbia University Press, New York, 1968).

SHEPPERD, Alan, *The Italian Campaign* (Arthur Barker, London, 1968).

SHERWOOD, Robert E., *Roosevelt and Hopkins* (Harper, New York, 1948).

SHERWOOD, Robert E., *The White House Papers of Harry L. Hopkins*, 2 vols. (Eyre & Spottiswoode, London, 1949).

SHONFIELD, Andrew, *British Economic Policy since the War* (Penguin Books, Harmondsworth, 1959).

SHONFIELD, Andrew, *Modern Capitalism: the Changing Balance of Public and Private Power* (Oxford University Press, Oxford and New York, 1965 and 1969).

SHTEMENKO, S. M., *General'nyi shtab v gody voiny*, 2 vols. (Moscow, Voennoe Izdatel'stvo, 1981).

SIRINTSEV, Yu. V., *I. V. Kurchatov i yadernaya energetika* (Nauka, Moscow, 1980).

SLADOVSKII, M. I., *Kitai i Angliya* (Nauka, Moscow, 1980).

SMYTH, H. D., *Atomic Energy: general account of the development of methods of using atomic energy for military purposes under the auspices of the United States Government, 1940–1945* (US Government Printing Office, Washington, DC, 1945).

SNOW, Edgar, *Journey to the Beginning* (Gollancz, London, 1959).

SOAMES, Mary, *Clementine Churchill: the Biography of a Marriage* (Cassell and Houghton Mifflin, London and New York, 1979).

STIMSON, Henry L. and BUNDY, McGEORGE, *On Active Service in Peace and War* (Harper and Hutchinson, New York and London, 1948 and 1949).

STUECK, William W., Jr., *The Road to Confrontation: American Policy Toward China and Korea, 1947–1950* (University of North Carolina Press, Chapel Hill, North Carolina, 1981).

TAYLOR, A. J. P., *A History of England, 1914–1945* (Clarendon Press, Oxford, 1965).

TAYLOR, A. J. P., *The Second World War* (Hamish Hamilton, London, 1979).

THORNE, Christopher, *Allies of a Kind* (Hamish Hamilton and Oxford University Press, Oxford and New York, 1979).

TOLSTOY, Nikolai, *Victims of Yalta* (Hodder & Stoughton, London, 1977).

TRUMAN, Harry S., *Year of Decisions, 1945* (Doubeday, New York, 1955).

TRUMAN, Harry S., *Years of Trial and Hope, 1946–53* (Doubleday, New York, 1956).

TRUMAN, Margaret, *Harry S. Truman* (Morrow, New York, 1973).

TURNER, Arthur C., *The Unique Partnership, Britain and the United States* (Pegasus, New York, 1971).

ULAM, Adam, *Expansion and Coexistence: Soviet Foreign Policy, 1917–1973* (Secker & Warburg and Holt, Rinehart & Winston, London and New York, 1968 and 1974 (2nd edn.)).

ULAM, Adam, *Dangerous Relations: the Soviet Union in World Politics, 1970–1982* (Oxford University Press, New York, 1984).

VALEO, Francis, *The China White Paper*, a summary of the State Department volume (*United States relations with China, with special reference to the period 1944–1949*, 5 August 1949) prepared for the Library of Congress Legislative Reference Service, Washington, DC, October 1949.

WASILIEWSKA, Wanda, *The Memoirs of Wanda Wasiliewska* (Archivum Ruchu Robotnichnego, Warsaw, 1982).

WATT, D. Cameron, *Succeeding John Bull, America in Britain's place, 1900–1975* (Cambridge University Press, Cambridge and New York, 1984).

WEBSTER, C. K., *The Congress of Vienna, 1814–1815* (Humphrey Milford, London, 1919).

WILLIAMS, Francis, *A Prime Minister Remembers* (Heinemann, London, 1961).

WILMOT, Chester, *The Struggle for Europe* (Collins, London and Westport, Conn., 1959 and 1952 (repr. 1972)).

WILSON, Harold, *A Prime Minister on Prime Ministers* (Weidenfeld & Nicolson and Michael Joseph, London, 1977).

WOODHOUSE, C. M., *Capodistria* (Oxford University Press, Oxford, 1973).

WOODHOUSE, C. M., *The Struggle for Greece, 1941–1949* (Hart-Davis, McGibbon and Beckman Publishers, London and Woodstock, N.Y., 1976 and 1979).

WU, Chengming, *Imperialist Investments in Old China* (People's Publishing Press, Peking, 1956).

ZAKHAROV, Marshal M. V. (ed.), *Finale* (Moscow, Progress, 1972).

ZHUKOV, G. K., *Vospominaniya i razmyshleniya* (Moscow, Novosti, 1969).

NEWSPAPERS

Chicago Daily News
Christian Science Monitor
Daily Mail
Daily Telegraph
Financial Times
Le Figaro
Le Monde
Manchester Guardian
New York Herald Tribune

New York Times
Observer
Pravda
St Louis Post-Despatch
Sunday Times
The Times
Wall Street Journal
Washington Post

PERIODICALS

Adelphi Papers
The Atlantic
Department of State Bulletin
The Economist
Foreign Affairs
Foreign Policy
Harpers Magazine
Historical Journal
International Affairs
International Security

Keesing's Contemporary Archives
Military Balance
New Statesman
Punch
Review of International Studies
RIIA Documents on International Affairs
Round Table
Soviet Studies
Survival

Index

Acheson, Dean G. 14, 19, 91, 92, 98, 161, 202, 208, 232, 278 n.; on Stimson's atomic memorandum (1945) 68; on American understanding of atomic issues 70; on the Washington atomic energy agreements 86; on multilateral trading system 96; 1949 British financial crisis and 108; view of Britain's role in Europe 111; on Middle East 123–4, 125; on US China policy options 141, 142, 302 n., 303 n.; rejects Pacific Pact proposal 146; on negotiating with Soviet Union 159–60; Marshall Plan and 162–3; North Atlantic Treaty and 177; US bases in Britain and 181; strengths and weaknesses 201; Hiss and 201; appointed Secretary of State (1949) 182, 200–1, 280 n., 314 n.; European integration and 183, 191–4; Bevin and 184; on American defence perimeter in the Pacific 204; NSC 68 report and 206–7; Schuman Plan and 210, 211, 212; Korean war and 217–18; German rearmament and 219–20; on the Washington summit meeting (Dec. 1950) 223–4; Middle East Command Plan and 235; 1962 speech on Britain's role 239–40; State Department's 'Policy Statement – United Kingdom' (June 1950) and 210, 252, 324 n.

Addison, Christopher 1st Viscount 282 n.

Aden 114, 115

Adenauer, Konrad 184, 186, 211, 219

Afghanistan 174

Africa, British Empire in 27

air forces, Anglo-American relationship and 13

Alexander, Field-Marshal Sir Harold (later 1st Earl Alexander of Tunis) 33

Alexander I, Tsar 1

Algeria 177, 232, 243

Allied Control Commission (Germany) 40

Allied Control Council (Germany) 61, 154, 178

Allied Council for Japan 73

Alphand, Hervé 299 n.

American bases in Britain 180–1, 218, 260; dual key arrangement 240; 'Attlee-Churchill' formula for use of 223, 230, 319 n.

American economy: post-war strength of 95, 98, 248, 249; 1946 exports 104; British dependence on 106; 1949 recessionary phase 107; economic union with Britain discussed within State Department 111; see also Bretton Woods agreements and Marshall Plan

American foreign policy: at time of Roosevelt's death 19; on Poland's frontiers 35; China the Achilles heel of 113, 234, 255; domestic politics and 113; Zionism and 113, 117, 126; anti-colonialism and 117, 123; post-war Middle East policy 123–4; East Asian policy after World War II: 138; China White Paper and (1949) 142; Truman doctrine and 156; see also Acheson, Dean G.; Byrnes, James F.; Dulles, John Foster; Hull, Cordell; Marshall, George C.; Stettinius, Edward R.

Anderson, Sir John (later 1st Viscount Waverley) 53, 79, 81, 85, 92, 93, 287 n.

Anglo-American relationship ix; post-war evolution 3; Churchill on 'special' nature of 5–6, 11; centre-piece of British foreign policy 6; in the inter-war period 12–14; common law allies 12, 279 n.; developed by Roosevelt and Churchill 11; in World War II 12, 15, 16; Quebec conference 17; Italian crisis and 35; Malta conference 36; Poland and, at Yalta 43, 48–9; withdrawal to agreed boundaries in Germany and (1945) 49–50; Truman-Attlee relationship and 22, 221–4; at Potsdam 50–66; at end of 1945: 72, 83, 94, 103; German reparations and 289; financial negotiations 77, 94–112, 257; post-war low points 103, 132–3; atomic negotiations and 77–93; relative political strengths in 1947: 88; American loan to Britain 94–112; China and 138–48; Palestine and 118–19, 124–6, 128–30, 132–3; effect of 1949 financial crisis on 110–12; revival of 151–60; remaking of Europe and 151–4, 253–5; occupation of Germany and 153; post-war high point 182; secret London meeting (June 1947) 183; Pentagon Talks (Oct. 1947) 129; European integration and 182–3, 189–94, 232, 253–5; effect of Korean War on 203, 205, 217–19; German rearmament and 219–20, 232; secret Washington talks (July 1950) 199; Washington summit meeting (Dec. 1950) 202, 221–4; Macmillan's Greco-Roman analogy 227, 319 n.; Churchill-

346 *Index*